Electronics and Microprocessors

Electronics and Microprocessors

By

P. MANIMEGALAI
BE, ME
Assistant Professor, EIE Department
Karunya University
Coimbatore **(Tamil Nadu)**

G. PABITHA
BE, ME
Assistant Professor, ECE Department
Karpagam College of Engineering and Technology
Coimbatore **(Tamil Nadu)**

UNIVERSITY SCIENCE PRESS
(An Imprint of Laxmi Publications Pvt. Ltd.)
An ISO 9001:2015 Company

BENGALURU • CHENNAI • GUWAHATI • HYDERABAD
KOCHI • KOLKATA • LUCKNOW • RANCHI
NEW DELHI

Printed and bound in India

First Edition: 2013; *Edition*: 2025

ISBN : 978-93-81159-45-3

Branches

Bengaluru	080-26 75 69 30
Chennai	044-24 34 47 26
Guwahati	0361-254 36 69
Hyderabad	040-27 55 53 83
Kochi	0484-405 13 03
Kolkata	033-40 04 77 79
Lucknow	0522-430 36 13
Ranchi	0651-224 24 64

PUBLISHED IN INDIA BY

UNIVERSITY SCIENCE PRESS

(An Imprint of Laxmi Publications Pvt. Ltd.)

An ISO 9001:2015 Company

113, GOLDEN HOUSE, GURUDWARA ROAD, DARYAGANJ,

NEW DELHI - 110002, INDIA

Corporate Office : B-33, Sector-5, Noida,

Gautambuddha Nagar (U.P.) - 201 301, INDIA

Telephone : +91-120-44 46 900

www.laxmipublications.com

info@laxmipublications.com

CONTENTS

PREFACE

This book is a response of the request of the students of Mechanical Engineering who prefer to have all the material on Electron Devices, Digital Electronics and Microprocessors in a single volume.

To make a reasonable sized single volume available to the students we have incorporated only some topics of electronics and microprocessors. These topics have been selected based on syllabus of Anna University.

This book contains five chapters. The first chapter introduces semiconductors and rectifiers. Chapter 2 deals with transistors and amplifiers and covers BJT, UJT, FET, UJT, SCR, DIAC, TRIACS transistors and feedback concept. Chapter 3 is on Digital Electronics Number Systems. Chapter 4 introduces 8085 microprocessors. Chapter 5 describes the details of interfacing and applications of microprocessor.

The language used in explaining the concepts is extremely simple. Since a proper understanding of the subject would involve a serious attempt to solve a variety of problems, a wide variety of problems with their step-by-step solutions has been provided.

We thank the publishers for encouraging us to prepare this book. We will be thankful to all users of this book for their criticism and suggestions.

Suggestions for improvement of the book will be gratefully acknowledged.

We hope that the teachers will find this book a useful aid to make their teaching interesting and purposeful to the students of engineering.

—Authors

1

SEMICONDUCTORS AND RECTIFIERS

1.1 INTRODUCTION

Materials whose properties lies between those of metals and insulators are called semiconductors. Their conductivity (ability to conduct) lies between that of a metal and insulator. Electronic devices such as diodes, transistors, field effect transistors, thyristors, integrated circuits etc., are made of semiconductors. The most popularly used semiconductor materials are silicon and germanium. The Gallum Arsenide (GaAs) and related compounds are used for high speed devices and optical devices. The conductivity of semiconductors depends on temperature. It increases with increase in temperature.

1.2 FUNDAMENTALS—PHYSICS

Any substance is made up of molecules and molecules are made up of atoms. The atoms contain tiny particles called protons, neutrons and electrons which are called as fundamental particles. Protons are positively charged, electrons are negatively charged and neutrons are electrically neutral. The arrangement of protons, neutrons and electrons will be different for different material.

Figure 1.1 (*a*) shows that an atom consists of a central 'nucleus' which is surrounded by orbiting electrons.

Nucleus : The nucleus of an atom consists of two types of particles — protons and neutrons. For a given atom, the number of protons in the nucleus is normally equally to the number of orbiting electrons.

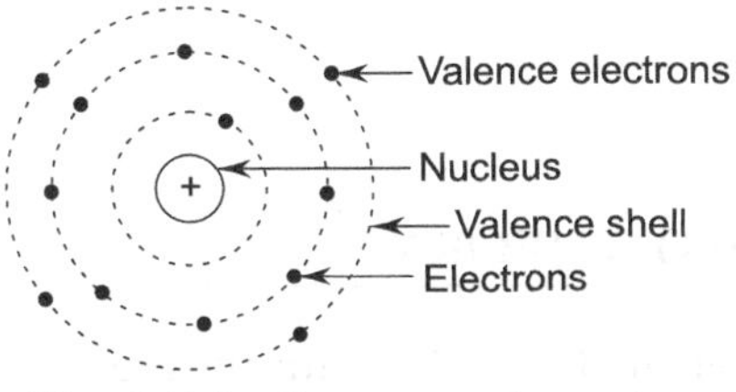

Fig. 1.1(*a*) Structure of an atom

Atoms : Since the protons and orbital electrons are equal in number, their equal and opposite charge will neutralize each other electrically. Therefore, atoms are electrically neutral.

If an atom loses an electron, then the number of protons becomes higher than the number of electrons. Therefore, the atom becomes positively charged and it is referred to as a positive ion. Similarly if an atom gains an additional electron then it becomes negatively charged and called as a negative ion.

Valence Shell and Valence Electrons

Electrons can occupy only certain orbital rings or shells. Each shell can contain only a particular number of electrons. First shell can contain 2 electrons, second one can contain 8 electrons and so on. The exception for this rule is that the outermost shell cannot contain more than eight electrons.

The outermost shell is known as the valence shell and the electrons in it are called as valence electrons. The valence shell may be completely filled or partially filled of valence electrons.

Energy Level

Each shell has an energy level associated with it. This energy level represents the amount of energy that is required to be supplied to extract an electron from shell.

An electron volt is defined as the amount of energy required to move one electron through a potential difference of one volt. The charge on one electron is 1.6×10^{-9} Coulomb. So the energy required by an electron to fall through a potential of 1 volt is called one electron volt.

$$\therefore \qquad 1 \text{ eV} = 1.6 \times 10^{-19} \text{ J}$$

Structure of Silicon and Germanium Atoms

A silicon atom consists of 14 protons and 14 neutrons inside the nucleus and 14 revolving electrons. These 14 electrons are distributed among various fields.

A germanium atom consists of 32 protons, 32 neutrons in the nucleus. Both silicon and germanium has 4 electrons in the valence shell.

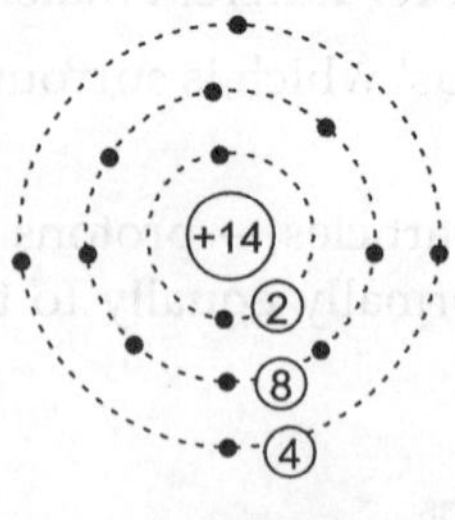

Fig. 1.1(*b*)

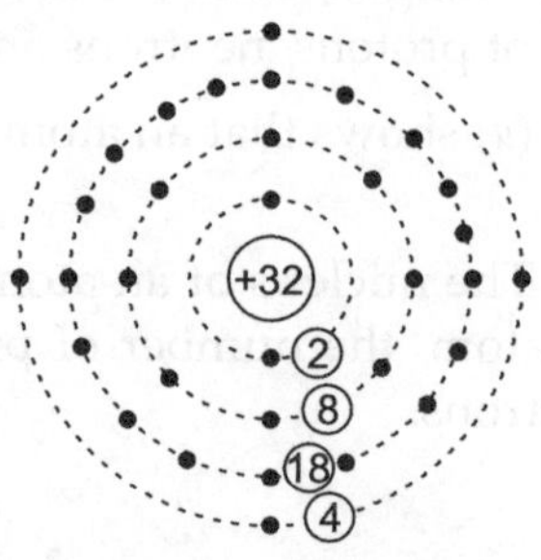

Fig. 1.1(*c*)

1.3 ENERGY BANDS (BAND THEORY)

The energy band diagram shown in Fig. 1.2 is simply graphical representation of the energy levels associated with electrons in different shells.

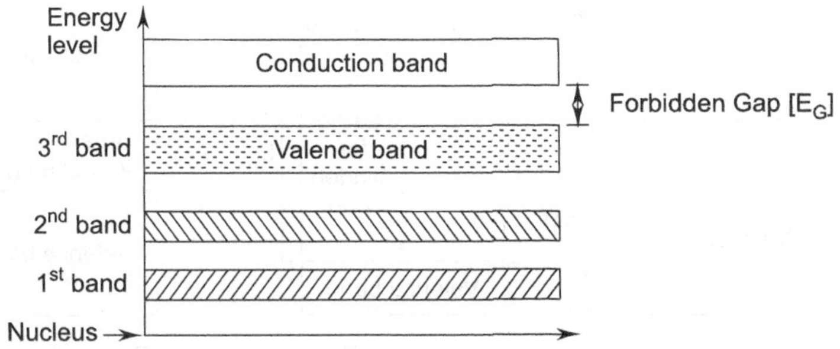

Fig. 1.2 Energy band diagram

Valence band corresponds to the valence electrons present in different atoms of the material. Energy associated with the valence band is the second highest.

Conduction band has the highest energy associated with it. The electrons in the conduction band are the 'free' electrons. Conduction band electrons are actually responsible for the flow of current.

Forbidden gap : The forbidden gap is the energy gap that separates the conduction and valence bands. For any given type of material the forbidden gap may be large, small or even non-existent. The materials are classified as conductors, insulators and semiconductors based on the relative width of the forbidden gap.

1.4 CLASSIFICATION OF MATERIALS BASED ON BAND THEORY

The materials can be broadly divided into three categories as conductors, insulators and semi-conductors depending on their ability to conduct the electric current.

Conductors

Conductors are the materials which allow current to flow very easily, due to the availability of large number of free electrons present in the conductors. For metals like copper, aluminium etc., there is no 'forbidden gap' present between the valence and conduction bands. They are some examples for conductors.

Insulators

The energy band diagram of an insulator is shown in Fig. 1.3. The forbidden gap between the conduction and valence band is extremely wide. Normally the valence electrons cannot jump that far and enter into conduction band. Hence conduction does not take place and these materials are known as insulators. Example of insulating materials are wood, glass, mica, paper etc.

Semiconductors

They have conduction properties which are in between those of conductors and insulators. The energy band diagram shows that the forbidden gap is very narrow as compared to that of the insulator. The forbidden region E_G in a semiconductor depends on the temperature. At room temperature (300°K) E_G = 1.1 eV for silicon and E_G = 0.72 eV for germanium. At very low temperatures of order of 0°K (–273°C), semiconductor act as insulators.

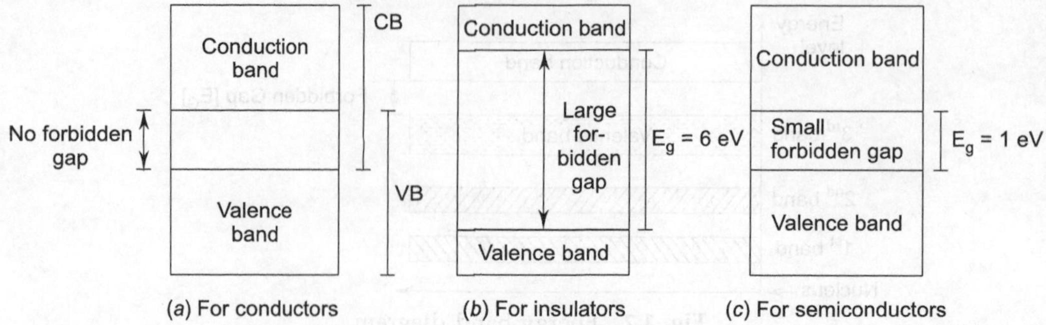

Fig. 1.3 Energy band diagram of conductors

1.5 COMPARISON OF CONDUCTORS, INSULATORS AND SEMICONDUCTORS

S.No.	Parameter	Conductors	Insulators	Semiconductors
1.	Conductivity	Very high	Very low	Moderate
2.	Resistivity	Very low	Vey high	Moderate
3.	Forbidden gap	No gap	Large gap (≈ 5 eV)	Medium (1 to 2 eV)
4.	Temperature coefficient of resistance	Positive	Negative	Negative
5.	Examples	Metals	Paper, Glass	Silicon, Germanium
6.	Applications	As conductors, wires	Capacitors, Insulation material	All electronic devices

1.6 INTRINSIC AND EXTRINSIC SEMICONDUCTORS

Semiconductors are classified into two categories as :

Intrinsic semiconductors and Extrinsic semiconductors

Intrinsic semiconductors : Intrinsic semiconductors are the purest form of semiconductors. The presence of impurity is as low as 1 part per 100 million parts of the semiconductors. At room temperature they are very poor conductors. Even at increase in temperature or by application of increased voltage. Conductivity would not increase significantly. Examples are Si and Ge.

Extrinsic semiconductors : Intrinsic means pure and extrinsic means impure, so we can obtain extrinsic by adding impurities to intrinsic semiconductors.

Doping : The process of adding impurities to intrinsic semiconductors is known as doping. Due to doping conductivity of the semiconductors increase.

Extrinsic semiconductors are of two types namely : Depending upon the impurity added, they can be classified as :

- *n*-type semiconductors
- *p*-type semiconductors

1.7 INTRINSIC SEMICONDUCTORS

Two dimensional representation of a silicon crystal :

In silicon atom there are four valence electrons. In an intrinsic silicon or germanium crystal these four valence electrons are bounded to four adjoining atoms. Each one of the four valence electrons in each atoms forms a bond with a valence electron from the adjoining atom. This bond is nothing but sharing of atoms. These bonds are known as covalent bonds.

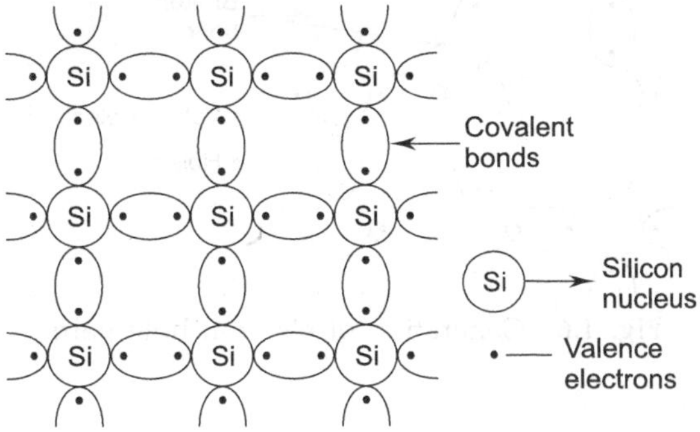

Fig. 1.4 2-D representation of silicon

Four of the electrons are their own and remaining four are due to the four covalent bonds. Due to this, the outermost shells of all the atoms are completely filled. This makes the outer atoms stable. Hence the intrinsic semiconductor act as insulator at absolute zero temperature 0°K or – 273°C.

Conduction in Intrinsic Semiconductors

The intrinsic semiconductor behave like perfect insulators at the absolute zero temperature. But when there is increase in temperature, its behaviour changes with increase in temperature, many valence electrons will absorb the thermal energy, break the covalent bonds and go into the conduction band. Thus they become free for conduction, and these electrons are called conduction electrons.

When an electron breaks a covalent bond and becomes free, a vacancy is created in the broken bond. This vacancy is called as 'hole'.

Therefore, the number of free electrons generated due to increased temperature is exactly equal to the number of holes. Both electrons and holes can act as charge carriers. The generation of electron hole pair is shown in Fig. 1.5. The effect of light on semiconductors is same as that of the heat.

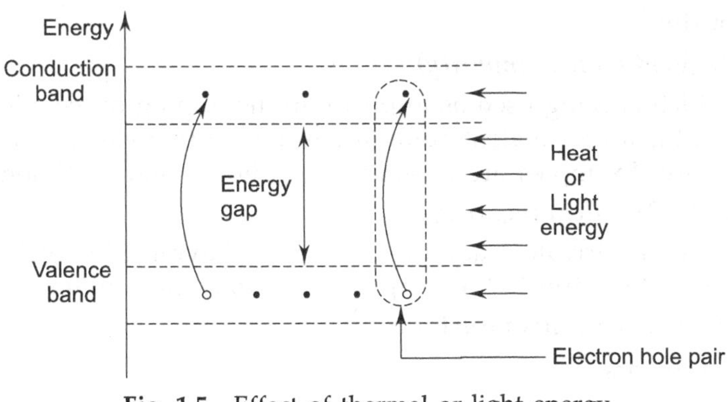

Fig. 1.5 Effect of thermal or light energy

In intrinsic semiconductors

$$n_e = n_p$$

n_e = number of electrons

n_p = number of holes.

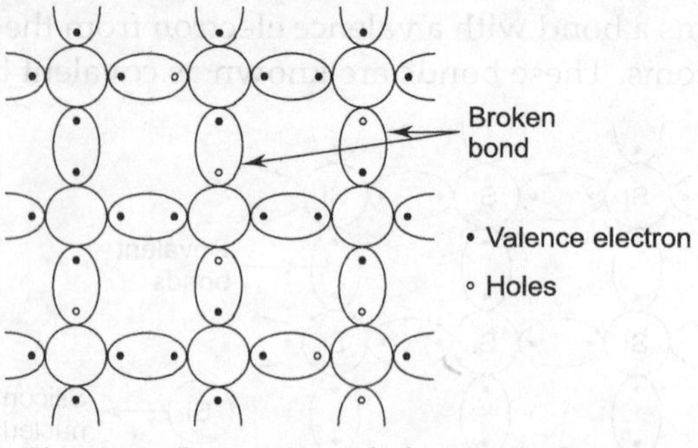

Fig. 1.6 Generation of electron hole pairs

Recombination

The free electrons in the conduction band, when come across the holes will jump into the holes. This process is called as the recombination process.

When a conduction band returns back to valence band due to external thermal energy, recombination takes place, it releases energy equal to energy gap between conduction and valence bands. This principle is used in Light Emitting Diode (LED).

1.8 EXTRINSIC SEMICONDUCTORS

Obtained by process called doping.

Doping

The process of adding impurities to pure silicon or germanium (intrinsic semiconductors) is known as doping.

The impurity materials can be of two types :

1. Donor impurity
2. Acceptor impurity

1. ***Donor impurity (Pentavalent impurity)***

   The material which is being used as impurity in the doping process is called as 'dopant'. When the dopant is a pentavalent atom (atom having five valence electrons), it is called as donor impurity. Examples are Arsenic, Phosphorus and Antimony.

2. ***Acceptor impurity (Trivalent impurity)***

   When the dopant is a trivalent atom *i.e.*, the atom having three valence electrons, then it is called as 'acceptor impurity'. Examples are Boron, Aluminium, Gallium and Indium.

Classification of Extrinsic Semiconductors :

1. *n*-type semiconductors
2. *p*-type semiconductors.

1.9 N-TYPE SEMICONDUCTORS

The *n*-type semiconductor is formed by adding a small amount of 'pentavalent', impurity to the pure Silicon or Germanium material. Examples for pentavalent electrons are Antimony, Arsenic and Phosphorus. The effect is shown in Fig. 1.7, when a pentavalent impurity such as Arsenic is added to the intrinsic semiconductors four valence electrons or Arsenic atom form four covalent bonds with neighbouring silicon atoms.

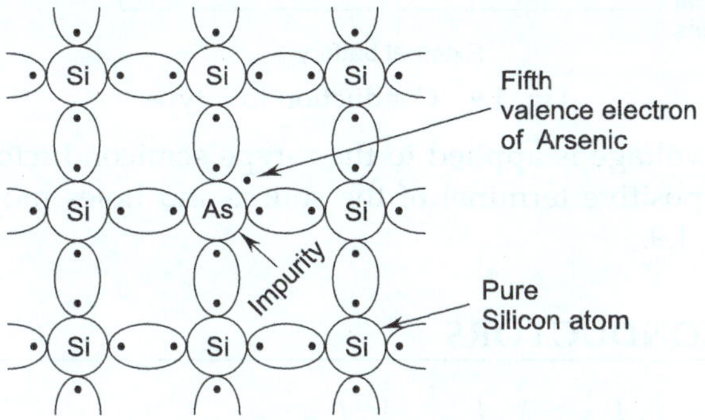

Fig. 1.7 Formation of *n*-type semiconductor

The fifth electron of the Arsenic atom does not form covalent bond. This additional valence electron can enter conduction band very easily to be a free electron. So, for each impurity atom an extra electron becomes available for conduction. As each impurity atom "donates" one free electron, pentavalent impurities are known as "donor impurities".

Energy band diagram of n-type semiconductor:

The doping process will alter the energy band diagram. The donor energy level (E_D) shown by the dotted line is very close to the conduction band energy level (E_C). Therefore, the impurity atoms can easily jump into the conduction band and become free electrons.

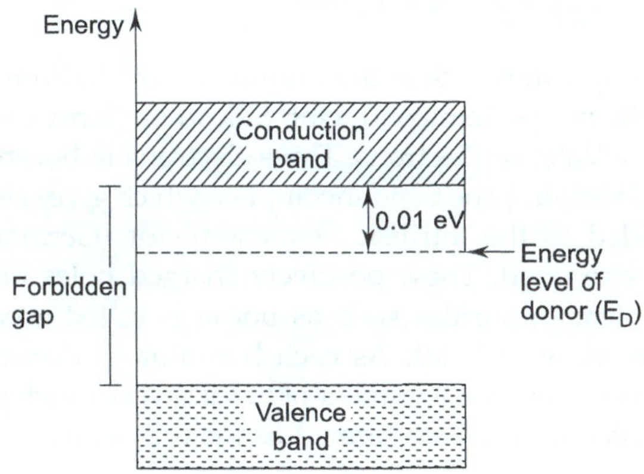

Fig. 1.8 Energy band diagram of *n*-type semiconductor

Majority and minority carriers in n-type semiconductor :

In *n*-type semiconductor, a large number of free electrons are present along with a small number or thermally generated holes in an *n*-type semiconductor. Therefore, free electrons are 'majority carriers' and holes are known as "minority carriers".

Conduction in n-type material :

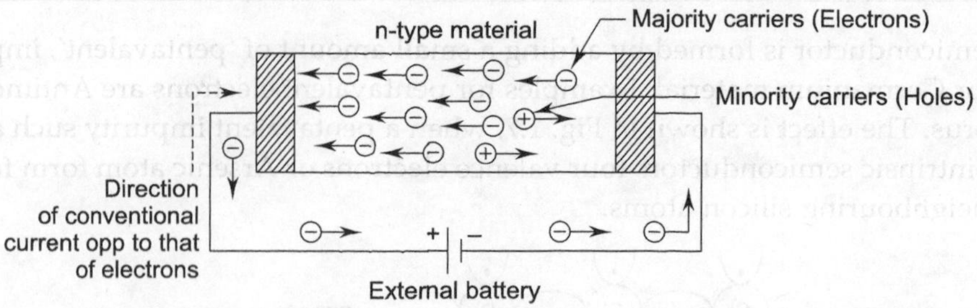

Fig. 1.9 Conduction to *n*-type

When an external DC voltage is applied to the *n*-type semiconductor material, the free electrons move towards the positive terminal of the source and holes move towards the negative terminal as shown in Fig. 1.9.

1.10 P-TYPE SEMICONDUCTORS

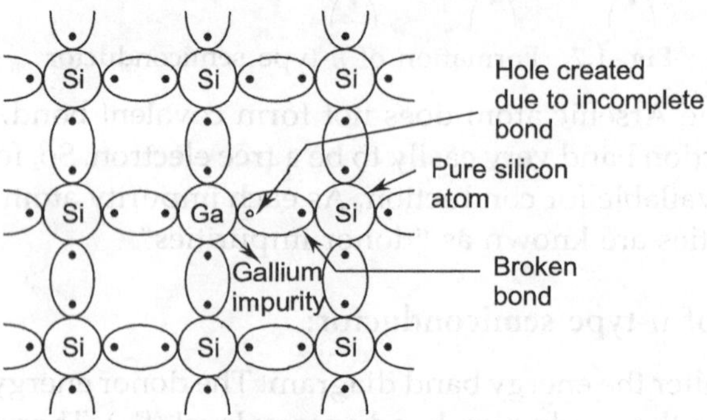

Fig. 1.10 *p*-type semiconductors

P-type Semiconductor

A small amount of invalent impurities such as aluminium, boron, indium or gallium is added to the pure semiconductor to get the *p*-type Semiconductor. Silicon or Germanium atom has four valence electrons and boron has three Valence Electrons. Three electrons in boron from covalent bond with four surrounding atoms of Ge leaving one bond incomplete which gives rise to a hole. Thus Trivalent impurity (boron) when added to the intrinsic Semiconductor (Germanium) introduces a large number of holes in the valence bond. These positively charged holes increase the conductivity of *p*-type semiconductor. Trivalent impurities such as boron is called acceptor impurity because it accepts free electrons in the place of holes. As each boron atom donates a hole for conduction, it becomes a negatively charged ion. As the number of holes is very much greater than the number of free electrons in a *p*-type material, holes are termed as majority carriers and electrons as minority carriers.

1.11 PN-JUNCTION

A *p*-type material consists of silicon atoms and trivalent atoms such as boron. The boron atom adds a hole when it bonds with the silicon atoms. However, since the number of protons and the number of electrons are equal throughout the material, there is no net charge in the material and so it is neutral.

An *n*-type silicon material consists of silicon atoms and pentavalent impurity atoms such as antimony. As you have seen, an impurity atom releases an electron when it bonds with four silicon atoms, since there is still an equal number of protons and electrons throughout the material, there is no net charge in the material and so it is neutral.

If a piece of intrinsic silicon is doped so that part is N-type and the other part is P-type, a *pn*-junction forms at the boundary between the two regions and a diode is created as shown in Fig. 1.11.

The *p*-region has many holes (majority carriers) from the impurity atoms and only a few thermally generated free electrons (minority carriers). The *n*-region has many free electrons (majority carriers) from the impurity atoms and only a few thermally generated holes (minority carriers).

A PN junction forms a very useful device and is called a semiconductor diode, PN junction diode or simply crystal diode.

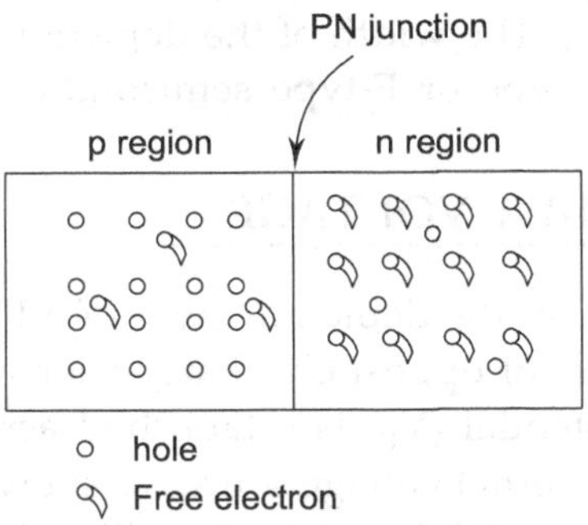

Fig. 1.11 PN Junction

Formation of the Depletion Region:

Figure 1.12(*a*) shows a semiconductor consisting of a PN junction, which has been just formed. It may be noted that the entire sample is a single crystal.

Fig. 1.12 Formation of depletion layer

We know that holes and electrons are mobile charges, and therefore are known as mobile charge carriers. On the otherhand, positive and negative ions are immobile charges and therefore do not take part in the conduction. The sample as a whole, is electrically neutral and so are the P-regions and N-regions, considered separately. Thus, in the P-region, the total positive charge on the hole is equal to the total negative charges on free electrons and immobile ions.

It may be noted that no external voltage has been connected to the PN junction. As soon as the junction is formed, the following takes place.

1. The holes, from the P-region diffuse into the N-region, where they combine with the free electrons.
2. The free electrons, from the N-region diffuse into the P-region, when they combine with the holes.

3. The diffusion of holes (from P to N region) and free electrons (from N to P region) takes place due to the reason that there is a difference of concentrations in the two regions *i.e.*, the P-region has more number of holes, whereas N-region has more number of free electrons. This difference in concentration creates a concentration gradient across the junction. It results in the diffusion of mobile charge carriers across the junction. Moreover, the electrons and holes move at random in all directions due to thermal energy.
4. The diffusion of holes and free electrons across the junction takes place for a short time. After a few recombination of holes and free electrons, a restraining force is automatically set up. This force is produced due to depletion region, which exists on either side of the junction. (*i.e.*, the additional holes, trying to diffuse into the N-region, are repelled by the uncovered positive charge of the donor ions. Similarly the electrons, trying to diffuse into the P-region, are repelled by the uncovered negative charge of the acceptor ions. Therefore, further diffusion is stopped). The width of the depletion layer depends upon, the doping level of the impurity in N-type or P-type semiconductor.

1.12 JUNCTION OR BARRIER VOLTAGE

In last topic we have discussed, that the depletion layer of a PN junction has no mobile charge carriers. But it contains fixed rows of oppositely charged ions on its two sides. Because of this charge separation, an electric potential (V_B) is established across the junction, even when the junction is not connected to any external voltage source as shown in Fig. 1.13. This electric potential is called junction or barrier potential. It exerts a repelling force on the mobile charge carriers trying to crossover the junction. This force steps the mobile charge carriers to crossover the junction unless the energy is supplied from an external source. At room temperature, the value of V_B is 0.3 V for Ge and 0.7 V for silicon.

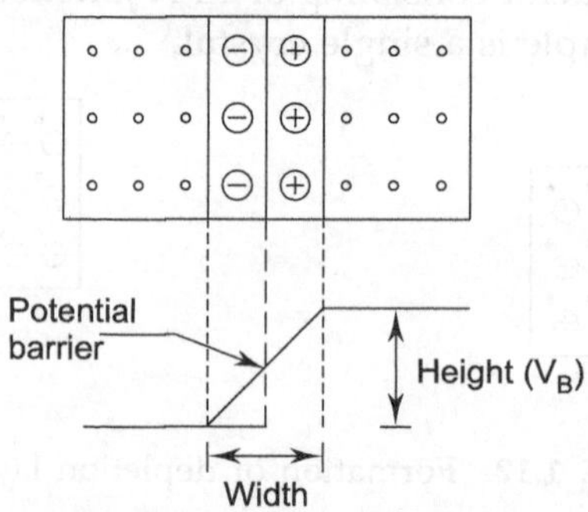

Fig. 1.13 Barrier potential

1.13 EFFECT OF TEMPERATURE ON BARRIER VOLTAGE

The barrier voltage depends on three factors namely density, electronic charge and temperature. For a given PN junction, the first two factors are constant, thus making the value of V_B dependent only one temperature. It has been observed that for both silicon and germanium, the value of V_B decreases by 2 mV/°C. Mathematically, the decrease in barrier voltage,

$$\Delta V_B = -0.002 \times \Delta t.$$

where Δt is the increase in temperature in °C.

1.14 BIASING THE JUNCTION

The unbiased PN junction is of no use in actual practice. Since there is no net flow of current across an unbiased PN junction (In unbiased PN junction, the majority carrier current and minority carrier current are equal in magnitude and flow in opposite directions. So obviously there is no net flow of current). A PN junction, connected to an external voltage source (*i.e.*, battery) is called a biased PN junction.

There are two ways of connecting a battery to a PN-junction.

1. **Forward bias :** In this case, positive terminal of the battery is connected to the P-side and negative terminal to the N-side as shown in Fig. 1.14(*a*). A large amount of current flows through the junction under this condition.
2. **Reverse bias :** In this case, positive terminal of the battery is connected to the N-side and negative terminal to the P-side as shown in Fig. 1.14(*b*). Practically no current flows through the junction under this condition.

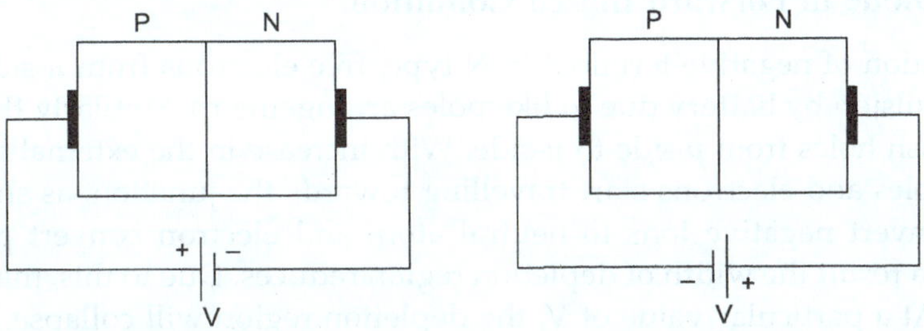

(*a*) Forward bias (*b*) Reverse bias

Fig. 1.14 Biasing the PN junction

1.15 PN-JUNCTION DIODE

The PN-junction itself forms the most basic semiconductor device called semiconductor diode or PN junction diode. The term diode means devices having "two electrodes" (di-ode), Cath*ode* and An*ode*.

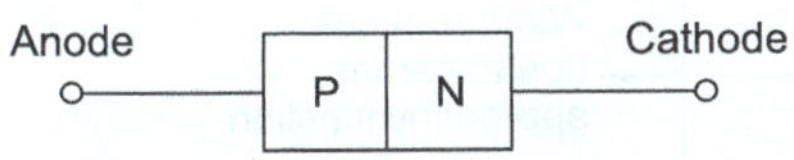

Fig. 1.15 PN-junction forms a diode

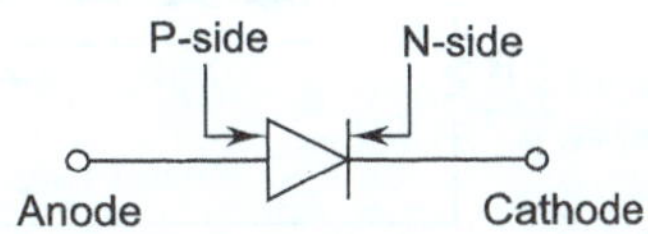

Fig. 1.16 Circuit symbol of diode

The symbol of a semiconductor diode is shown in Fig. 1.16. The two electrodes are named as anode and cathode. The arrow head points in the direction of conventional current through the device.

Biasing of a PN-Junction Diode:

Biasing is the process of applying external DC voltage to the semiconductor diode. When external voltage is not applied to the diode, the PN-junction will remain in the equilibrium state and hence no current flows through it. To make the current to flow, it is necessary to 'bias' the diode. Biasing can be of two types forward and reverse bias.

1.16 FORWARD BIASING A DIODE

If the *p*-region is connected to positive terminal or external DC source or battery and *n*-region to negative terminal, then it is said to be "forward biasing". Generally a resistance is connected in series with the diode to limit current flow. It is shown in Fig. 1.17 with its symbolic representation. The current I_F is a conventional current due to biasing.

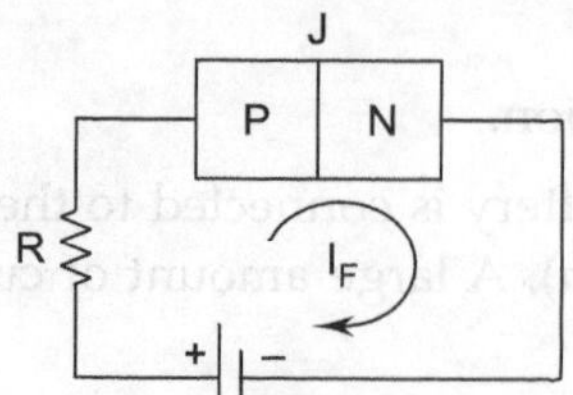

Fig. 1.17 Forward biasing

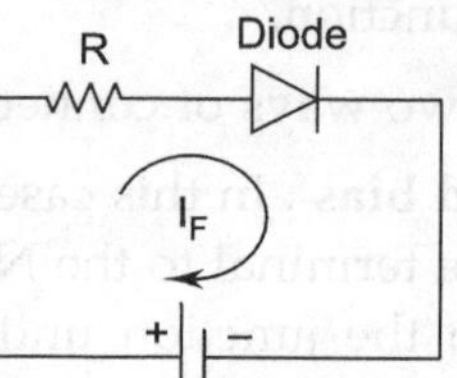

Fig. 1.18 Symbolic representation

Operation of a Diode in Forward Biased Condition:

Due to the connection of negative terminal to N type, free electrons from *n*-side are pushed towards *p*-side (Repulsion by battery due to like poles arrangement). Similarly the positive end of the supply will push holes from *p*-side to *n*-side. With increase in the external supply voltage V, more and more holes and electrons start travelling towards the junction, as shown in Fig. 1.19. The holes will convert negative ions to neutral atom and electron convert positive ions into neutral atoms. As a result the width of depletion region reduces. Due to this, the barrier potential will also reduce. At a particular value of V, the depletion region will collapse. A large majority carriers will cross the junction constituting a current called forward current (which is in opposite direction to conventional current).

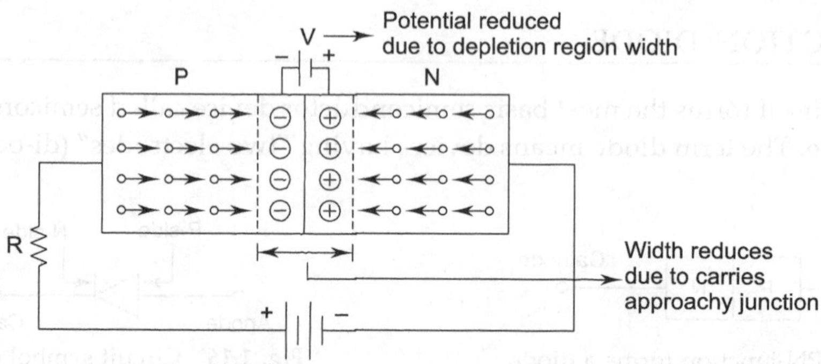

Fig. 1.19 Effect of forward bias

Current Flow in Forward Biased Diode:

The forward current through a diode flows due to the majority carrier and its direction of flow (conventional) is always from anode to cathode.

Voltage Drop:

The forward voltage drop V_F is due to the two factors namely drop due to barrier potential and drop due to internal resistance R. Forward voltage drop is equal to 0.7 V for silicon and 0.3 V for germanium.

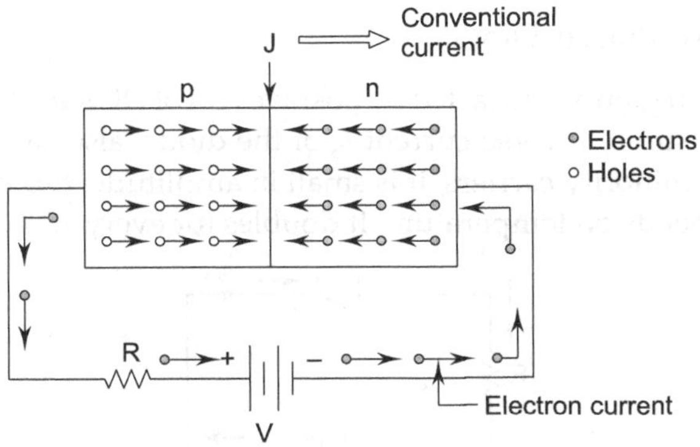

Fig. 1.20 Current flow in forward biased diode

1.17 REVERSE BIASING A DIODE

If the *p*-region is connected to negative terminal of DC source and n-region to positive terminal of external DC supply then it is reverse biasing. Figure 1.21 shows reverse biasing. The reverse current I_S flows exactly opposite to the forward current.

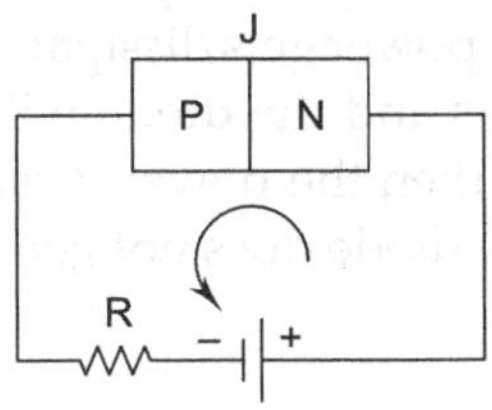

Fig. 1.21 Reverse biasing

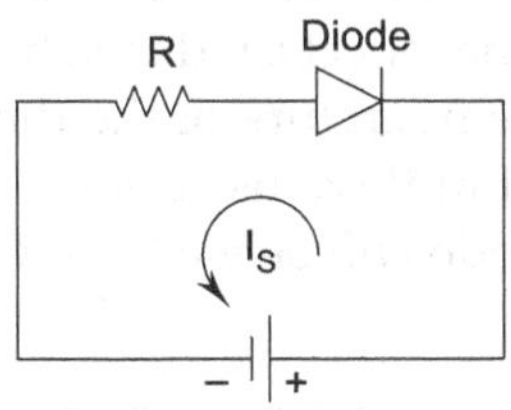

Fig. 1.22 Symbolic representation

Operation of Reverse Biased Diode:

When a diode is reverse biased, holes in the *p*-region are attracted towards the negative terminal of the supply and electrons in *n*-side are attracted towards positive terminal of the supply as shown in Fig. 1.23

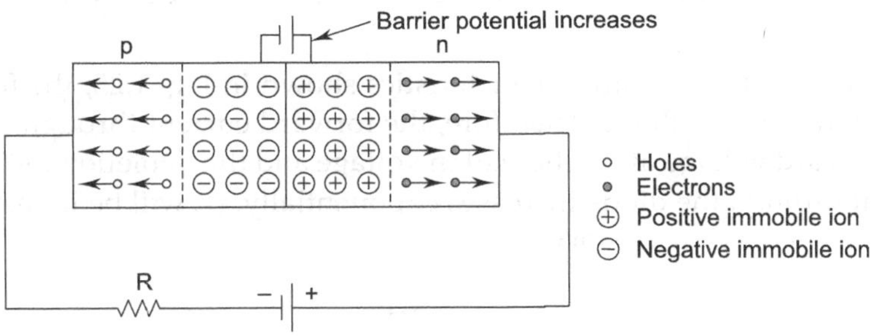

Fig. 1.23

Due to the movement of electrons and holes away from junction, width of the depletion region increases. (Due to the creation of more number of positive and negative immobile ions). Due to the presence of more number of ions the barrier potential increases. The process does not continue for a long time, since there is no steady flow of holes.

Current Flow in Reverse Biased Diode:

The minority electrons in *p*-region are attracted by positive end of DC supply. Hence the electrons will cross the junction and constitute reverse current I_S of the diode, also called as "reverse saturation current". As this is due to minority carriers, it is small in amplitude, typically a few μA for Ge and few nA for Si diode. It depends on temperature. It doubles for every 10°C rise in temperature.

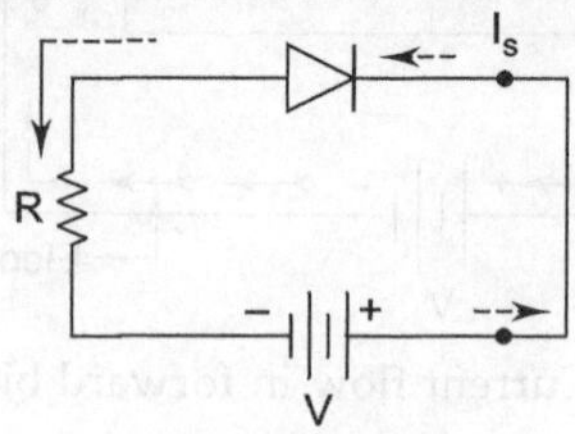

Fig. 1.24 Current flow

Breakdown in Reverse Biased Diode:

The reverse saturation current flowing in the reverse biased diode is dependent only on the temperature and independent of the reverse voltage applied externally. But due to very large reverse voltage, breakdown occurs (avalanche effect). Reverse breakdown can take place due to another effect called zener effect *i.e.*, due to heavy doping. Due to this large power gets dissipated in the diode. The junction temperature of the diode may exceed its safe limits and the diode will be damaged permanently. But if we use external current limiting resistance, then the power dissipation taking place in the diode will be well below the dangerous level and the diode does not get damaged.

1.18 V-I CHARACTERISTICS OF A DIODE

The behaviour of a *pn* junction diode is indicated by its Volt-Ampere characteristics (V-I characteristics). It can be divided into two parts : 1. Forward characteristics 2. Reverse characteristics.

Forward Characteristics of PN Junction Diode

It is the graph of the anode to cathode forward voltage V_F versus the forward current through the diode (I_F).

In the region A to B of the forward characteristics shown in Fig. 1.25, the forward voltage is small and less than the cut in voltage. Therefore, the forward current through the diode is small. With increase in forward voltage, it reaches cut in voltage and the depletion region width goes on decreasing, current through the diode increases exponentially. (It will be in mA range).

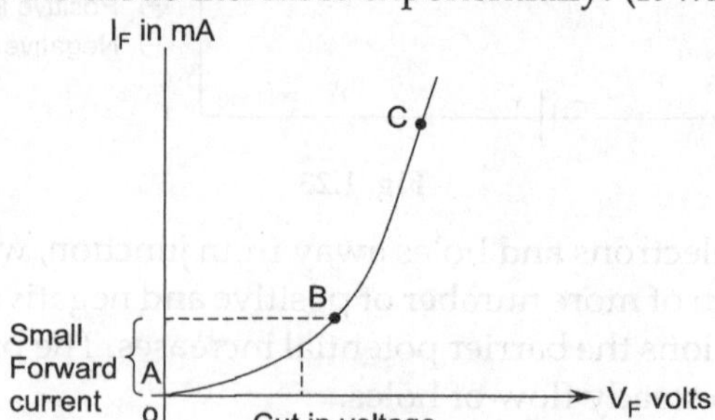

Fig. 1.25 Forward characteristics

The voltage at which the forward diode current starts increasing rapidly is known as "cut-in" voltage of a diode (V_γ) for silicon $V_\gamma = 0.6$ V and for germanium $V_\gamma = 0.2$ V. R_F will be very small in order of few ohms.

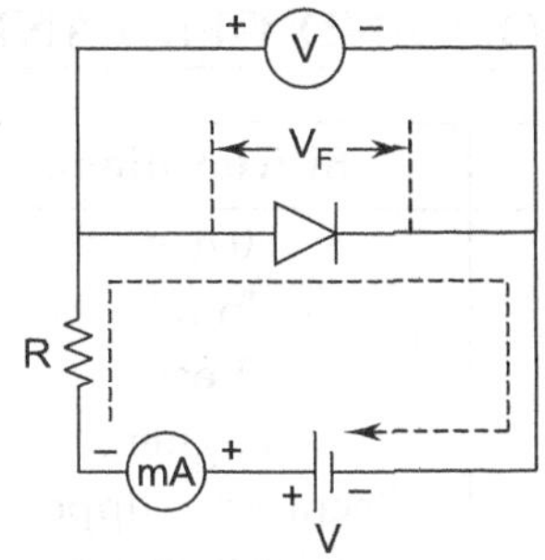

Fig. 1.26 Set up to get forward characteristics

Reverse Characteristics of a Diode:

It is graph of reverse voltage (V_R) versus reverse current (I_S) as shown in Fig. 1.27. Current flowing through the diode is the reverse saturation current which flows due to the minority carriers.

As the voltage is increased, the reverse saturation current remains constant equal to I_o if the temperature is constant (Reverse current depends on temperature only). Operation in the breakdown region should be avoided, the diode may get damaged due to large power dissipation. Reverse breakdown voltage for diode is in range of 50 to 100 volts. Reverse resistance will be very high in order of few hundred kΩ.

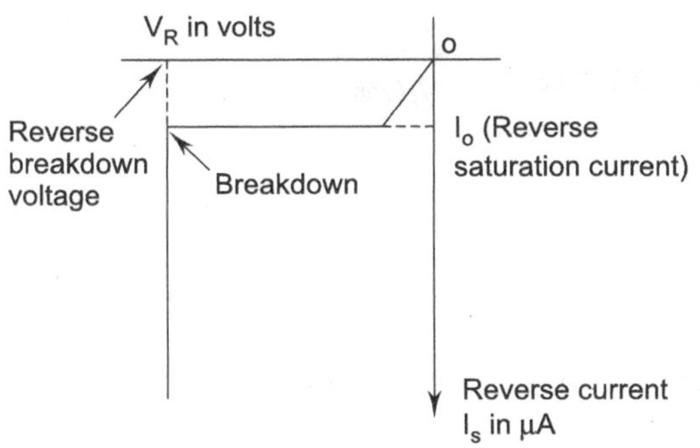

Fig. 1.27 Reverse characteristics of a diode

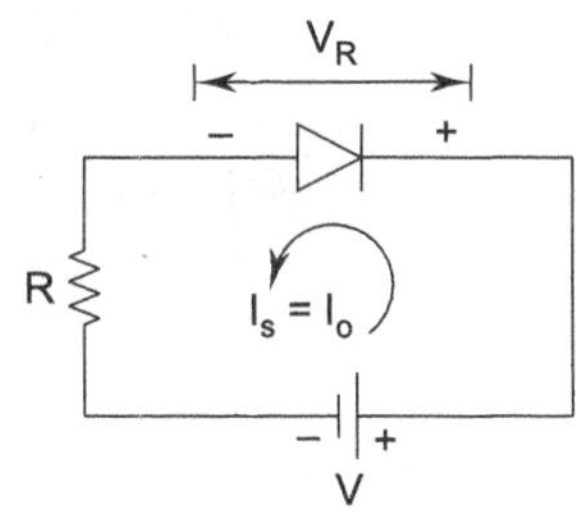

Fig. 1.28 Arrangement to get reverse bias a diode

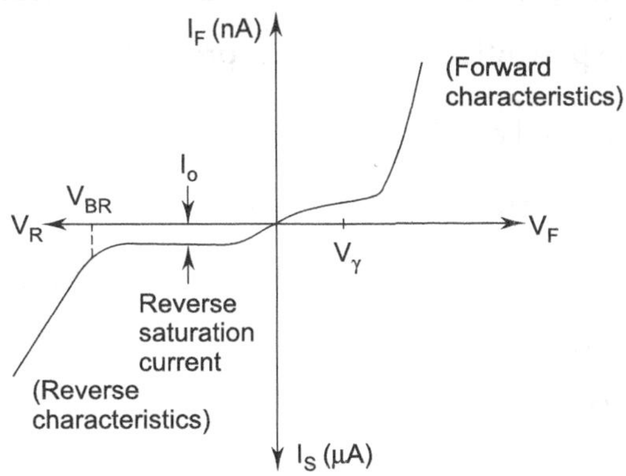

Fig. 1.29 V-I characteristics of a diode

Complete V-I Characteristics of a Diode:

It is obtained by combining the forward and reverse characteristic of a diode.

1.19 COMPARISON OF SILICON AND GERMANIUM DIODES

S.No.	Parameter	Silicon diode	Germanium diode
1.	Cut-in voltage	0.6 V	0.2 V
2.	Reverse saturation current	In nA	In μA
3.	Effect of temperature	Less	More
4.	Breakdown voltage	Higher	Lower
5.	Application	Rectifiers clippers, clampers etc.	Low voltage, low temperature applications

1.20 V-I CHARACTERISTIC OF SILICON AND GERMANIUM DIODE

1. Cut-in voltage for Si and Ge are 0.6 and 0.2 V respectively.
2. Breakdown voltage of Si is higher than Ge diode.
3. Reverse saturation current for Ge is few μA and Si is in nA.

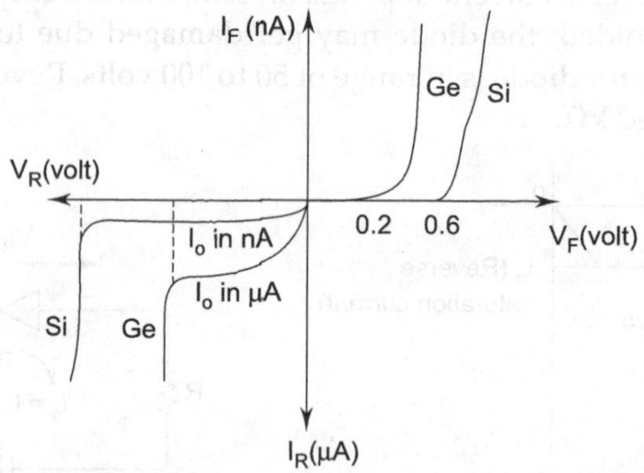

Fig. 1.30 V-I characteristic of silicon and germanium

1.21 MATHEMATICAL EXPRESSION FOR DIODE CURRENT

The diode current I_D has an exponential shape. It is expressed as :

$$I_D = I_o \left[e^{\frac{V}{\eta . V_T}} - 1 \right] \quad ...(1)$$

where I_o = Reverse saturation current

$V_T = \frac{T}{11.600}$ (Volt equivalent of temperature)

η = 1 for Ge and 2 for Si (Emission coefficient)

T = Temperature in °K

V = Diode voltage

Forward current :

Substituting $I_D = I_F$ and $V = V_F$ in equation (1), we get

$$I_F = I_o \left[e^{\frac{V_F}{\eta V_T}} - 1 \right].$$

$$I_F \; e^{\frac{V_F}{\eta V_T}} >> 1, \text{ then}$$

$$I_F \approx I_o \, e^{\frac{V_F}{\eta V_T}}.$$

Reverse current :

Substituting $I_o = I_S$ and $V = -V_R$ in equation (1), we get

$$I_S = I_o \left[e^{-\frac{V_R}{\eta V_T}} - 1 \right]$$

But $\quad e^{-\frac{V_R}{\eta V_T}} << 1$

$\therefore \quad I_S \approx I_o$

Applications of Diode

Used in:

1. Rectifier circuits
2. Clipper and clamper circuits
3. Voltage multipliers
4. AM detection
5. Free wheeling diodes.

Example 1.1. *The forward current through a silicon diode is 10 mA at room temperature (27°C). The corresponding forward voltage is 0.75 volts. Calculate the reverse saturation current I_o.*

Solution. Given $\quad I_F = 10$ mA, $V_F = 0.75$ volt, $T = 27°C = 300°K$, $\eta = 2$ for Si.

$$\therefore \quad V_T = \frac{T}{11600} = \frac{300}{11600} = -\,26 \text{ mV}$$

$$I_F = I_o \left(e^{\frac{V_F}{\eta V_T}} - 1 \right)$$

Substituting, we get $10 \times 10^{-3} = I_o \left[e^{\frac{0.75}{2 \times 26 \times 10^{-3}}} - 1 \right] = 5.446$ nA.

1.22 ZENER DIODE

'Zener diode' is a special type of pn junction diode. Its construction is similar to that of a conventional p-n junction diode. But is constructing zener diode, the reverse breakdown voltage is adjusted between 3 V to 200 V. That is done by adjusting the doping level of the impurity.

Circuit symbol :

The circuit symbol of a zener diode is shown in Fig. 1.31. It has two terminals, anode and cathode. The arrow-head in the symbol point towards conventional direction of current through the zener diode, when it is forward biased.

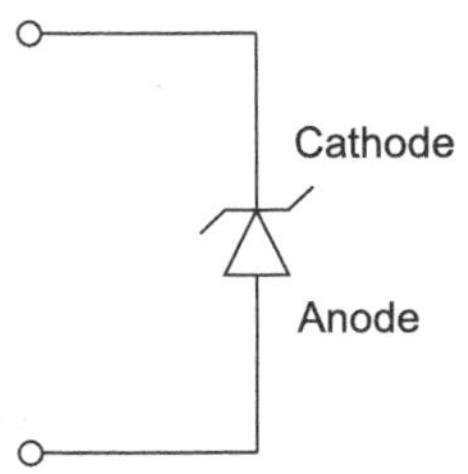

Fig. 1.31 Circuit symbol of zener diode

1.23 FORWARD BIASING OF ZENER DIODE

When the anode of the zener diode is connected to the positive terminal of the battery or external DC source and cathode to the negative terminal, the zener diode is said to be forward biased. Generally, the zener diode is not used in forward biased condition. Forward biasing is shown in Fig. 1.32.

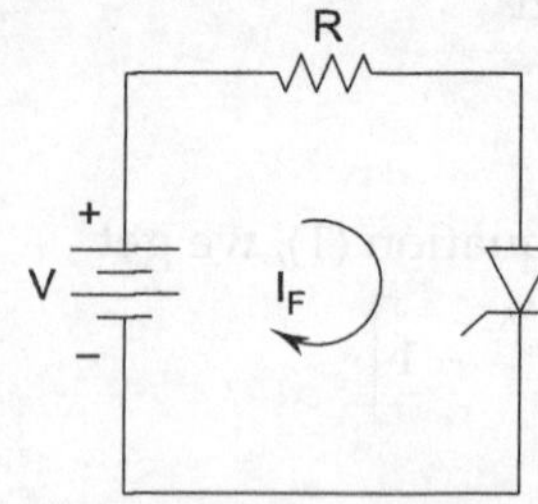

Fig. 1.32 Forward biasing

1.24 REVERSE BIASING OF ZENER DIODE

When the cathode is connected to positive terminal and anode is connected to the negative terminal of the DC source, the zener diode is said to be reverse biased. It is shown in Fig. 1.33. Zener in reverse biased condition is used as a voltage regulator.

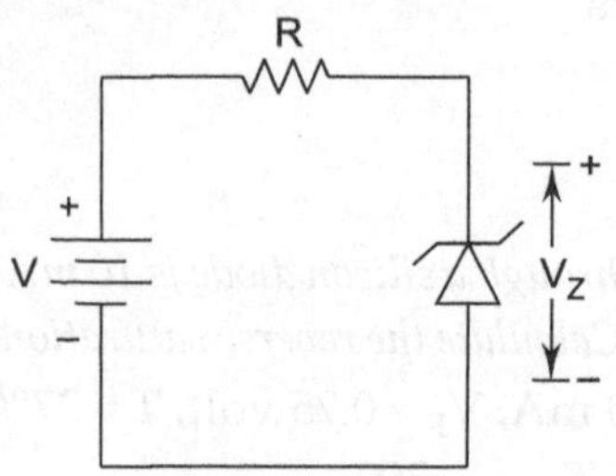

Fig. 1.33 Reverse biasing

1.25 V-I CHARACTERISTICS OF A ZENER DIODE

Forward Characteristics : The forward characteristics of a zener diode is shown in Fig. 1.34. It is almost identical with *pn* diode forward characteristics.

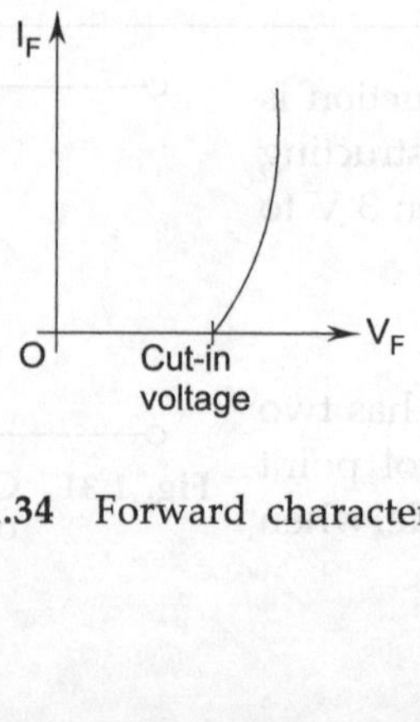

Fig. 1.34 Forward characteristics

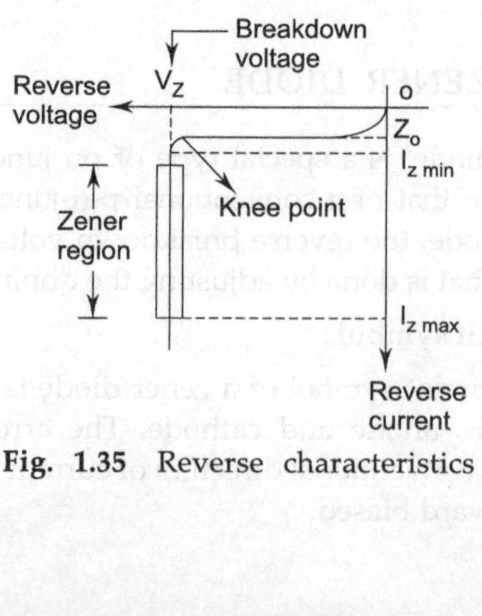

Fig. 1.35 Reverse characteristics

Reverse Characteristics : The reverse characteristics of a zener diode is different from that of *p-n* junction diode. It is shown in Fig. 1.35. As we increase the reverse voltage, initially a small reverse saturation current 'I_o' will flow (in μA). This is due to minority carriers, which are generated thermally. At a certain value of reverse voltage, the reverse current will increase suddenly and sharply. Here the breakdown occurs. This voltage is called as zener breakdown voltage or zener voltage (V_Z). This value V_Z depend on the doping level of impurity. After breakdown, the voltage remains constant equal to V_Z. The zener current can be increased by increasing source voltage. Zener diode is operated in zener region when it is being used as a voltage regulator.

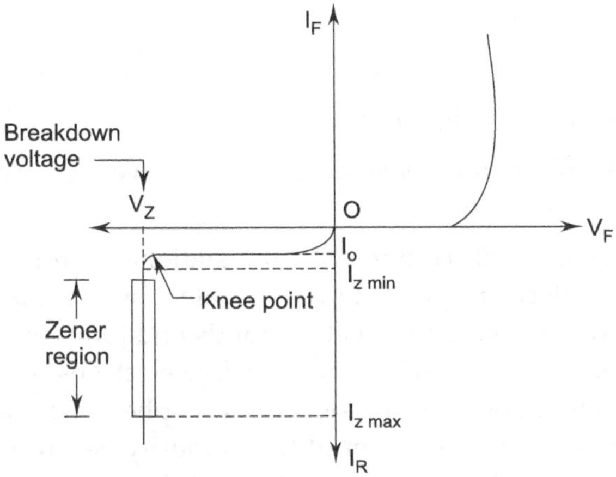

Fig. 1.36 Complete V-I characteristics

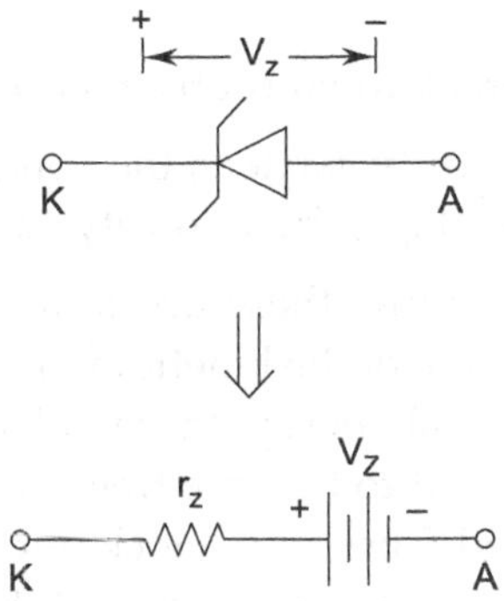

Fig. 1.37 Equivalent circuit of zener diode

Figure 1.37 shows the zener diode is equivalent to a series combination of a small resistor 'r_z' and a battery, 'V_z'.

1.26 BREAKDOWN MECHANISMS IN ZENER DIODE

In zener, there are two different breakdown mechanisms. They are :

- Zener breakdown
- Avalanche breakdown

1.27 ZENER BREAKDOWN

Zener breakdown is observed in the zener diode having V_z less than 5 V or between 5 to 8 volts.

When a reverse voltage (5 V or less) is applied to a zener diode, it causes a very intense electric field to appear across a narrow depletion region. The field intensity will be in order of 3×10^5 V/cm. Such an intense electric field is strong enough to pull some of the valence electron into the conduction band by breaking their covalent bonds. These electrons will become free electrons. A large number of free electrons will constitute a large reverse current through the zener diode and breakdown is said to have occurred due to the "zener effect". In this type of breakdown, breakdown voltage depends on temperature. It decreases with increase in junction temperature. Its characteristics is shown in Fig. 1.38.

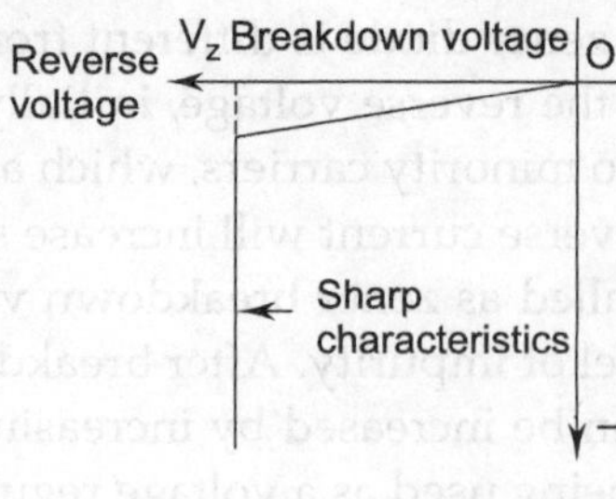

Fig. 1.38 Zener breakdown

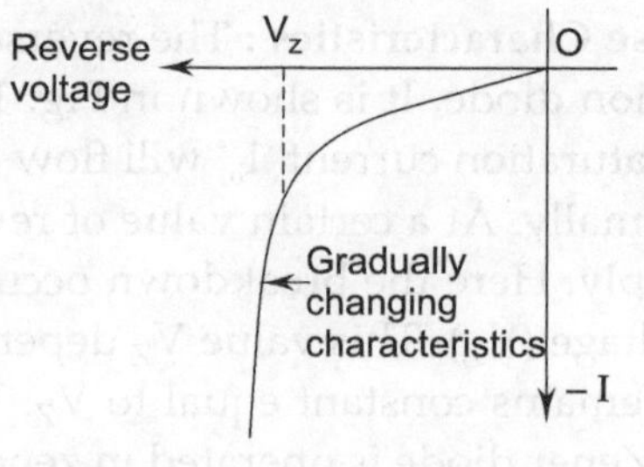

Fig. 1.39 Avalanche breakdown

1.28 AVALANCHE BREAKDOWN

Avalanche breakdown is observed in zener diode having V_z higher than 8 V.

Due to minority carriers, the conductor will take place in reverse biased case. As we increase the reverse voltage, the minority carriers tends to accelerate.

While travelling, these accelerated minority carriers will collide with the stationary atoms, and impart some of the kinetic energy to the valence electrons present in the covalent bonds. Due to this additional energy, these valence electrons will break their covalent bonds and jump into conduction band to become free for conduction. Now these newly generated free electrons will get accelerated. They will knock out some more valence electrons by collision. This phenomenon is called as 'carrier multiplication'. In very short time, a large number of free minority electrons and holes will be available for conduction. This is called "Avalanche effect". A large reverse current starts flowing through zener diode. The breakdown voltage increases with increase in temperature.

1.29 COMPARISON OF ZENER AND AVALANCHE BREAKDOWN

S.No.	Zener Breakdown	Avalanche Breakdown
1.	Observed in zener diode having V_z between 5 to 8 volts	Observed when V_z is greater than 8 volts
2.	Valence electrons are pulled to conduction band due to very intence electric field	Valence electrons are pushed into conduction band due to energy imparted by colliding accelerated minority carriers
3.	Sharp V-I characteristics	Gradually increasing V-I characteristics
4.	Breakdown voltage decreases with temperature increase	Breakdown voltage increases with increase in temperature

1.30 APPLICATION OF ZENER DIODE

1. As a regulated power supply
2. As voltage reference in regulators
3. In clipping circuits
4. In MOSFET protection circuits.

1.31 VOLTAGE REGULATION

A voltage regulator is an electronic circuit that provides a stable DC voltage independent of the load current, temperature and AC line voltage regulations.

Line regulation: Line regulation is defined as the change in output voltage for a change in line supply voltage keeping the load current and temperature constant.

Line regulation is given by

$$\text{Line regulation} = \frac{\text{Change in output voltage}}{\text{Change in input voltage}} = \frac{\Delta V_o}{\Delta V_{in}}$$

Load regulation: Load regulation is expressed as

$$\text{Load regulation} = \frac{\left(V_{\text{no load}} - V_{\text{full load}}\right)}{V_{\text{no load}}}$$

or

$$\text{Load regulation} = \frac{\left(V_{\text{no load}} - V_{\text{full load}}\right)}{V_{\text{full load}}}$$

where $V_{\text{no load}}$ is the output voltage at zero load current and $V_{\text{full load}}$ is the output voltage at rated load current. This is usually denoted in percentage. The plot of the output voltage V_o versus load current I_L for a full wave rectifier is given in figure. The drop in the characteristics is a measure of the internal resistance of the power supply.

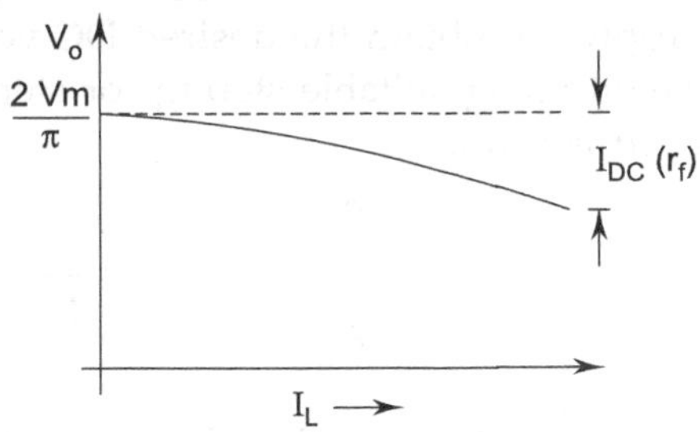

Fig. 1.40 Load regulation characteristics

Zener diode shunt regulator: A zener diode, under reverse bias breakdown condition, can be used to regulate the voltage across a load, irrespective of the supply or load current variations. A simple zener voltage regulator circuit is shown in Fig. 1.41. The zener diode is selected with V_Z equal to the voltage desired across the load. The zener diode has a characteristic that under reverse bias condition, the voltage across it practically remains constant, even if the current through it changes by a large extent. Under normal condition, $I_i = I_L + I_Z$ flows through resistor R. The input voltage V_i can be written as $V_i = I_i R + V_Z = (I_L + I_Z) R + V_Z$.

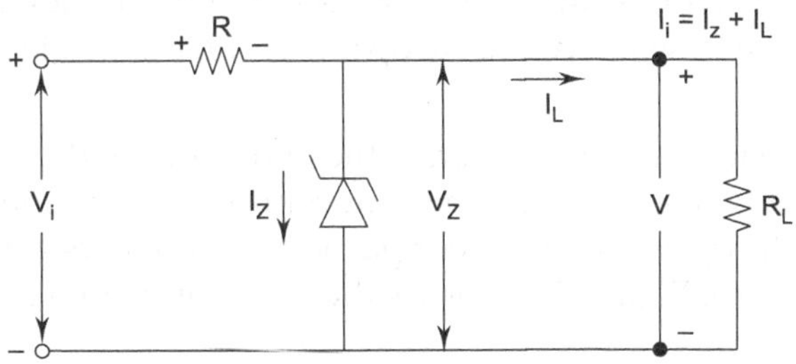

Fig. 1.41 Zener voltage regulator

When the input voltage V_i increases (due to supply voltage variations), as the voltage across zener diode remains constant, the drop across resistor R will increase with a corresponding increases in $I_L + I_Z$. As V_Z is a constant, the voltage across the load will also remain constant and hence, I_L will be a constant. Therefore, an increase in $I_L + I_Z$ will result in an increase in I_Z which will not alter the voltage across the load. It must be ensured that the reverse voltage applied to the zener diode never exceeds PIV of the diode and the same time, the applied input voltage must be greater than the breakdown voltage of the zener diode for its operation.

1.32 RECTIFIERS

Rectifier is defined as an electronic device used for converting AC voltage into unidirectional voltage. A rectifier utilizes unidirectional, conduction devices like a vacuum diode or PN junction diode. Rectifiers are classified depending upon the period of conduction as Half-wave rectifier, Full-wave rectifier and Bridge rectifier.

Half-Wave Rectifier

Half-wave rectifier converts an AC voltage into a pulsating DC voltage using only one half of the applied AC voltage. The rectifying diode conducts only during positive half of the AC cycle. The negative half cycles of AC supply are eliminated from the output.

Figure 1.42 shows the circuit diagram. It consists of resistive load, rectifying element, *i.e.*, *p-n* junction diode, and the source of AC voltage, all connected in series. Usually, the rectifier circuits are operated from AC mains supply. To obtain the desired DC voltage across the load, the AC voltage is applied to rectifier circuit using suitable step-up or step-down transformer, mostly a step-down one, with necessary turns ratio.

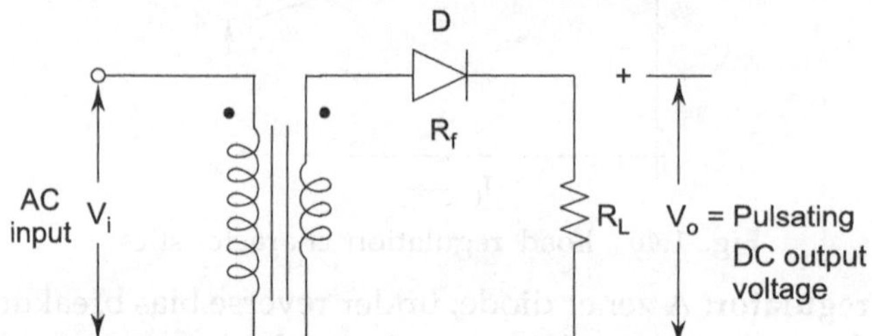

Fig. 1.42 Circuit diagram of a half-wave rectifier

Let V_i be the voltage to the primary of the transformer and given by the equation

$$V_i = V_m \sin \omega t \; V_m >> V_r$$

where V_r is the cut-in voltage of the diode. During the positive half-cycle of the input signal, the anode of the diode becomes more positive with respect to the cathode and hence, diode D conducts. For an ideal diode, the forward voltage drop is zero. So the whole input voltage will appear across the load resistance, R_L.

During negative half-cycle of the input signal, the anode of the diode becomes negative with respect to the cathode and hence, diode D does not conduct. For an ideal diode, the impedance offered by the diode is infinity. So the whole input voltage appears across diode D. Hence, the voltage drop across R_L is zero. Figure 1.43 shows the waveforms of a half-wave rectifier.

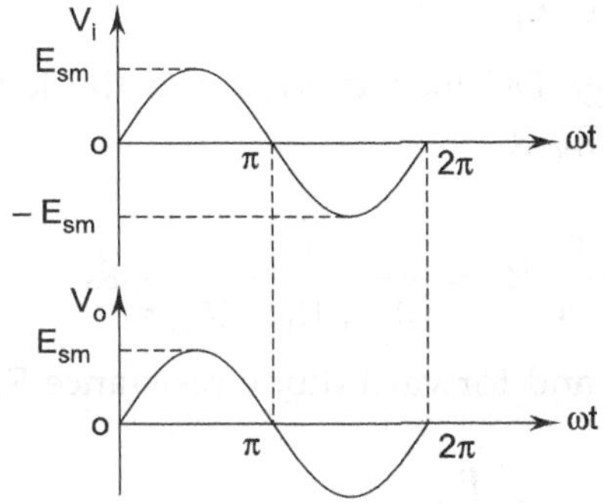

Fig. 1.43 Input and output waveforms of half-wave rectifier

The DC output waveform is expected to be a straight line but the half-wave rectifier gives output in the form of positive sinusoidal pulses. Hence the output is called pulsating DC. It is discontinuous in nature. Hence it is necessary to calculate the average value of load current and average value of output voltage.

(*a*) **Average DC Load Current (I_{DC}) :**

The average or DC value of alternating current is obtained by integration. For finding out the average value of an alternating waveform, we have to determine the area under the curve over one complete cycle *i.e.*, from 0 to 2π and then dividing it by the base *i.e.*, 2π. Mathematically, current waveform can be described as

$$i_L = I_m \sin \omega t \qquad \text{for } 0 \le \omega t \le \pi$$

$$i_L = 0 \qquad \text{for } \pi \le \omega t \le 2\pi$$

where I_m = Peak value of load current

$$\therefore \quad I_{DC} = \frac{1}{2\pi}\int_0^{2\pi} i_L \, d(\omega t) = \frac{1}{2\pi}\int_0^{2\pi} I_m \sin(\omega t)\, d(\omega t)$$

As no current flows during negative half cycle of AC input voltage, *i.e.*, between $\omega t = \pi$ to $\omega t = 2\pi$, we change the limits of integration.

$$I_{DC} = \frac{1}{2\pi}\int_0^{\pi} I_m \sin(\omega t)\, d(\omega t) = \frac{I_m}{2\pi}\left[-\cos(\omega t)\right]_0^{\pi}$$

$$= -\frac{I_m}{2\pi}\left[\cos(\pi) - \cos(0)\right] = -\frac{I_m}{2\pi}\left[-1-1\right] = \frac{I_m}{\pi}$$

$$I_{DC} = \frac{I_m}{\pi} = \text{average value}$$

Applying Kirchhoff's voltage law we can write,

$$I_m = \frac{E_{sm}}{R_f + R_L + R_s}$$

where E_{sm} = the peak value of the secondary AC voltage

R_s = resistance of secondary winding of transformer

R_f = forward resistance of the diode

R_L = load resistance

I_m = peak value of load current.

If R_s is not given, it should be neglected while calculating I_m.

(*b*) **Average DC Load Voltage (E_{DC}) :**

It is the product of average DC load current and the load resistance R_L.

$$E_{DC} = I_{DC} R_L$$

Substituting value of I_{DC},

$$E_{DC} = \frac{I_m}{\pi} R_L = \frac{E_{sm}}{(R_f + R_L + R_s)\pi} R_L$$

The winding resistance R_s and forward diode resistance R_f are practically very small compared to R_L.

$$\therefore \quad E_{DC} = \frac{E_{sm}}{\pi\left[\frac{R_f + R_s}{R_L} + 1\right]}$$

But R_f and R_s are small compared to R_L, $\frac{(R_f + R_s)}{R_L}$ is negligibly small compared to 1. So neglecting it, we get

$$E_{DC} \approx \frac{E_{sm}}{\pi}$$

(*c*) **R.M.S. Value of Load Current (I_{RMS}) :**

The R.M.S. means squaring, finding mean and then finding square root. Hence RMS value of load current can be obtained as

$$I_{RMS} = \sqrt{\frac{1}{2\pi}\int_0^{\pi} (I_m \sin \omega t)^2 \, d(\omega t)}$$

$$= \sqrt{\frac{1}{2\pi}\int_0^{\pi} \left(I_m^2 \sin^2 \omega t\right) d(\omega t)}$$

$$= I_m \sqrt{\frac{1}{2\pi}\int_0^{\pi} \frac{[1 - \cos(2\omega t)]\, d(\omega t)}{2}}$$

$$= I_m \sqrt{\frac{1}{2\pi}\left\{\frac{\omega t}{2} - \frac{\sin(2\omega t)}{4}\right\}_0^{\pi}}$$

$$= I_m \sqrt{\frac{1}{2\pi}\left(\frac{\pi}{2}\right)} \quad \text{as } \sin(2\pi) = \sin(0) = 0$$

$$= \frac{I_m}{2}$$

$$I_{RMS} = \frac{I_m}{2}$$

(*d*) **DC Power Output (P_{DC}) :**

The DC power output can be obtained as,

$$P_{DC} = E_{DC} I_{DC} = I_{DC}^2 R_L$$

$$\text{DC power output} = I_{DC}^2 R_L = \left[\frac{I_m}{\pi}\right]^2 R_L = \frac{I_m^2}{\pi^2} R_L$$

$$P_{DC} = \frac{I_m^2}{\pi^2} R_L$$

where $$I_m = \frac{E_{sm}}{R_f + R_L + R_s}$$

$\therefore$ $$P_{DC} = \frac{E_{sm}^2 R_L}{\pi^2 \left[R_f + R_L + R_s\right]^2}$$

(*e*) **AC Power Input (P_{AC}) :**

The power input taken from the secondary of transformer is the power supplied to three resistances namely load resistance R_L, the diode resistance R_f and winding resistance R_s. The AC power is given by

$$P_{AC} = I_{RMS}^2 \left[R_L + R_f + R_s\right]$$

but $$I_{RMS} = \frac{I_m}{2}$$

$\therefore$ $$P_{AC} = \frac{I_m^2}{4}\left[R_L + R_f + R_s\right]$$

(*f*) **Efficiency (η) :**

The ratio of DC output power to AC input power is known as rectifier efficiency (η).

$$\eta = \frac{\text{DC output power}}{\text{AC input power}} = \frac{P_{DC}}{P_{AC}}$$

$$= \frac{\frac{I_m^2}{\pi^2} R_L}{\frac{I_m^2}{4}\left[R_f + R_L + R_s\right]} = \frac{(4/\pi^2) R_L}{(R_f + R_L + R_s)}$$

$$= \frac{0.406}{1 + \left(\frac{R_f + R_s}{R_L}\right)}$$

If $(R_f + R_s) \ll R_L$ as mentioned earlier, we get the maximum theoretical efficiency of half wave rectifier as

$$\%\,\eta_{max} = 0.406 \times 100 = 40.6\%.$$

(*g*) **Ripple Factor (Γ) :**

The ratio of rms value of AC component to the DC component in the output is known as ripple factor (Γ)

$$\Gamma = \frac{\text{rms value of AC component of output}}{\text{Average or DC value of component of output}}$$

$$= \frac{I_{AC}}{I_{DC}}$$

where $$I_{AC} = \sqrt{I_{RMS}^2 - I_{DC}^2}$$

$$\Gamma = \sqrt{\left(\frac{I_{RMS}}{I_{DC}}\right)^2 - 1}$$

Substituting the value of I_{RMS} and I_{DC}, we get

$$\Gamma = \sqrt{\left(\frac{I_m/2}{I_m/\pi}\right)^2 - 1}$$

$$= \sqrt{\frac{\pi^2}{4} - 1}$$

$$= \sqrt{1.4674}$$

$$\Gamma = 1.211.$$

From this expression it is clear that the amount of AC present in the output is 121% of the DC voltage. So, the half-wave rectifier is not practically useful in converting AC into DC.

(*h*) **Peak Inverse Voltage (PIV) :**

It is defined as the maximum reverse voltage that a diode can withstand without destroying the junction. The peak inverse voltage across a diode is the peak of the negative half cycle. For half-wave rectifier, PIV is E_{sm}.

(*i*) **Transformer Utilization Factor (TUF) :**

The factor which indicates how much is the utilization of the transformer in the circuit is called transformer utilization factor (TUF).

The TUF is defined as the ratio of DC power delivered to the load to the AC power rating of the transformer.

The TUF for half-wave rectifier can be obtained as,

AC power rating of transformer $= E_{RMS}\, I_{RMS}$

$$= \frac{E_{sm}}{\sqrt{2}} \cdot \frac{I_m}{2}$$

$$= \frac{E_{sm}\, I_m}{2\sqrt{2}}$$

DC power delivered to the load $= I_{DC}^2 R_L$

$$= \left(\frac{I_m}{\pi}\right)^2 R_L$$

$$\therefore \quad TUF = \frac{(I_m/\pi)^2 R_L}{\left(E_{sm}\, I_m/2\sqrt{2}\right)}$$

Neglecting the drop across R_f and R_s we can write

$$E_{sm} = I_m R_L$$

$$TUF = \frac{I_m^2}{\pi^2} \cdot \frac{R_L \cdot 2\sqrt{2}}{I_m^2 R_L} = \frac{2\sqrt{2}}{\pi^2} = 0.287$$

The TUF for a half-wave rectifier is 0.287.

The value of TUF is flow which shows that in half-wave circuit, the transformer is not fully utilized.

(*j*) **Form Factor :**

$$\text{Form factor} = \frac{\text{rms value}}{\text{average value}}$$

$$= \frac{I_m/2}{I_m/\pi} = \frac{\pi}{2} = 1.57$$

(*k*) **Peak Factor :**

$$\text{Peak factor} = \frac{\text{peak value}}{\text{rms value}}$$

$$= \frac{I_m}{I_m/2} = 2$$

(*l*) **Voltage Regulation :**

The voltage regulation is the factor which tells us about the change in the DC output voltage as load from no load to full load condition.

if $(V_{DC})_{NL}$ = DC voltage on no load

$(V_{DC})_{FL}$ = DC voltage on full load

then voltage regulation is defined as,

$$\text{Voltage regulation} = \frac{(V_{DC})_{NL} - (V_{DC})_{FL}}{(V_{DC})_{FL}}$$

Less the value of voltage regulation, better is the performance of rectifier circuit.

For a half-wave circuit,

$$(V_{DC})_{NL} = \frac{E_{sm}}{\pi}$$

while

$$(V_{DC})_{FL} = I_{DC} R_L = \frac{I_m}{\pi} R_L = \frac{E_{sm}}{\pi\left[R_f + R_s + R_L\right]} \times R_L$$

$$\% R = \frac{\dfrac{E_{sm}}{\pi} - \dfrac{E_{sm}}{\pi} \cdot \dfrac{R_L}{\left[R_f + R_s + R_L\right]}}{\dfrac{E_{sm}}{\pi} \cdot \dfrac{R_L}{R_f + R_s + R_L}}$$

$$= \frac{1 - \dfrac{R_L}{R_f + R_s + R_L}}{\dfrac{R_L}{R_f + R_s + R_L}} \times 100 = \frac{R_f + R_s}{R_L} \times 100$$

Neglecting winding resistance,

$$\% R = \frac{R_f}{R_L} \times 100, \quad \text{where } R_f = \text{Diode forward resistance.}$$

Example 1.2. *Voltage $V = 250 \cos 75t$ is applied to a half wave rectifier, with $R_L = 8\ k\Omega$. The rectifier may be represented by ideal diode in series with a resistance of 2 kΩ. Calculate,*

(i) I_m *(ii)* I_{RMS} *(iii) DC power* *(iv) AC power*

(v) Efficiency and *(vi) Ripple factor*

Solution. The diode circuit is as shown in Fig. 1.44

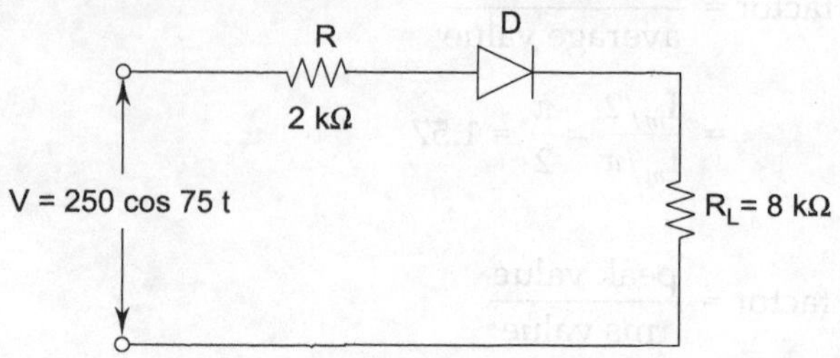

Fig. 1.44

The given voltage $V = 250 \cos 100t$ volts

Compare with $V_o = E_{sm} \sin \omega t$

$E_{sm} = 250$ volts

$R = 2\ k\Omega$ = resistance in series with diode

$R_L = 8\ k\Omega,\ R_s = R_f = 0\ \Omega$

(i)
$$I_m = \frac{E_{sm}}{R + R_L + R_f + R_s}$$

$$= \frac{250}{(2 + 8 + 0 + 0) \times 10^3}$$

$$= \frac{250}{10 \times 10^3}$$

$= 0.025$ A

$= 25$ mA

(ii)
$$I_{RMS} = \frac{I_m}{2}$$

$$= \frac{25 \times 10^{-3}}{2}$$

$= 12.5$ mA

(iii)
$$I_{DC} = \frac{I_m}{\pi} = \frac{25 \times 10^{-3}}{\pi} = 7.958 \text{ mA}$$

$\therefore$ $P_{DC} = I_{DC}^2 R_L$

$= (7.958 \times 10^{-3})^2 \times 8 \times 10^3$

$= 0.507$ watts.

(iv)
$$P_{DC} = I_{RMS}^2 \left(R + R_L + R_s + R_f\right)$$

$= (12.5 \times 10^{-3})^2\ (2 + 8 + 0 + 0) \times 10^3$

$= 1.5625$ watts.

(v) $$\eta = \frac{P_{DC}}{P_{AC}} \times 100$$

$$= \frac{0.507}{1.5625} \times 100$$

$$= 32.448\%.$$

(vi) $$\text{Ripple factor} = \sqrt{\left(\frac{I_{RMS}}{I_{DC}}\right)^2 - 1}$$

$$= \sqrt{\left(\frac{12.5 \times 10^{-3}}{7.958 \times 10^{-3}}\right)^2 - 1}$$

$$= 1.211.$$

1.33 FULL-WAVE RECTIFIER

The full-wave rectifier conducts during both positive and negative half-cycles of input AC supply. In order to rectify both the half cycles of AC input, two diodes are used in this circuit. The diodes feed a common load R_L with the help of a center tap transformer. The AC voltage is applied through a suitable power transformer with proper turns ratio. The full-wave rectifier circuit is shown in Fig. 1.45. For the proper operation of the circuit, a center-tap on the secondary winding of the transformer is essential. Figure 1.46 shows the waveforms of full-wave rectifier.

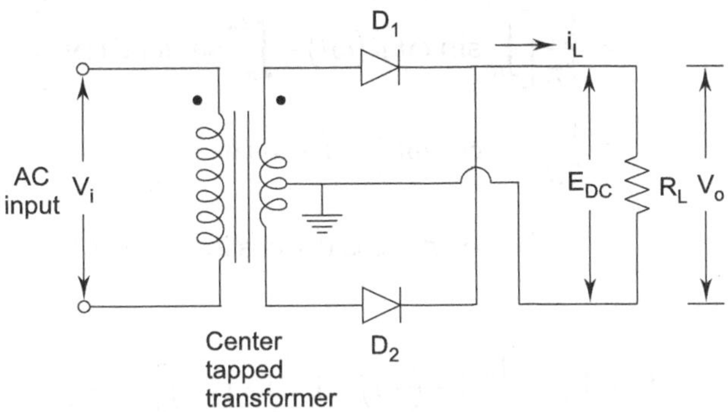

Fig. 1.45 Full-wave rectifier

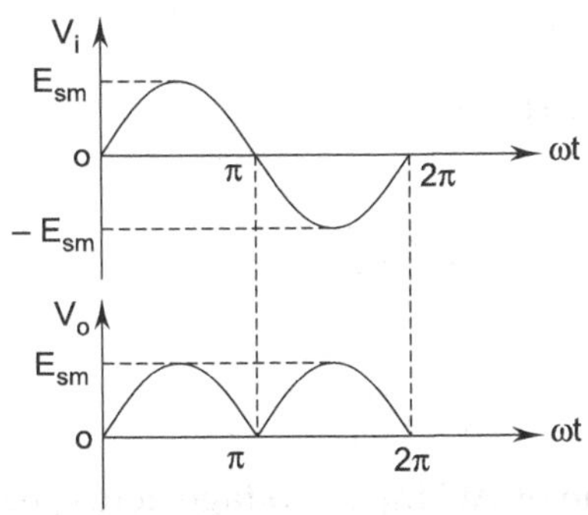

Fig. 1.46 Input and output waveforms of full-wave rectifier

During positive half of the input signal, anode of diode D_1 becomes positive and at the same time the anode of diode D_2 becomes negative. Hence, D_1 conducts and D_2 does not conduct. The load current flows through D_1 and the voltage drop across R_L will be equal to the input voltage.

During negative half-cycle of the input, the anode of D_1 becomes negative and the anode of D_2 becomes positive. Hence, D_1 does not conduct and D_2 conducts. The load current flows through D_2 and the voltage drop across R_L will be equal to the input voltage.

Thus the full wave rectifier circuit essentially consists of two half-wave rectifier circuits working independently (working in alternate half-cycles of AC) of each other but feeding a common load. The output load current is still pulsating DC and not pure DC.

(*a*) **Average DC Load Current (I_{DC}) :**

Consider one cycle of load current i_L from 0 to 2π to obtain the average value which is DC value of load current.

$$i_L = I_m \sin \omega t \quad 0 \leq \omega t \leq \pi$$

But for π to 2π, the current i_L is again positive while $\sin \omega t$ term is negative during π to 2π. Hence in the region π to 2π the positive i_L can be represented as negative of $I_m \sin \omega t$

$$\therefore \quad i_L = -I_m \sin \omega t \quad \pi \leq \omega t \leq 2\pi$$

$$\therefore \quad I_{av} = I_{DC} = \frac{1}{2\pi}\int_0^{2\pi} i_L \, d(\omega t)$$

$$= \frac{1}{2\pi}\left[-\int_0^{\pi} I_m \sin \omega t \, d(\omega t) + \int_{\pi}^{2\pi} -I_m \sin \omega t \, d(\omega t)\right]$$

$$= \frac{I_m}{2\pi}\left[\int_0^{\pi} \sin \omega t \, d(\omega t) - \int_{\pi}^{2\pi} \sin \omega t \, d(\omega t)\right]$$

$$= \frac{I_m}{2\pi}\left[(-\cos \omega t)_0^{\pi} - (-\cos \omega t)_{\pi}^{2\pi}\right]$$

$$= \frac{I_m}{2\pi}\left[-\cos \pi + \cos 0 + \cos 2\pi - \cos \pi\right]$$

but $\qquad \cos \pi = -1$

$$I_{av} = I_{DC} = \frac{I_m}{2\pi}\left[-(-1) + 1 + 1 - (-1)\right] = \frac{4I_m}{2\pi}$$

$$I_{DC} = \frac{2I_m}{\pi}$$

(*b*) **Average DC Load Voltage (E_{DC}) :**

The DC load voltage is,

$$E_{DC} = I_{DC} R_L = \frac{2I_m R_L}{\pi}$$

where $\qquad I_m = \dfrac{E_{sm}}{R_s + R_f + R_L}$

where E_{sm} = maximum value of AC input voltage across each half of secondary winding

R_s = winding resistance of each half of secondary

R_f = forward resistance of diodes

R_L = load resistance

I_m = maximum value of load current i_L

Substituting the value of I_m in the above equation, we get

$$E_{DC} = \frac{2E_{sm} R_L}{\pi\left[R_s + R_f + R_L\right]} = \frac{2E_{sm}}{\pi\left[\frac{R_s + R_f}{R_L} + 1\right]}$$

But as R_f and $R_s << R_L$, hence $\frac{R_f + R_s}{R_L} << 1$

$$E_{DC} = \frac{2E_{sm}}{\pi}$$

(*c*) **RMS Value of Load Current (I_{RMS}) :**

The RMS means squaring, finding mean and then finding square root. Hence RMS value of load current can be obtained as

$$I_{RMS} = \sqrt{\frac{1}{2\pi}\int_0^{2\pi} i_L^2 \, d(\omega t)}$$

Since two half-wave rectifier are similar in operation we can write,

$$I_{RMS} = \sqrt{\frac{2}{2\pi}\int_0^{\pi}\left[I_m \sin \omega t\right]^2 d(\omega t)}$$

$$= I_m\sqrt{\frac{1}{\pi}\int_0^{\pi}\left[\frac{1-\cos 2\omega t}{2}\right] d(\omega t)} \qquad \left[\because \quad \sin^2 \omega t = \frac{1-\cos 2\omega t}{2}\right]$$

$$= I_m\sqrt{\frac{1}{2\pi}\left[(\omega t)_0^{\pi} - \left(\frac{\sin 2\omega t}{2}\right)_0^{\pi}\right]}$$

$$= I_m\sqrt{\frac{1}{2\pi}\left[\pi - 0\right]}$$

$$= I_m\sqrt{\frac{1}{2\pi}(\pi)} \qquad [\because \quad \sin(2\pi) = \sin(0) = 0]$$

$$\therefore \qquad I_{RMS} = \frac{I_m}{\sqrt{2}}.$$

(*d*) **DC Power Output (P_{DC}) :**

The DC power output can be obtained as,

$$P_{DC} = E_{DC} I_{DC} = I_{DC}^2 R_L$$

$$\therefore \qquad P_{DC} = I_{DC}^2 R_L = \left(\frac{2I_m}{\pi}\right)^2 R_L$$

$$\therefore \qquad P_{DC} = \frac{4}{\pi^2} I_m^2 R_L$$

Substituting value of I_m, we get

$$P_{DC} = \frac{4}{\pi^2} \cdot \frac{E_{sm}^2}{(R_s + R_f + R_L)^2} \times R_L$$

(*e*) **AC Power Input (P_{AC}) :**

The AC power input is given by,

$$P_{AC} = I_{RMS}^2 (R_f + R_s + R_L) = \left(\frac{I_m}{\sqrt{2}}\right)^2 (R_f + R_s + R_L)$$

$$\therefore \quad P_{AC} = \frac{I_m^2 (R_f + R_s + R_L)}{2}$$

Substituting value of I_m, we get

$$\therefore \quad P_{AC} = \frac{E_{sm}^2}{(R_f + R_s + R_L)^2} \times \frac{1}{2} \times (R_f + R_s + R_L)$$

$$\therefore \quad P_{AC} = \frac{E_{sm}^2}{2(R_f + R_s + R_L)}$$

Efficiency (η):

$$\eta = \frac{P_{DC} \text{ output}}{P_{AC} \text{ input}}$$

$$\therefore \quad \eta = \frac{\frac{4}{\pi^2} I_m^2 R_L}{\frac{I_m^2 (R_f + R_s + R_L)}{2}}$$

$$\therefore \quad \eta = \frac{8R_L}{\pi^2 (R_f + R_s + R_L)}$$

But if $R_f + R_s << R_L$, neglecting it from denominator

$$\eta = \frac{8R_L}{\pi^2 (R_L)} = \frac{8}{\pi^2}$$

$$\therefore \quad \% \eta_{max} = \frac{8}{\pi^2} \times 100 = 81.2\%.$$

This is the maximum theoretical efficiency of full wave rectifier.

(*f*) **Ripple Factor (γ) :**

Ripple factor, $$\gamma = \sqrt{\left(\frac{I_{RMS}}{I_{DC}}\right)^2 - 1}$$

For full-wave, $I_{RMS} = \frac{I_m}{\sqrt{2}}$ and $I_{DC} = \frac{2I_m}{\pi}$. So, substituting in the above equation,

$$\text{Ripple factor} = \sqrt{\left[\frac{I_m/\sqrt{2}}{2I_m/\pi}\right]^2 - 1}$$

$$= \sqrt{\frac{\pi^2}{8} - 1}$$

∴ Ripple factor, $\gamma = 0.48$

(*g*) **Peak Inverse Voltage (PIV) :**

We know that each diode in a full-wave rectifier, is alternately forward and reverse biased. The maximum value of reverse voltage, which each diode must withstand, is equal to the maximum secondary voltage (*i.e.*, $2E_{sm}$).

∴ Peak inverse voltage of each diode in a center tapped full-wave rectifier,

$$\text{PIV} = 2E_{sm}.$$

(*h*) **Transformer Utilization Factor (TUF) :**

In full-wave rectifier, the secondary current flows through each half separately in every half-cycle. While the primary of transformer carries current continuously. Hence, TUF is calculated for primary and secondary windings separately and then the average TUF is determined. The average TUF for full-wave circuit will be

$$\text{Average TUF for full-wave rectifier circuit} = \frac{\text{TUF of primary} + \text{TUF of secondary}}{2}$$

$$\text{TUF for primary winding} = 2 \times \text{TUF of half wave circuit}$$

$$= 2 \times 0.287$$

$$= 0.574.$$

$$\text{TUF for secondary winding} = \frac{\text{DC power to the load}}{\text{AC power rating of secondary}}$$

$$= \frac{I_{DC}^2 R_L}{E_{RMS}\, I_{RMS}} = \frac{\left(\frac{2}{\pi} I_m\right)^2 R_L}{\frac{E_{sm}}{\sqrt{2}} \times \frac{I_m}{\sqrt{2}}}$$

Neglecting forward resistance R_f of diode, $E_{sm} \approx I_m R_L$

$$\therefore \quad \text{TUF for secondary winding} = \frac{\frac{4}{\pi^2} \times I_m^2 R_L}{\frac{I_m^2 R_L}{2}}$$

$$= \frac{8}{\pi^2}$$

$$= 0.812$$

Therefore, average TUF for full-wave rectifier circuit becomes

$$\therefore \quad \text{Average TUF for full-wave rectifier circuit} = \frac{0.574 + 0.812}{2}$$

∴ Average TUF for full-wave rectifier = 0.693.

Thus in full-wave circuit transformer gets utilized more than the half-wave rectifier circuit.

(*i*) **Form Factor:**

$$\text{Form factor} = \frac{\text{rms value}}{\text{average value}}$$

$$= \frac{I_m/\sqrt{2}}{2I_m/\pi} = \frac{\pi}{2\sqrt{2}} = 1.11$$

(*j*) **Peak Factor:**

$$\text{Peak factor} = \frac{\text{peak value}}{\text{rms value}} = \frac{I_m}{I_m/\sqrt{2}} = \sqrt{2}$$

(*k*) **Voltage Regulation:**

For a full-wave circuit,

$$(V_{DC})_{NL} = \frac{2E_{sm}}{\pi}$$

and $(V_{DC})_{FL} = I_{DC} R_L$

The regulation can be expressed as,

$$\% R = \frac{(V_{DC})_{NL} - (V_{DC})_{FL}}{(V_{DC})_{FL}} \times 100$$

$$\% R = \frac{\frac{2E_{sm}}{\pi} - I_{DC} R_L}{I_{DC} R_L} \times 100$$

Now, $$I_m = \frac{E_{sm}}{R_f + R_L + R_s}$$

$\therefore$ $$E_{sm} = I_m (R_f + R_L + R_s)$$

and $$I_{DC} = \frac{2I_m}{\pi}$$

$$\% R = \frac{\frac{2I_m}{\pi}\left[R_f + R_L + R_s\right] - \frac{2I_m}{\pi} R_L}{\frac{2I_m}{\pi} R_L} \times 100$$

$$= \frac{R_f + R_L + R_s - R_L}{R_L} \times 100$$

$$\% R = \frac{R_f + R_s}{R_L} \times 100$$

Neglecting winding resistance R_s, the regulation can be expressed as

$$\% R \approx \frac{R_f}{R_L} \times 100$$

Advantages

1. The DC output voltage and load current values are twice than those of a half-wave rectifier.
2. The ripple factor is much less (0.482) than that of half-wave rectifier (1.21)

3. The efficiency is twice that of half-wave rectifier. For a full-wave rectifier, the maximum possible value of efficiency is 81.2%, while that of half-wave rectifier is 40.6%.

Disadvantages

1. The output voltage is half of the secondary voltage.
2. The Peak Inverse Voltage (PIV) of a diode is twice that of the diode used in the half-wave rectifier.
3. It is expensive to manufacture a center-tapped transformer.

Example 1.3. *A full-wave rectifier circuit is fed from a transformer having a center-tapped secondary winding. The rms voltage from either end of secondary to center tap is 50 V. If the diode forward resistance is 5 Ω and that of the half secondary is 10 Ω, for a load of 2 kΩ, calculate*

(a) power delivered to load
(b) % regulation at full load
(c) efficiency of rectification
(d) TUF of secondary
(e) form factor
(f) ripple factor

Solution. (*a*) **Power delivered to the load** = $I_{DC}^2 R_L$

$$I_{DC} = \frac{2}{\pi} I_m$$

where $$I_m = \frac{E_{sm}}{R_f + R_L + R_s}$$

where $$E_{sm} = E_{RMS}\sqrt{2}$$

where $$E_{RMS} = 50 \text{ V}$$

$$\therefore \quad E_{sm} = 50\sqrt{2} = 70.7 \text{ V}$$

$$\therefore \quad I_m = \frac{70.7}{5 + 2000 + 10} = 35 \text{ mA}$$

$$\therefore \quad I_{DC} = \frac{2}{\pi} \times 35 \times 10^{-3} = 22.28 \text{ mA}$$

$$\therefore \quad \text{Power delivered to the load} = I_{DC}^2 R_L = \left(22.28 \times 10^{-3}\right)^2 \times 2000 = 0.993 \text{ W.}$$

(*b*) **% Regulation at full load** $= \dfrac{(V_{DC})_{NL} - (V_{DC})_{FL}}{(V_{DC})_{FL}} \times 100$

$$(V_{DC})_{NL} = \frac{2}{\pi} E_{sm} = \frac{2}{\pi} \times 70.7 = 45 \text{ V}$$

$$(V_{DC})_{FL} = I_{DC} R_L = 22.28 \times 10^{-3} \times 2000 = 44.56$$

$$\therefore \quad \% \text{ Regulation} = \frac{45 - 44.56}{44.56} \times 100$$

$$= 0.987\%.$$

(*c*) **Efficiency of rectification** $= \dfrac{P_{DC} \text{ output}}{P_{AC} \text{ input}}$

$$= \frac{8}{\pi^2} \times \frac{1}{1 + \dfrac{R_f + R_s}{R_L}}$$

$$= \frac{8}{\pi^2} \times \frac{1}{1 + \frac{5+10}{2000}}$$

$$= \frac{8}{\pi^2} \times 0.9925$$

$$= 0.804 \text{ i.e., } 80.4\%.$$

(*d*) **Transformer secondary rating,**

$$E_{RMS}\, I_{RMS} = (50\text{ V})\left(\frac{35\, mA}{\sqrt{2}}\right)$$

$$= 1.237 \text{ W}$$

$$\text{TUF} = \frac{\text{DC power output}}{\text{AC rating}} = \frac{0.993}{1.237} = 0.803.$$

(*e*) **Form factor**

$$= \frac{\text{rms value}}{\text{average value}}$$

$$= \frac{I_m/\sqrt{2}}{2I_m/\pi}$$

$$= \frac{\pi}{2\sqrt{2}} = 1.11.$$

(*f*) **Ripple factor**

$$= \sqrt{\frac{\pi^2}{8} - 1}$$

Ripple factor, $\gamma = 0.48$.

1.34 BRIDGE RECTIFIER

The need for a centre tapped transformer in a full-wave rectifier is eliminated in the bridge rectifier. As shown in figure, the bridge rectifier has four diodes connected to form a bridge. The AC input voltage is applied to the diagonally opposite ends of the bridge. The load resistance is connected between the other two ends of the bridge.

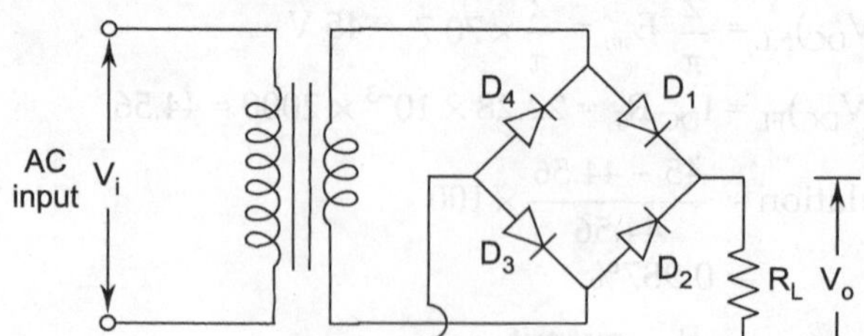

Fig. 1.47 Bridge rectifier

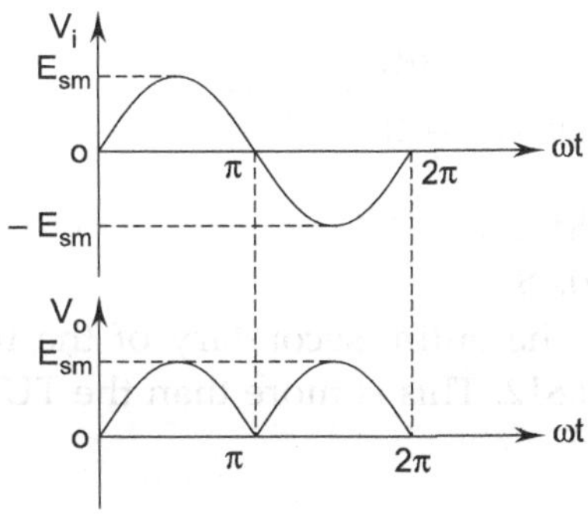

Fig. 1.48 Input and output waveform of bridge rectifier

For the positive half-cycle of the input AC voltage, diodes D_1 and D_3 conduct, whereas diode D_2 and D_4 do not conduct. The conducting diodes will be in series through the load resistance R_L. So the load current flows through R_L.

During the negative half-cycle of the input AC voltage, diodes D_2 and D_4 conduct, whereas diodes D_1 and D_3 do not conduct. The conducting diode D_2 and D_4 will be in series through the load R_L and the current flows through R_L in the same direction as in the previous half-cycle. Thus a bidirectional wave is converted into an unidirectional one.

The bridge rectifier circuit, being basically a full-wave rectifier circuit, all the characteristic discussed previously for a full-wave circuit using two diodes, are the characteristic of a bridge rectifier circuit.

The relation between I_m the maximum value of load current and I_{DC}, I_{RMS} remains same as derived earlier for the full wave rectifier circuit.

$$I_{DC} = \frac{2I_m}{\pi} \text{ and } I_{RMS} = \frac{I_m}{\sqrt{2}}.$$

The expression for I_m will change slightly. In this, the two diodes conduct simultaneously for each half cycle. Hence maximum value of load current is,

$$I_m = \frac{E_{sm}}{R_s + 2R_f + R_L}$$

The remaining expressions identical to those derived for two diode full-wave rectifier and reproduced for the convenience of the reader

$$E_{DC} = I_{DC} R_L = \frac{2E_{sm}}{\pi}$$

$$P_{DC} = I_{DC}^2 R_L = \frac{4}{\pi^2} I_m^2 R_L$$

$$P_{AC} = I_{RMS}^2 \left(R_s + 2R_f + R_L\right)$$

$$= \frac{I_m^2 \left(2R_f + R_s + R_L\right)}{2}$$

$$\eta = \frac{8R_L}{\pi^2 (R_s + 2R_f + R_L)}$$

$$\% \eta_{max} = 81.2\%$$

$$\gamma = 0.48.$$

As the current flows through the entire secondary of the transformer for all the time, the transformer utilization factor is 0.812. This is more than the TUF for full-wave rectifier circuit.

Peak Inverse Voltage (PIV)

The reverse voltage appearing across the reverse biased diodes is $2E_{sm}$ but two diodes are sharing it. Hence PIV rating of the diode is E_{sm} and not $2E_{sm}$ as in case of full-wave rectifier.

Advantages

1. It can be used, with advantage, in applications allowing floating output terminals *i.e.*, no output terminal is grounded.
2. The transformer is less costly as it is required to provide only half the voltage of an equivalent center-tapped transformer used in a full-wave rectifier circuit.
3. No center-tap is required on the transformer.

Disadvantages

The only disadvantage of bridge rectifier is the use of four diodes as compared to two diodes in normal full-wave rectifier. This causes additional voltage drop as indicated by term $2R_f$ present in expression of I_m instead of R_f. This reduces the output voltage.

Example 1.4. *The four semiconductor diodes used in bridge rectifier circuit each having a forward resistance of 0.2 Ω and infinite reverse resistance, feed a DC current of 15 A to a resistive load from a sinusoidally varying alternating supply of 35 V (r.m.s.). Determine the resistance of the load and the efficiency of the circuit.*

Solution. **(*a*) To find the Load resistance, R_L**

$$I_m = \frac{E_{sm}}{2R_f + R_s + R_L}$$

w.r.t., $$I_m = \frac{\pi I_{DC}}{2} = \frac{\pi \times 15}{2} = 23.56 \text{ A}$$

$$E_{sm} = E_{sm} \text{ (r.m.s.)} \times \sqrt{2} = 35 \times \sqrt{2} = 49.497 \text{ V}$$

Substituting in the above equation, we get

$$23.56 = \frac{49.497}{2 \times 0.2 + R_L}$$

∴ $$23.56 = \frac{49.497}{0.4 + R_L}$$

$$9.424 + 23.56 R_L = 49.497$$

$$23.56 R_L = 49.497 - 9.424$$

$$R_L = \frac{40.073}{23.56}$$

$$R_L = 1.7\ \Omega$$

(*b*) **To find efficiency of the circuit**

$$\%\ \eta = \frac{P_{DC}}{P_{AC}} \times 100$$

Now $$P_{DC} = I_{DC}^2\ R_L = (15)^2 \times 1.7$$

$$= 382.5\ \text{W}$$

$$P_{AC} = I_{RMS}^2 \left(2R_f + R_s + R_L\right)$$

and $$I_{RMS} = \frac{I_m}{\sqrt{2}} = \frac{23.56}{\sqrt{2}} = 16.66\ \text{A}$$

$\therefore$ $$P_{AC} = (16.66)^2\ (2 \times 0.2 + 1.7)$$

$$= (277.556)\ (2.1)$$

$$= 582.8676\ \text{W}$$

$\therefore$ $$\%\ \eta = \frac{382.5}{582.8676} \times 100 = 65.62\%.$$

Example 1.5. *A 230 V, 50 Hz voltage is applied to the primary of a 4 : 1 step-down transformer used in a bridge rectifier having a load resistance of 800 Ω. Assuming the diodes to be ideal, determine (a) DC output voltage (b) DC power delivered to the load (c) PIV and (d) output frequency.*

Solution. (*a*) **DC output voltage,** $E_{DC} = \dfrac{2E_{sm}}{\pi}$

$$E_{sm} = E_{RMS} \times \sqrt{2}$$

$$E_{RMS} = \frac{230}{4} = 57.5\ \text{V}$$

$\therefore$ $$E_{sm} = 57.5 \times \sqrt{2} = 81.3\ \text{V}$$

$\therefore$ $$E_{DC} = \frac{2 \times 81.3}{\pi} = 52\ \text{V}$$

(*b*) **DC power delivered to the load**

$$P_{DC} = \frac{E_{DC}^2}{R_L} = \frac{(52)^2}{800} = 3.38\ \text{W}$$

(*c*) **PIV across each diode** $= E_{sm} = 81.3$ V

(*d*) **Output frequency** $= 2 \times 50 = 100$ Hz.

1.35 COMPARISON OF RECTIFIERS

S.No.	Parameter	Half-wave Rectifier	Full-wave Rectifier	Bridge Rectifier
1.	Number of diodes	1	2	4
2.	Average DC Current (I_{DC})	$\frac{I_m}{\pi}$	$\frac{2I_m}{\pi}$	$\frac{2I_m}{\pi}$
3.	Average DC Voltage (E_{DC})	$\frac{E_{sm}}{\pi}$	$\frac{2E_{sm}}{\pi}$	$\frac{2E_{sm}}{\pi}$
4.	RMS Current (I_{RMS})	$\frac{I_m}{2}$	$\frac{I_m}{\sqrt{2}}$	$\frac{I_m}{\sqrt{2}}$
5.	DC Power Output (P_{DC})	$\frac{I_m^2 R_L}{\pi^2}$	$\frac{4}{\pi^2} I_m^2 R_L$	$\frac{4}{\pi^2} I_m^2 R_L$
6.	AC Power Input (P_{AC})	$\frac{I_m^2 (R_L + R_f + R_s)}{4}$	$\frac{I_m^2 (R_f + R_L + R_s)}{2}$	$\frac{I_m^2 (2R_f + R_s + R_L)}{2}$
7.	Efficiency	40.6%	81.2%	81.2%
8.	Ripple factor (γ)	1.21	0.482	0.482
9.	Maximum Load Current (I_m)	$\frac{E_{sm}}{R_s + R_f + R_L}$	$\frac{E_{sm}}{R_s + R_f + R_L}$	$\frac{E_{sm}}{R_s + 2R_f + R_L}$
10.	PIV rating of diode	E_{sm}	$2E_{sm}$	E_{sm}
11.	Output frequency	f	$2f$	$2f$
12.	TUF	0.287	0.693	0.812
13.	Form factor	1.57	1.11	1.11
14.	Peak factor	2	$\sqrt{2}$	$\sqrt{2}$

2

TRANSISTORS AND AMPLIFIERS (BIPOLAR JUNCTION TRANSISTOR)

2.1 INTRODUCTION TO BJT

A Bipolar Junction Transistor (BJT) is a three terminal semiconductor device in which the operation depends on the interaction of both majority and minority carriers and hence the name Bipolar. The BJT is analogous to a vacuum triode and is comparatively smaller in size. It is used in amplifier and oscillator circuits, and as a switch in digital circuits. It has wide applications in computers, satellites and other modern communication systems.

The BJTs are of two types :

1. *n-p-n* type
2. *p-n-p* type

Construction

When a transistor is formed by sandwiching a single *p*-region between two *n*-regions, as shown in Fig. 2.1(*a*), it is an *n-p-n* type transistor. The *p-n-p* type transistor has a single *n*-region between two *p*-regions, as shown in Fig. 2.1(*b*).

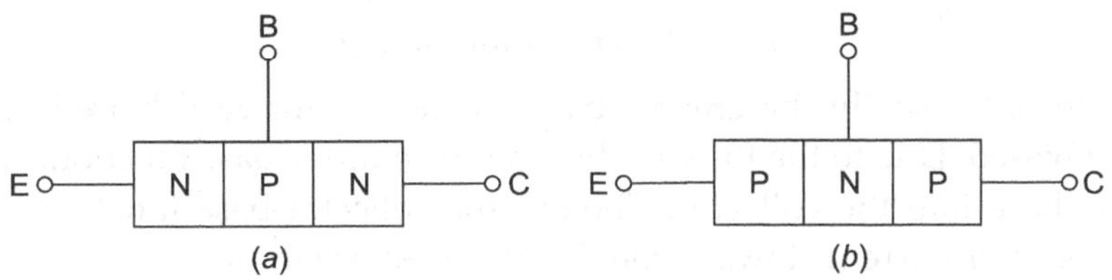

Fig. 2.1 Bipolar transistor construction

The middle region of each transistor type is called the base of the transistor. This region is very thin and lightly doped. The remaining two regions are called emitter and collector. The emitter and collector are heavily doped. But the doping level in emitter is slightly greater than that of collector and the collector region-area is slightly more than that of emitter.

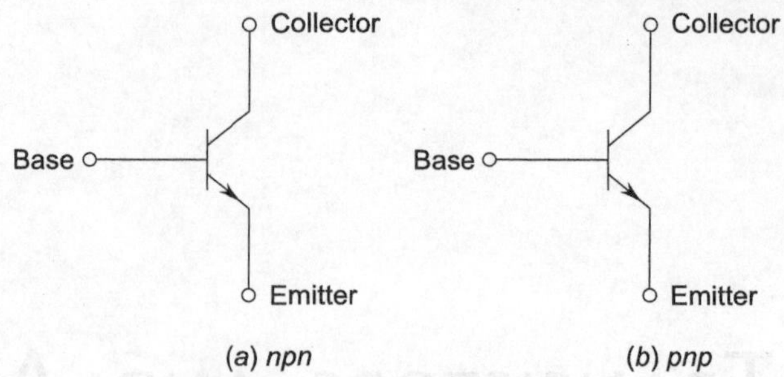

Fig. 2.2 Transistor symbols

Figures 2.2(*a*) and (*b*) shows the symbols of *npn* and *pnp* transistors. A transistor has two *p-n* junctions. One junction is between the emitter and the base, and is called the emitter-base junction, or simply the emitter junction. The other junction is between the base and the collector, and is called collector-base junction, or simply collector junction. Thus, transistor is like two *pn* junction diodes connected back to back as shown in Figs 2.3(*a*) and (*b*).

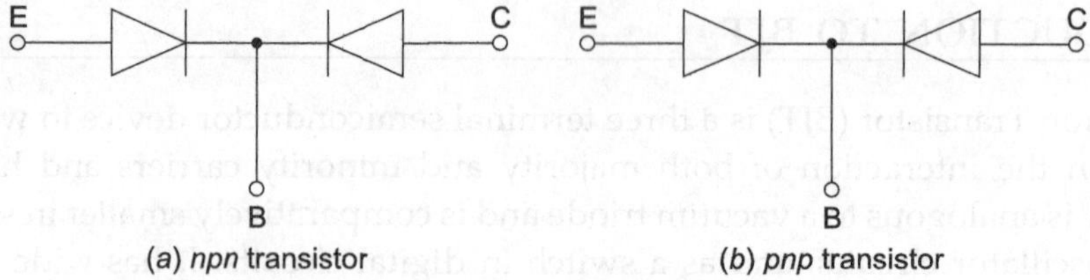

Fig. 2.3 Two-diode transistor analogy

Transistor Biasing:

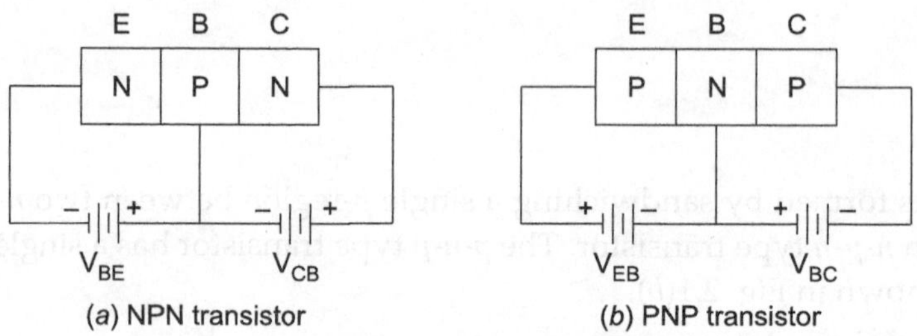

Fig. 2.4 Transistor biasing

As shown in Fig. 2.4, usually the emitter base junction is forward biased and collector-base junction is reverse biased. Due to the forward bias on the emitter-base junction, an emitter current flows through the base into the collector. Though the collector-base junction is reverse biased, almost the entire emitter current flows through the collector circuit.

2.2 WORKING OF NPN AND PNP TRANSISTOR

Working of NPN Transistor:

As shown in Fig. 2.5, the forward bias applied to the emitter base junction of an NPN transistor causes a lot of electrons from the emitter region to crossover to the base region. As the base is lightly doped with *p*-type impurity, the number of holes in the base region is very small and hence the number of electrons that combine with holes in the *p*-type base region is also very small. Hence a few electrons combine with holes to constitute a base current I_B. The remaining electrons (more than 95%) crossover into the collector region to constitute a collector current I_C. Thus the base and collector current summed up gives the emitter current, *i.e.*, $I_E = -(I_C + I_B)$.

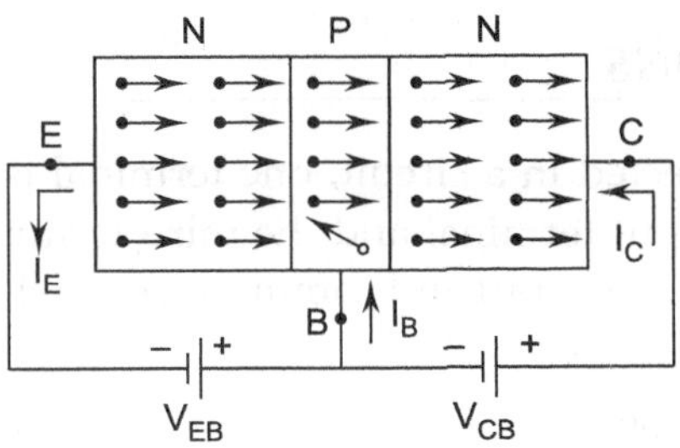

Fig. 2.5 Current in NPN transistor

In the external circuit of the NPN bipolar junction transistor, the magnitudes of the emitter current I_E, the base current I_B and the collector current I_C are related by

$$I_E = I_C + I_B$$

Since the base current is very small, therefore,

$$I_E \approx I_C$$

Working of PNP Transistor:

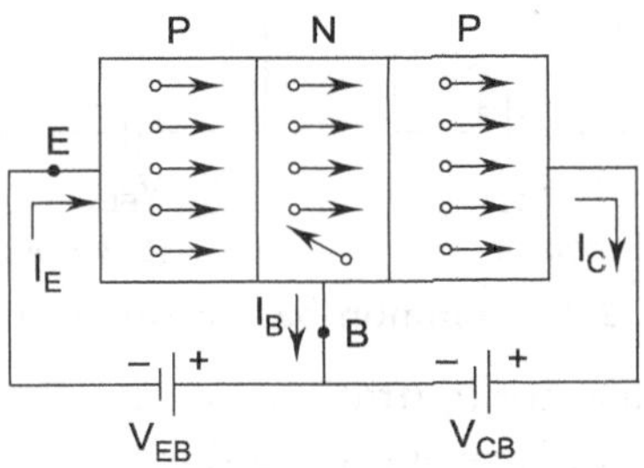

Fig. 2.6 Current in PNP transistor

As shown in Fig. 2.6, the forward bias applied to the emitter-base junction of a PNP transistor causes a lot of holes from the emitter region to crossover to the base region as the base is lightly doped with N-type impurity. The number of electrons in the base region is very small and hence the number of holes combined with electrons in the N-type base region is also very small. Hence a few holes combined with electrons to constitute a base current I_B. The remaining holes (more than 95%) crossover into the collector region to constitute a collector current I_C. Thus the collector and base current when summed up gives the emitter current, *i.e.*, $I_E = -(I_C + I_B)$.

In the external circuit of the PNP bipolar junction transistor, the magnitude of the emitter current I_E, the base current I_B and the collector current I_C are related by

$$I_E = I_C + I_B$$

Example 2.1. *In a certain transistor, the emitter current is 15 mA and its collector current is 14.8 mA. Calculate the value of the base current.*

Solution. Given $I_E = 15$ mA

and $I_C = 14.8$ mA

w.r.t., $I_E = I_C + I_B$

Substituting, we get

$$15 = 14.8 + I_B$$

$$\therefore \quad I_B = 15 - 14.8 = 0.2 \text{ mA.}$$

2.3 BJT CONFIGURATIONS

When a transistor is to be connected in a circuit, one terminal is used as an input terminal, the other terminal is used as an output terminal and the third terminal is common to the input and output. Depending upon the input, output and common terminal, a transistor can be connected in three configurations. They are

1. Common base configuration
2. Common emitter configuration
3. Common collector configuration

1. Common Base Configuration

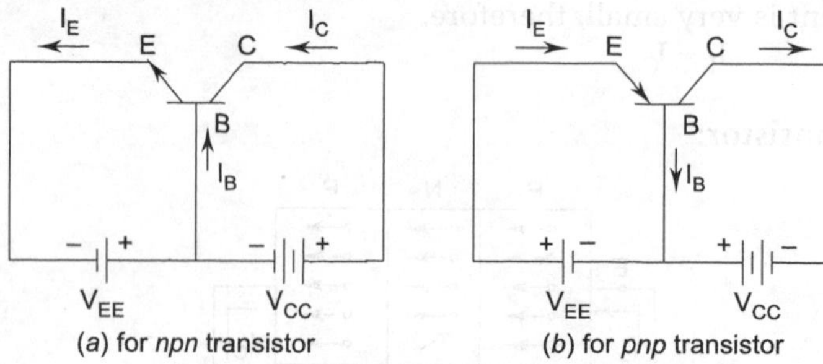

Fig. 2.7 Common base configuration

This is also called grounded base configuration. As shown in Fig. 2.7, in this configuration, input is applied between emitter and base and output is taken from the collector and base. Here base of the transistor is common to both input and output circuits and hence the name common base configuration. Common base configurations for both *npn* and *pnp* transistors are shown in Figs. 2.7(*a*) and (*b*) respectively.

Common Base Characteristics

To understand complete electrical behaviour of a transistor it is necessary to study the interrelation of the various currents and voltages. These relationships can be plotted graphically which are commonly known as the characteristics of transistor. The most important characteristics of transistor in any configuration are input and output characteristics.

Input Characteristics

To determine the input characteristics, the collector-base voltage V_{CB} is kept constant at zero volt and the emitter current I_E is increased from zero in suitable equal steps by increasing V_{EB}. This is repeated for higher fixed values of V_{CB}. A curve is drawn between emitter current I_E and emitter-base voltage V_{EB} at constant collector-base voltage V_{CB}. The input characteristics thus obtained are shown in Fig. 2.8.

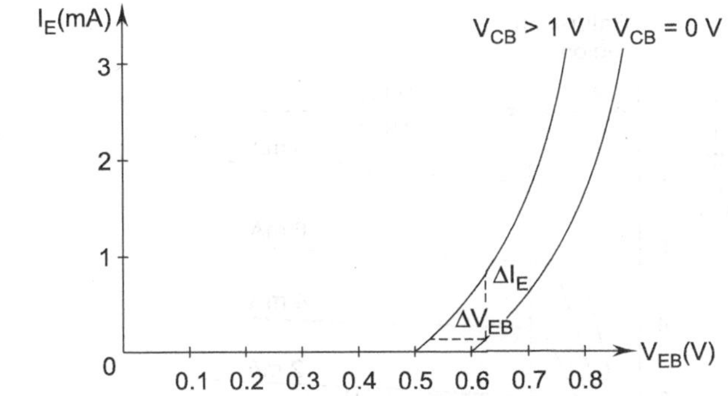

Fig. 2.8 Input characteristics of transistor in CB configuration

When V_{CB} is equal to zero and the emitter-base junction is forward biased as shown in the characteristics, the junction behaves as a forward biased diode. So that emitter current I_E increases rapidly with small increase in emitter-base voltage V_{EB}. When V_{CB} is increased keeping V_{EB} constant, the width of the base region will decrease. This effect results in an increase of I_E. Therefore, the curves shift towards the left as V_{CB} is increased. The input characteristics may be used to determine the value of dynamic input resistance. Its value at any point on the curve is given by the ratio of a change in emitter to base voltage (ΔV_{EB}) to the resulting change in emitter current (ΔI_E) for a constant collector to base voltage (V_{CB}). Mathematically, the dynamic input resistance,

$$R_i = \frac{\Delta V_{EB}}{\Delta I_E}$$

It may be noted that input characteristic is linear in the upper region, but non-linear (*i.e.*, curved) in the lower region. Therefore, the dynamic input resistance depends upon the location of the operating point selected along the curve. Its value, in the linear region of the curve, is about 50 ohms.

Early effect or base width modulation

As the collector voltage V_{CC} is made to increase the reverse bias, the space charge width between collector and base tends to increase, with the result that the effective width of the base decreases. This dependency of base-width on collector to emitter voltage is known as the early effect. This decrease in effective base-width has three consequences :

(*i*) There is less chance for recombination within the base region. Hence, the current gain increases with increasing $|V_{CB}|$.

(*ii*) The charge gradient is increased within the base, and consequently, the current of minority carriers injected across the emitter junction increases.

(*iii*) For extremely large voltages, the effective base-width may be reduced to zero, causing voltage breakdown in the transistor. This phenomenon is called the punch through.

Output Characteristics

It is the curve between collector current I_C and collector base voltage V_{CB} at constant emitter current I_E. The collector current is taken along Y-axis and collector-base voltage magnitude along X-axis. Figure 2.9 shows the output characteristics of a typical transistor in common base configuration.

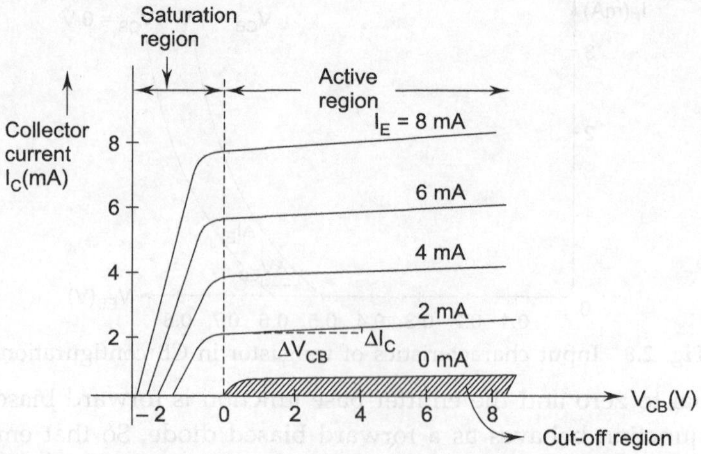

Fig. 2.9 Output characteristics of transistor in CB configuration

From this characteristics, we observe following points :

1. The output characteristics has three basic regions. Active, cutoff and saturation.

State	Emitter base junction	Collector base junction
Active	Forward biased	Reverse biased
Cut-off	Reverse biased	Reverse biased
Saturation	Forward biased	Forward biased

2. For the operation in the active region, the emitter base junction is forward biased while collector base junction is reverse biased. In this region, collector current I_C is approximately equal to the emitter current (I_E) and transistor works as an amplifier.
3. A small collector current flows even when emitter current (I_E) is zero. This current is called collector leakage current and is designated as I_{CBO}. The region below the curve $I_E = 0$ is known as cut-off region, where the collector current is nearly zero and the collector-base and emitter-base junctions of a transistor are reverse biased.
4. The saturation region is that region of the characteristic which is to the left of $V_{CB} = 0$. It may be noted that in this region, collector to base voltage (V_{CB}) is negative for a NPN transistor. It means that collector-base junction of a transistor is also forward biased in the saturation region. In this region, a small change in V_{CB} results in a large value of current.
5. In the active region, the collector current is essentially almost constant, and the graph is almost parallel to X-axis. The collector current I_C is almost independent on collector base voltage V_{CB} and the transistor can be said to work as constant-current source. This provides very high dynamic output resistance. Dynamic output resistance is the ratio of change in collector-base voltage (ΔV_{CB}) to the resulting change in collector current (ΔI_C) at constant emitter current (I_E). Its typical value is about 500 kΩ (very high)

$$R_o = \left.\frac{\Delta V_{CB}}{\Delta I_C}\right|_{I_E = \text{constant}}$$

6. As I_E increases I_C also increases. Thus, I_C depends upon Input current I_E but not on collector voltage. Hence, input current controls output current. Since transistor requires some current to device it, it is called current operating device.

7. The current gain of transistor in CB configuration is given as

$$\alpha_{DC} = \frac{I_C}{I_B}$$

The value of α_{DC} is less than unity. It ranges from 0.95 to 0.995 (typical value is 0.98). If we take small change in I_C, ΔI_C, to small change I_E, ΔIE, we get

$$\alpha_{AC} = \frac{\Delta I_C}{\Delta I_E}$$

2. Common Emitter Configuration

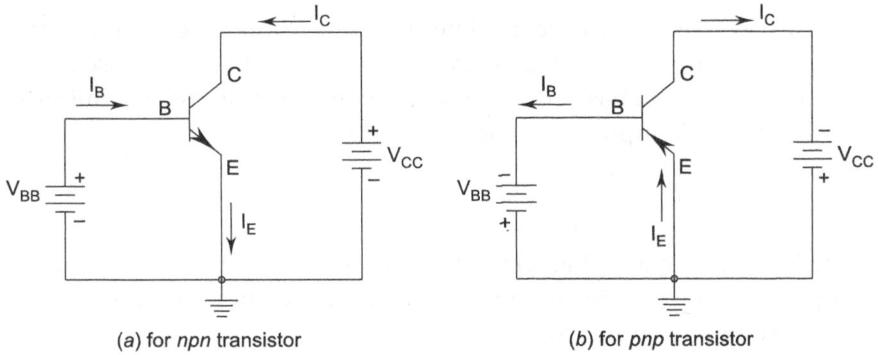

Fig. 2.10 Common emitter configuration

This is also called grounded emitter configuration.

As shown in Fig. 2.10, in this configuration, input is applied between base and emitter and output is taken from collector and emitter. Here, emitter of the transistor is common to both, input and output circuits, and hence the name common emitter configuration. Common emitter configurations for both *npn* and *pnp* transistors are shown in Figs. 2.10(*a*) and (*b*) respectively.

Common Emitter Characteristics

Input Characteristics

It is the curve between input current I_B (base current and input voltage V_{BE} (base-emitter voltage) at constant collector-emitter voltage, V_{CE}. The base current is taken along Y-axis and base emitter voltage V_{BE} is taken along X-axis. Figure 2.11 shows the input characteristics of a typical transistor in common-emitter configuration.

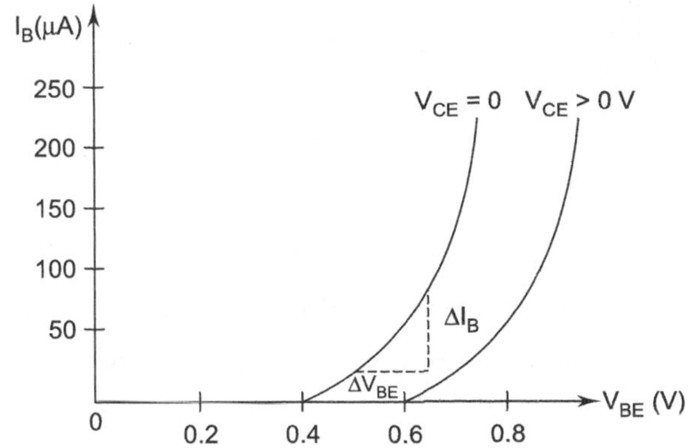

Fig. 2.11 Input characteristics of transistor in CE configuration

To determine the input characteristics, the collector to emitter voltage is kept constant at zero volt and base current is increased from zero in equal steps by increasing V_{BE}. The value of the V_{BE} is noted for each setting of I_B. This procedure is repeated for higher fixed values of V_{CE}, and the curves of I_B versus V_{BE} are drawn.

When $V_{CE} = 0$, the emitter-base junction is forward biased and the junction behaves as a forward biased diode. Hence the input characteristic for $V_{CE} = 0$ is similar to that of a forward biased diode. When V_{CE} is increased, the width of the depletion region at the reverse biased collector-base junction will increase. Hence the effective width of the base will decrease. This effect cause a decrease in the base current I_B. Hence to get the same value of I_B as that for $V_{CE} = 0$, V_{BE} should be increased. Therefore, the curve shifts to the right as V_{CE} increases.

The input characteristic may be used to determine the value of common-emitter transistor dynamic input resistance (R_i). Its value is given by the ratio of change in base to emitter voltage (ΔV_{BE}) to the resulting change in base current (I_B) at a constant collector to emitter voltage (V_{CE}). Mathematically, the dynamic input resistance,

$$R_i = \frac{\Delta V_{BE}}{\Delta I_B}$$

It may be noted that the input characteristic is not linear in the lower region of the curve. Therefore, the input resistance varies with the location of the operating point. The value of dynamic input resistance ranges from 600 Ω to 4000 Ω.

Output Characteristics

This characteristics shows the relation between the collector current I_C and collector voltage, V_{CE}, for various fixed values of I_B. Figure 2.12 shows the output characteristics of a typical transistor in common emitter configuration.

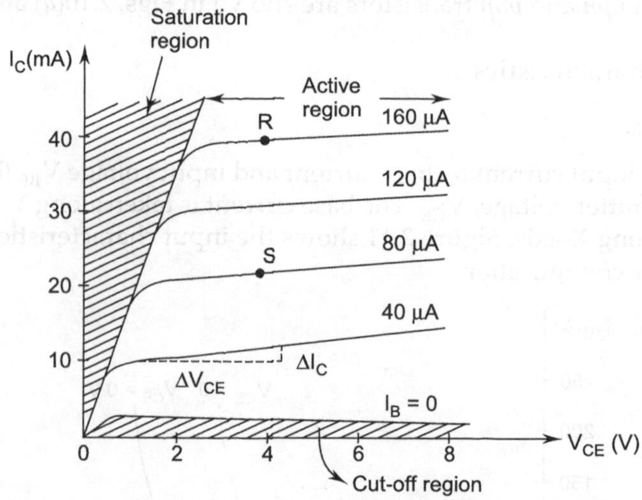

Fig. 2.12 Output characteristics of transistor in CE configuration.

The output characteristics give us the information about the following important points.

1. The output characteristics may be divided into three important regions — Active region, Saturation region and Cut-off region.

(i) **Active region :** The region where the curves are approximately horizontal is the "active" region of the CE configuration. In the active region, the collector region is reverse biased. As V_{CE} is increased, reverse bias increases. This causes depletion region to spread more in base than in collector, reducing the changes of recombinations in the base. This increases the value of α_{DC}. This early effect causes collector current to rise more sharply with increasing V_{CE} in the linear region of output characteristics of CE transistor.

(ii) **Saturation region :** If V_{CE} is reduced to a small value such as 0.2 V, then collector-base junction becomes forward biased, since the emitter base junction is already forward biased by 0.7 V. The input junction in CE configuration is base to emitter junction, which is always forward biased to operate transistor in active region. Thus, input characteristics of CE configuration is similar to forward characteristics of *p-n* junction diode. When both the junctions are forward biased, the transistor operates in the saturation region, which is indicated on the output characteristics. The saturation value of V_{CE}, designated $V_{CE\,(sat)}$, usually ranges between 0.1 V to 0.3 V.

(iii) **Cut-off region :** When the input base current is made equal to zero, the collector current is the reverse leakage current I_{CEO}. The region below $I_B = 0$ is the cut-off region of operation for the transistor. In this region, both the junctions of the transistor are reverse biased.

2. In the active region, the collector base junction is reverse biased. For every transistor there is limit on the maximum value for this reverse bias voltage. If this limit is exceeded, the breakdown occurs in the transistor. This effect is commonly known as punch through effect. This large current may damage transistor. Hence, in practice, maximum collector emitter voltage rating, $V_{CE\,(max)}$ should never be exceeded for the safe operation of the transistor.
3. From the output characteristics we can see that change in collector-emitter voltage (ΔV_{CE}) causes the little change in the collector current (ΔI_C) for constant base current I_B. Thus, the output dynamic resistance is high in CE configuration.

$$R_o = \left.\frac{\Delta V_{CE}}{\Delta I_C}\right|_{I_B = \text{constant}}$$

4. The characteristic may be used to determine the common emitter current gain or DC beta (β_{DC}) of a transistor. This can be done by selecting two points R and S on the characteristics and note down the corresponding valued of I_C and I_B

$$\beta_{DC} = \frac{I_C}{I_B}$$

Relationship between α_{DC} and β_{DC}

w.r.t.,
$$\beta_{DC} = \frac{I_C}{I_B} \qquad \text{...(1)}$$

or
$$\alpha_{DC} = \frac{I_C}{I_E} \qquad \text{...(2)}$$

Now, $I_E = I_C + I_B$

or $I_B = I_E - I_C$

Substituting the value of I_B in equation (1), we get

$$\beta_{DC} = \frac{I_C}{I_E - I_C} \qquad \text{...(3)}$$

Dividing the numerator and denominator of R.H.S. of equation (3) by I_E, we get

$$\beta_{DC} = \frac{I_C/I_E}{I_E/I_E - I_C/I_E}$$

w.r.t., $\alpha_{DC} = \frac{I_C}{I_B}$ from equation (2)

$$\therefore \qquad \beta_{DC} = \frac{\alpha_{DC}}{1-\alpha_{DC}} \qquad \text{...(4)}$$

dividing the R.H.S. and L.H.S. by $1+\beta_{dc}$, we get

$$\frac{\beta_{DC}}{1+\beta_{DC}} = \frac{\frac{\alpha_{DC}}{1-\alpha_{DC}}}{1+\beta_{DC}}$$

By substituting value of β_{DC} from equation (4) in R.H.S., we get

$$\frac{\beta_{DC}}{1+\beta_{DC}} = \frac{\frac{\alpha_{DC}}{1-\alpha_{DC}}}{1+\frac{\alpha_{DC}}{1-\alpha_{DC}}}$$

$$\frac{\beta_{DC}}{1+\beta_{DC}} = \frac{\frac{\alpha_{DC}}{1-\alpha_{DC}}}{\frac{1-\alpha_{DC}+\alpha_{DC}}{1-\alpha_{DC}}}$$

Cancelling common denominator terms, we get

$$\frac{\beta_{DC}}{1+\beta_{DC}} = \frac{\alpha_{DC}}{1-\alpha_{DC}+\alpha_{DC}} = \alpha_{DC}$$

$$\therefore \qquad \alpha_{DC} = \frac{\beta_{DC}}{1+\beta_{DC}}$$

3. Common Collector Configuration

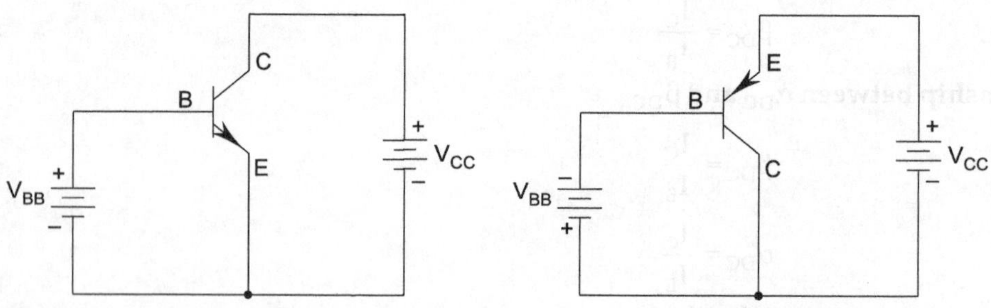

(*a*) for *npn* transistor (*b*) for *pnp* transistor

Fig. 2.13 Common collector configuration

This is also called grounded collector configuration. As shown in Fig. 2.13, in this configuration input is applied between base and collector, and output is taken from emitter and collector. Here, collector of the transistor is common to both input and output circuits, and hence the name common collector configuration. Common collector connections for both *npn* and *pnp* transistors are shown in Fig. 2.13(*a*) and (*b*) respectively.

The common collector configuration is same as the common emitter configuration, with the exception that the load resistance is in the emitter circuit rather than in the collector circuit and the output is taken from emitter lead instead of the collector.

Common Collector Characteristics

Input characteristics

It is the graph of input current I_B (base current) versus input voltage V_{CB} (collector base voltage) at constant V_{CE}. The base current is taken along Y-axis and collector base voltage V_{CB} is taken along X-axis. Figure 2.14 shows the input characteristics of a typical transistor in common-collector configuration.

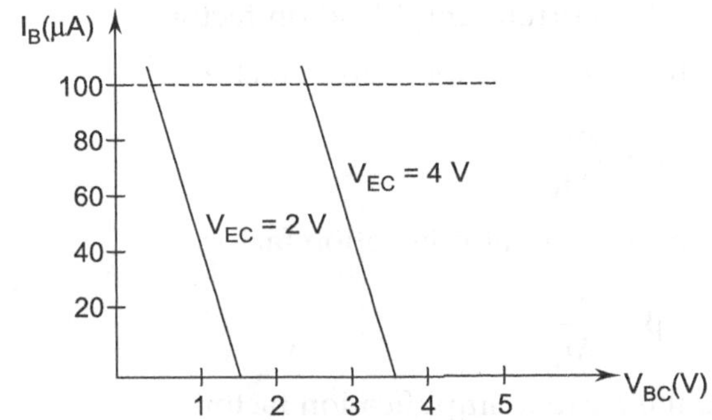

Fig. 2.14 Input characteristics of transistor in CC configuration

To determine the input characteristics, V_{EC} is kept at a suitable fixed value. The base-collector voltage V_{BC} is increased in equal steps and the corresponding increase in I_B is noted. This is repeated for different fixed values of V_{EC}. This configuration has very high input resistance (R_i).

Output characteristics

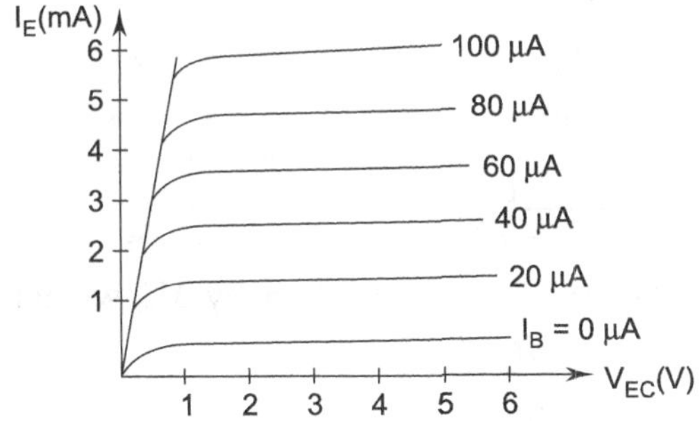

Fig. 2.15 Output characteristics of transistor in CC configuration

It is the curve between emitter current I_E and collector to emitter voltage V_{CE} at constant base current I_B. The emitter current is taken along Y-axis and collector to emitter voltage along X-axis. Figure 2.15 shows the output characteristics of a typical transistor in common collector configuration.

Since, I_C is approximately equal to I_E, the common collector output characteristics are practically similar to those of the common emitter output characteristics.

This characteristic may be used to determine the common collector current gain or DC gamma (γ_{DC}) of a transistor. The value of γ_{DC} of the transistor can be found at any point on the characteristics by taking the ratio of I_E to I_B at that point, *i.e.*, $\gamma_{DC} = \frac{I_E}{I_B}$. This is known as DC gamma for the transistor.

The common collector configuration has very low output impedance. This configuration is commonly used for impedance matching. As its current gain is high and voltage gain is nearly unity it is also commonly used as a buffer.

Current Amplification Factor

In a transistor amplifier with AC input signal, the ratio of change in output current to the change in input current is known as the current amplification factor.

In the CB configuration, the current amplification factor

$$\alpha = \frac{\Delta I_C}{\Delta I_E}$$

In the CE configuration, the current amplification factor,

$$\beta = \frac{\Delta I_C}{\Delta I_B}$$

In the CC configuration, the current amplification factor,

$$\gamma = \frac{\Delta I_E}{\Delta I_B}$$

Relation among α, β and γ

In the CC transistor amplifier circuit, I_B is the input current and I_E is the output current.

w.r.t., $$\gamma = \frac{\Delta I_E}{\Delta I_B}$$

Substituting, $\Delta I_B = \Delta I_E - \Delta I_C$, we get

$$\gamma = \frac{\Delta I_E}{\Delta I_E - \Delta I_C}$$

Dividing the numerator and denominator on R.H.S. by ΔI_E, we get

$$\gamma = \frac{\frac{\Delta I_E}{\Delta I_E}}{\frac{\Delta I_E}{\Delta I_E} - \frac{\Delta I_C}{\Delta I_E}} = \frac{1}{1-\alpha}$$

Therefore, $$\gamma = \frac{1}{1-\alpha} = (\beta + 1)$$

Comparison of Transistor Configurations

S.No.	Characteristic	Common base	Common emitter	Common collector
1.	Input resistance	Very low	Low	High
2.	Output resistance	Very high	High	Low
3.	Current gain	1	High	High
4.	Voltage gain	About 150	About 500	Less than 1
5.	Phase shift between input and output voltages	0° or 360°	180°	0° or 360°
6.	Input current	I_E	I_B	I_B
7.	Output current	I_C	I_C	I_E
8.	Input voltage applied between	Emitter and base	Base and emitter	Base and collector
9.	Output voltage taken between	Collector and base	Collector and emitter	Emitter and collector
10.	Current amplification factor	$\alpha_{DC} = \frac{I_C}{I_E}$	$\beta_{DC} = \frac{I_C}{I_B}$	$\gamma = \frac{I_E}{I_B}$
11.	Applications	As a input stage of multistage amplifier	For audio signal amplifications	For impedance matching

2.4 BIASING CIRCUITS

In order to produce distortion – free output in amplifier circuits, the supply voltages and resistances in the circuit must be suitably chosen. These voltages and resistances establish a set of DC voltage V_{CEQ} and current I_{CQ} to operate the transistor in the active region. These voltages and currents are called quiescent values which determine the operating point or Q-point for the transistor. The process of giving proper supply voltages and resistances for obtaining the desired Q-point is called biasing. The circuits used for getting the desired and proper operting point are known as biasing circuits.

DC Load Line

Referring to the biasing circuit of Fig. 2.16 (*a*), the values of V_{CC} and R_C are fixed and I_C and V_{CE} are dependent on R_B.

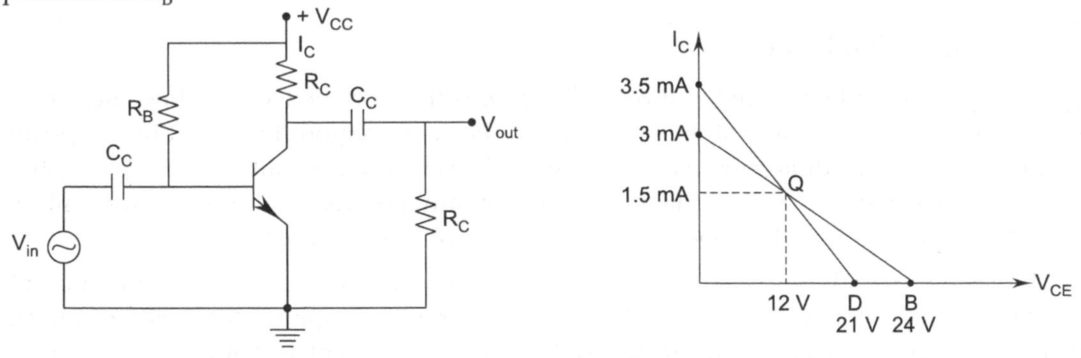

(*a*) Biasing circuit (*b*) CE output characteristics and load line

Fig. 2.16

Applying Kirchhoff's voltage law to the collector circuit in Fig. 2.16(*a*), we get $V_{CC} = I_C R_C + V_{CE}$.

The straight line represented by AB in Fig. 2.16(*b*) is called the DC load line. The co-ordinates of the end point A are obtained by substituting $V_{CE} = 0$ in the above equation. Then $I_C = \frac{V_{CC}}{R_C}$.

Therefore, the co-ordinates of A are $V_{CE} = 0$ and $I_C = \frac{V_{CC}}{R_C}$.

The co-ordinates of B are obtained by substituting $I_C = 0$ in the above equation. Then $V_{CE} = V_{CC}$. Therefore, the co-ordinates of B are $V_{CE} = V_{CC}$ and $I_C = 0$. Thus, the DC load line AB can be drawn if the values of R_C and V_{CC} are known.

As shown in Fig. 2.16 (*b*), the optimum Q-point is located at the mid-point of the DC load line. AB between the saturation and cut-off regions, *i.e.*, Q is exactly midway between A and B. In order to get faithful amplification, the Q-point must be well within the active region of the transistor.

AC load line

After drawing the DC load line, the operating point Q is properly located at the center of the DC load line. This operating point is chosen under zero input signal condition of the circuit. Hence, the AC load line should also pass through the operating point Q. The effective AC load resistance, R_{AC}, is the combintion of R_C parallel to R_L, *i.e.*, $R_{AC} = R_C \| R_L$. So the slope of the AC load line CQD will be $\left(-\frac{1}{R_{AC}}\right)$.

To draw an AC load line, two end points, *viz.*, maximum V_{CE} and maximum I_C when the signal is applied are required.

Maximum $V_{CE} = V_{CEQ} + I_{CQ} R_{AC}$, which locates the point D (OD) on the V_{CE} axis.

Maximum $I_C = I_{CQ} + \frac{V_{CEQ}}{R_{AC}}$, which locates the point C (OC) on the I_C axis.

By joining points C and D, AC load line CD is constructed. As $R_C > R_{AC}$, the DC load line is less step than the AC load line.

When the signal is zero, we have the exact DC conditions, from the Fig. 2.16(*b*), it is clear that the interaction of DC and AC load lines is the operating point Q.

Selection of Operating Point

The operating point can be selected at three different positions on the DC load line ; near saturation region, near cut-off region or at the centre, *i.e.*, in the active region. The selection of operating point will depend on its application. When transistor is used as an amplifier, the Q point should be selected at the center of the DC load line to prevent any possible distortion in the amplified output signal. This is well understood by going through following cases.

Case 1. Biasing circuit is designed to a fix a Q-point at point P, as shown in Fig. 2.17. Point P is very near to the saturation region. In this, the collector current is clipped at the positive half cycle. So, even though base current varies sinusoidally, collector current is not a useful sinusoidal waveform, *i.e.*, distortion is present at the output. Therefore, point P is not a suitable operating point.

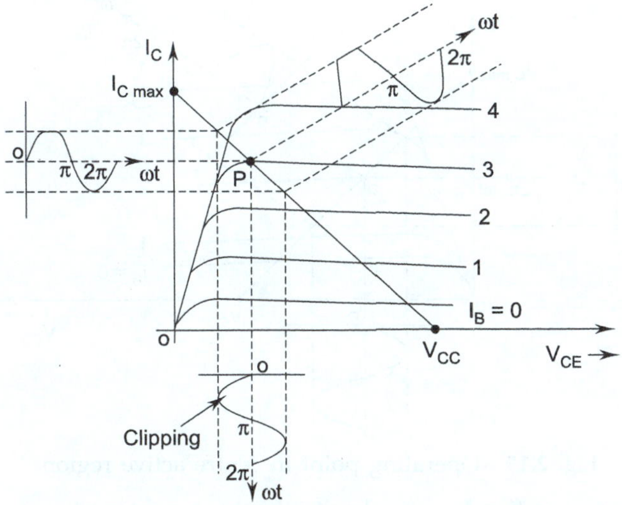

Fig. 2.17 Operating point near saturation region gives clipping at the positive peaks

Case 2. Biasing circuit is designed to fix a Q-point at point R as shown in Fig. 2.18. Point R is very near to the cut-off region. In this, the collector current is clipped at the negative half cycle. So. point R is also not a suitable operating point.

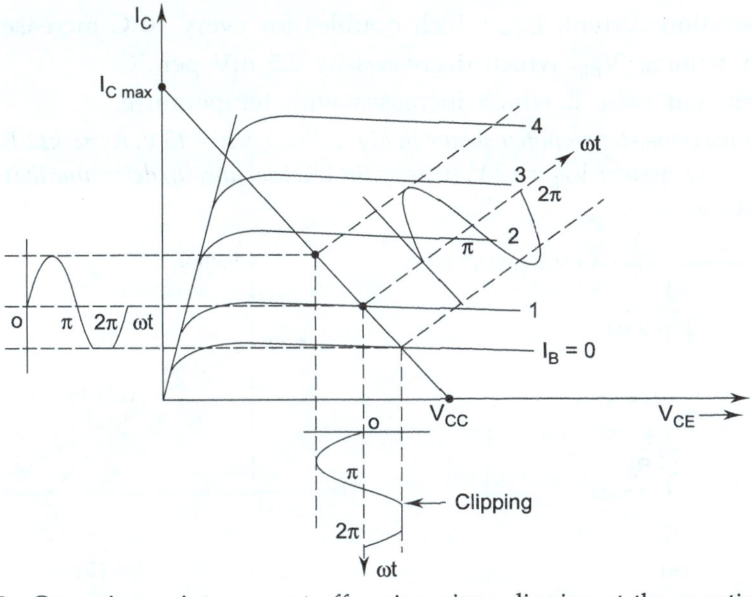

Fig. 2.18 Operating point near cut-off region gives clipping at the negative peaks

Case 3. Biasing circuit is designed to fix a Q-point at point Q as shown in Fig. 2.19. The output signal is sinusoidal waveform without any distortion. These point Q is the best operating point.

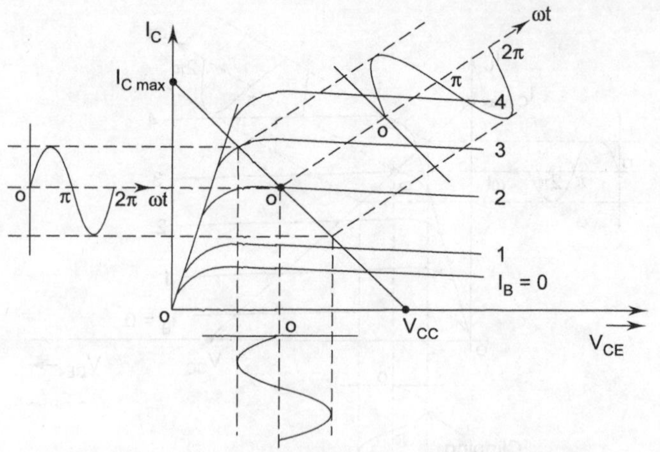

Fig. 2.19 Operating point in centre active region

Even though the Q-point is fixed properly, it is very important to ensure that the operating point remains stable where it is originally fixed. If the Q-point shifts nearer to either A or B, the output voltage and current get clipped, thereby output signal is distorted.

In practice, the Q-point tends to shift its position due to any or all of the following three main factors.

(*i*) Reverse saturation current, I_{CO}, which doubles for every 10°C increase in temperature.

(*ii*) Base-emitter voltage, V_{BE}, which decreases by 2.5 mV per °C

(*iii*) Transistor current gain, β which increases with temperature.

Example 2.2. *For the transistor amplifier shown in Fig. 2.20 (a), $V_{CC} = 15$ V, $R_1 = 9$ kΩ, $R_2 = 5$ kΩ, $R_C = 2$ kΩ, $R_E = 1.5$ kΩ and $R_L = 2$ kΩ. Assume $V_{BE} = 0.7$ V, (i) draw the DC load line, (ii) determine the operating point, and (iii) draw the AC load line.*

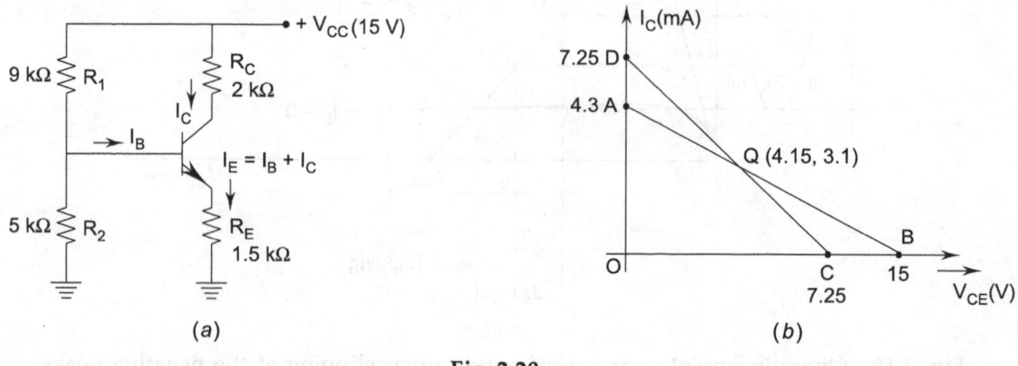

Fig. 2.20

Solution.

(*i*) **DC load line :** Referring to Fig. 2.20(*a*), we have $V_{CC} = V_{CE} + I_C (R_C + R_E)$.

To draw the DC load line, we need to end points, *viz.*, maximum V_{CE} point at ($I_C = 0$) and maximum I_C point (at $V_{CE} = 0$) maximum $V_{CE} = V_{CC} = 15$ V, which locates the point B (OB = 15 V) of the DC load line.

$$\text{Maximum} \quad I_C = \frac{V_{CC}}{R_C + R_E} = \frac{15}{(2+1.5)\times 10^3} = 4.3 \text{ mA}$$

This locates the point A (OA = 4.3 mA) of the DC load line. Figure 2.20(*b*) shows the DC load line AB, with (15 V, 4.3 mA).

(*ii*) **Operating point Q :**

The voltage across R_2 is $V_2 = \frac{R_2}{R_1 + R_2} V_{CC}$

Therefore, $$V_2 = \frac{5 \times 10^3}{(9+5) \times 10^3} \times 15 = 5.36 \text{ V}$$

$$V_2 = V_{BE} + I_E R_E$$

Therefore, $$I_E = \frac{V_2 - V_{BE}}{R_E} = \frac{5.36 - 0.7}{1.5 \times 10^3} = 3.1 \text{ mA}$$

$$I_C \approx I_E = 3.1 \text{ mA}$$

$$V_{CE} = V_{CC} - I_C (R_C + R_E)$$
$$= 15 - 3.1 \times 10^{-3} \times (2 + 1.5) \times 10^3$$
$$= 4.15 \text{ V}.$$

Therefore, the operating point Q is at 4.15 V and 3.1 mA, which is shown on the DC load line.

(*iii*) **AC load line :** To draw the AC load line, we need two end points, *viz.*, maximum V_{CE} and maximum I_C when signal is applied

AC load, $$R_{AC} = R_C \| R_L = \frac{2 \times 2}{4} = 1 \text{ k}\Omega$$

Therefore, maximum $V_{CE} = V_{CEQ} + I_{CQ} R_{AC}$
$$= 4.15 + 3.1 \times 10^{-3} \times 1 \times 10^3$$
$$= 7.25 \text{ V}$$

This locates the point C (OC = 7.25 V) on the V_{CE} axis

Maximum $$I_C = I_{CQ} + \frac{V_{CEQ}}{R_{AC}}$$
$$= 3.1 \times 10^{-3} + \frac{4.15}{1 \times 10^3}$$
$$= 7.25 \text{ mA}.$$

This locates the point D (OD = 7.25 mA) on the I_C axis. By joining points C and D, AC load line CD is constructed.

Thermal Runaway

The collector current for the CE circuit of figure is given by $I_C = \beta I_B + (1 + \beta) I_{CO}$. The three variables in the equation, β, I_B and I_{CO} increase with rise in temperature. In particular, the reverse saturation current or leakage current I_{CO} changes greatly with temperature. Specifically, it doubles for every 10°C rise in temperature. The collector current I_C causes the collector-base junction temperature to rise which, in turn, increase I_{CO}, as a result I_C will increase still further, which will further rise the temperature at the collector-base junction. This process will become cumulative leading to "thermal runaway". Consequently, the ratings of the transistor are exceeded which may destroy the transistor itself.

The collector is normally made larger in size than the emitter in order to help dissipate the heat developed at the collector junction.

However, if the circuit is designed such that the base current I_B is made to decrease automatically with rise in temperature, then the decrease in βI_B will compensate for the increase in $(1 + \beta) I_{CO}$, keeping I_C almost constant.

In power transistors, the heat developed at the collector junction may be removed by the use of heat sink, which is a metal sheet fitted to the collector and whose surface radiates heat quickly.

Stability Factor (S)

The stability factor may be defined as the rate of change of collector current with respect to the reverse saturation current keeping the common-emitter current gain (β) and base current (I_B) as constant. Mathematically, the stability factor

$$S = \frac{dI_C}{dI_{CO}}$$

The stability factor is a measure of bias stability of a transistor circuit. It will be interesting to know that a higher value of stability factor indicates poor stability, whereas a lower value indicates good stability. In this sense, the stability factor should be called as an instability factor.

It may be noted that if the rate of change of collector current (dI_C) is equal to the rate of change of reverse saturation current (dI_{CO}), then the value of stability factor is unity. The unity is the lowest value of stability factor. It may be noted that if the value of stability factor is closer to the unity, the variation of collector current with the temperature will be less. Therefore, there will be less variation in the Q-point.

The stability factor may also be expressed alternatively by using the relationship between base current (I_B) collector current (I_C) and reverse saturation current (I_{CO}). We know that the value of collector current in a transistor is given by the relation,

$$I_C = \beta I_B + (1 + \beta) I_{CO}.$$

Differentiating the above expression, with respect to I_C,

$$1 = \frac{d(\beta \cdot I_B)}{dI_C} + \frac{d(1+\beta) I_{CO}}{dI_C}$$

$$= \beta \times \frac{dI_B}{dI_C} + (1+\beta)\frac{dI_{CO}}{dI_C} \qquad \text{(Assuming } \beta \text{ as constant)}$$

$$= \beta \times \frac{dI_B}{dI_C} + (1+\beta)\frac{1}{S}$$

$$S = \frac{1+\beta}{1-\beta\left(\dfrac{dI_B}{dI_C}\right)} \qquad \text{...(1)}$$

The above expression is a general expression, which is very useful for determining the stability factor (S) of any biasing circuit. It can be done first by finding the relationship between the base current and collector current and then using the above equation to determine the value of stability factor.

Stability Factor S′ and S″

The stability factor S′ is defined as the rate of change of I_C with V_{BE}, keeping I_{CO} and β constant

$$S' = \frac{\partial I_C}{\partial V_{BE}} \approx \frac{\Delta I_C}{\Delta V_{BE}}$$

The stability factor S″ is defined as the rate of change of I_C with respect to β, keeping I_{CO} and V_{BE} constant.

$$S'' = \frac{\partial I_C}{\partial \beta} \simeq \frac{\Delta I_C}{\Delta \beta}$$

Methods of Transistor Biasing

The stability factor for some commonly used biasing circuits are discussed below :

1. Fixed bias or base resistor method

A common emitter amplifier using fixed bias circuit as shown in Fig. 2.21. The DC analysis of the circuit yields the following equation.

$$V_{CC} = I_B R_B + V_{BE}$$

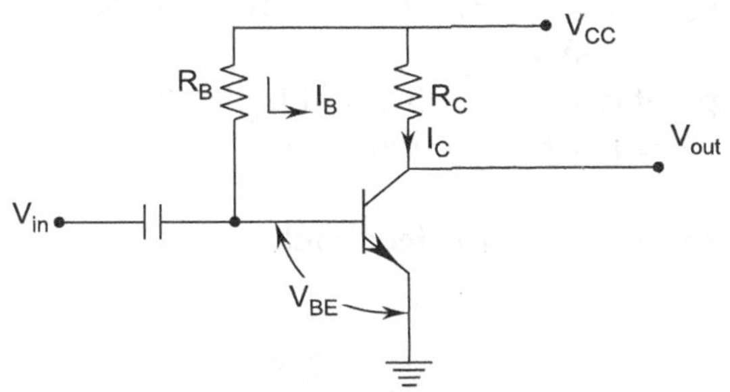

Fig. 2.21 Fixed bias circuit

Therefore, $$I_B = \frac{V_{CC} - V_{BE}}{R_B}$$

Since this equation is independent of current I_C, $\frac{dI_B}{dI_C} = 0$ and the stability factor given in equation (1) reduces to

$$S = 1 + \beta$$

Since β is a large quantity, this is a very poor bias stable circuit. Therefore, in practice, this circuit is not used for biasing the base.

The advantage of this method are : (*a*) Simplicity (*b*) Small number of components required, and (*c*) If the supply voltage is very large as compared to V_{BE} of the transistor, then the base current becomes largely independent of the voltage V_{BE}.

Example 2.3. *In a fixed bias compensation method (refer to Fig. 2.22) a silicon transistor with $\beta = 100$ is used. $V_{CC} = 8$ V, $R_C = 3.5$ kΩ, $R_B = 550$ kΩ. Draw the DC load line and determine the operating point and the stability factor.*

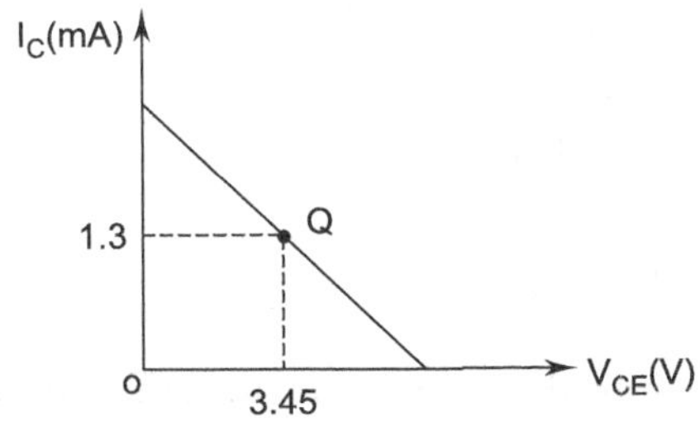

Fig. 2.22 DC load line

Solution.

(*i*) **DC load line :**

$$V_{CE} = V_{CC} - I_C R_C$$

when $I_C = 0$, $\quad V_{CE} = V_{CC} = 8$ V

when $V_{CE} = 0$, $\quad I_C = \frac{V_{CC}}{R_C} = \frac{8}{3.5 \times 10^3} = 2.3$ mA

(*ii*) **Operating point Q :**

For silicon transistor, $V_{BE} = 0.7$ V

$$V_{CC} = I_B R_B + V_{BE}$$

$$\therefore \quad I_B = \frac{V_{CC} - V_{BE}}{R_B} = \frac{8 - 0.7}{550 \times 10^3} = 13\ \mu A$$

$$\therefore \quad I_C = \beta I_B = 100 \times 13 \times 10^{-6} = 1.3 \text{ mA}$$

$$V_{CE} = V_{CC} - I_C R_C = 8 - 1.3 \times 10^{-3} \times 3.5 \times 10^3 = 3.45 \text{ V}$$

Therefore, operating point is $V_{CEQ} = 3.45$ V and $I_{CQ} = 1.3$ mA

(*iii*) **Stability factor :** $\quad S = 1 + \beta = 1 + 100 = 101.$

2. Collector to base bias or biasing with feedback resistor

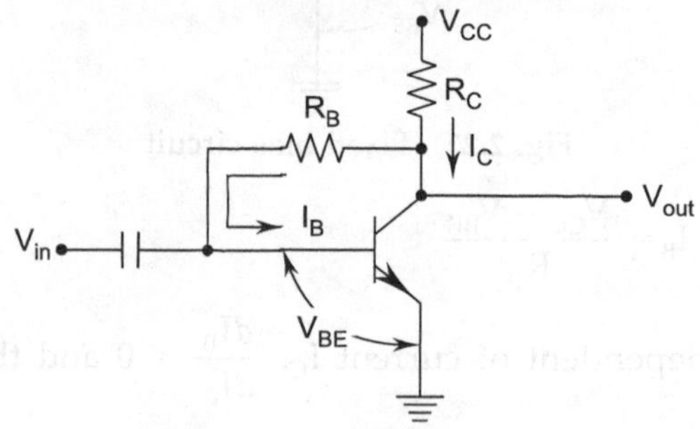

Fig. 2.23 Collector to base circuit

A common emitter amplifier using collector to base bias circuit is shown in Fig. 2.23. This circuit is the simplest way to provide some degree of stabilization to the amplifier operating point.

If the collector current I_C tends to increase due to either increase in temperature or the transistor has been replaced by the one with a higher β, the voltage drop across R_C increases, thereby reduces the value of V_{CE}. Therefore, I_B decreases which in turn, compensates the increase in I_C. Thus, greater stability is obtained.

The loop equation for this circuit is,

$$V_{CC} = (I_B + I_C) R_C + I_B R_B + V_{BE}$$

$$V_{CC} = I_B (R_C + R_B) + I_C R_C + V_{BE}$$

i.e.,

$$I_B = \frac{V_{CC} - I_C R_C - V_{BE}}{R_C + R_B}$$

$$\therefore \quad \frac{dI_B}{dI_C} = \frac{-R_C}{R_C + R_B}$$

Substituting the value of $\frac{dI_B}{dI_C}$ into the stability factor equation (1), we get

$$S = \frac{1+\beta}{1+\beta\left(\frac{R_C}{R_C+R_B}\right)}$$

As can be seen, this value of the stability factor is smaller than the value obtained by fixed bias circuit. Also, S can be made small and the stability can be improved by making R_B small or R_C large.

If R_C is very small, then $S = (\beta + 1)$, *i.e.*, stability is very poor. Hence, the value of R_C must be quite large for good stabilization. Thus, collector to base bias arrangement is not satisfactory for the amplifier circuits like transformer coupled amplifier where the DC load resistance in collector circuits is very small. For such amplifiers, emitter bias or self bias will be the most satisfactory transistor biasing for stabilization.

Example 2.4. *In the biasing with feedback resistor method, a silicon transistor with feedback resistor is used. The operating point is at 8 V, 2 mA and V_{CC} = 14 V.*

Assume β = 100. Determine (a) the value of R_B, (b) Stability factor, and (c) What will be the new operating point if β = 50 with all other circuit values are same.

Solution. Refer to Fig. 2.23, we know that for a silicon transistor, $V_{BE} = 0.7$ V

(*a*) **To determine R_B :**

The operating point is at $V_{CE} = 8$ V and $I_C = 2$ mA

Here,
$$R_C = \frac{V_{CC} - V_{CE}}{I_C} = \frac{14-8}{2\times10^{-3}} = 3\ k\Omega$$

$$I_B = \frac{I_C}{\beta} = \frac{2\times10^{-3}}{100} = 20\ \mu A$$

Using the relation,

$$R_B = \frac{V_{CC} - V_{BE} - I_C R_C}{I_B} = \frac{14 - 0.7 - 2\times10^{-3}\times10^{3}}{20\times10^{-6}}$$
$$= 365\ k\Omega$$

(*b*) **Stability factor :**

$$S = \frac{1+\beta}{1+\beta\left(\frac{R_C}{R_C+R_B}\right)}$$
$$= \frac{1+100}{1+100\left(\frac{3\times10^3}{(3+365)\times10^3}\right)}$$
$$= 55.6$$

(*c*) **To determine new operating point when β = 50**

$$V_{CC} = \beta I_B R_C + I_B R_B + V_{BE}$$
$$= I_B(\beta R_C + R_B) + V_{BE}$$

i.e.,
$$14 = I_B(50\times3\times10^3 + 365\times10^3) + 0.7$$

$$I_B = \frac{14-0.7}{50\times3\times10^3 + 365\times10^3} = 25.8\ \mu A$$

Therefore,
$$I_C = \beta I_B = 50\times25.8\times10^{-6} = 1.29\ mA$$
$$V_{CE} = V_{CC} - I_C R_C = 14 - 1.29\times10^{-3}\times3\times10^3$$
$$= 10.13\ V$$

Therefore, the co-ordinates of the new operating point are $V_{CEQ} = 10.13$ V and $I_{CQ} = 1.29$ mA.

3. Self bias or emitter bias or voltage divider bias

A simple circuit used to establish a stable, operating point is the self-biasing configuration. The self bias, also called as emitter bias, or emitter resistor and (potential) voltage divider circuit, that can be used for low collector resistance, is shown in Fig. 2.24. The current in the emitter resistor R_E causes a voltage drop which is in the direction to reverse bias the emitter junction. For the transistor to remain in the active region, the base-emitter junction has to be forward biased. The required base bias is obtained from the power supply through the potential divides network of the resistances R_1 and R_2.

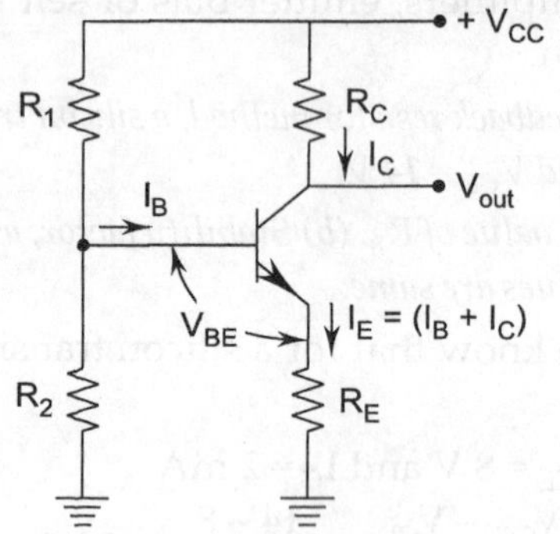

Fig. 2.24 Self bias

If I_C tends to increase, say, due to increase in I_{CO} with temperature, the current in R_E increases. Hence, the voltage drop across R_E increases thereby decreasing the base current. As a result, I_C is maintained almost constant.

Applying Thevenin's theorem to the circuit of figure, for finding the base current, we have

$$V_T = \frac{R_2 V_{CC}}{R_1 + R_2} \text{ and } R_B = \frac{R_1 R_2}{R_1 + R_2}$$

The loop equation around this base circuit can be written as

$$V_T = I_B R_B + V_{BE} + (I_B + I_C) R_E$$

Differentiating this equation with respect to I_C, we get

$$\frac{dI_B}{dI_C} = -\frac{R_E}{R_E + R_B}$$

Substituting this equation in stability factor equation, we get

$$S = \frac{1+\beta}{1+\beta\left(\dfrac{R_E}{R_E + R_B}\right)}$$

$$\therefore \qquad S = (1+\beta)\frac{1+\dfrac{R_B}{R_E}}{1+\beta+\dfrac{R_B}{R_E}}$$

As can be seen, the value of S is equal to one if the ratio $\frac{R_B}{R_E}$ is very small as compared to 1. As this ratio becomes comparable to unity, and beyond towards infinity, the value of the stability factor goes on increasing till $S = 1 + \beta$.

This improvement is the stability up to a factor equal to 1 is achieved at the cost of power dissipation. To improve the stability, the equivalent resistance R_B must be decreased, forcing more current in the voltage divider network of R_1 and R_2.

Often, to prevent the loss of gain due to negative feedback, R_E is shunted by a capacitor C_E. The capacitive reactance X_{CE} must be equal to about one-tenth of the value of the resistance R_E at the lowest operating frequency.

Example 2.5. *In a CE germanium transistor amplifier circuit, the bias is provided by self bias, i.e., emitter resistor and potential divides arrangement (refer to Fig. 2.24). The various parameters are :* $V_{CC} = 15$ V, $R_C = 4$ kΩ, $R_E = 3$ kΩ, $R_1 = 58$ kΩ, $R_2 = 25$ kΩ *and* $\alpha = 0.985$. *Determine (a) the coordinates of the operating point, and (b) the stability factor S.*

Solution. (*a*) **For a germanium transistor,** $V_{BE} = 0.3$ V. As $\alpha = 0.985$

$$\beta = \frac{\alpha}{1-\alpha} = \frac{0.985}{1-0.985} = 66$$

To find the co-ordinates of the operating point

Thevenin's voltage, $V_T = \frac{R_2}{R_1 + R_2} V_{CC}$

$$= \frac{25 \times 10^3}{(58+25) \times 10^3} \times 15$$

$$= 4.51 \text{ V}$$

Thevenin's resistance, $R_B = \frac{R_1 R_2}{R_1 + R_2} = \frac{58 \times 10^3 \times 25 \times 10^3}{(58+25) \times 10^3} = 17.469 \text{ k}\Omega$

The Loop equation around the base circuit is

$$V_T = I_B R_B + V_{BE} + (I_B + I_C) R_E$$

$$= \frac{I_C}{\beta} R_B + V_{BE} + \left(\frac{I_C}{\beta} + I_C\right) R_E$$

$$4.51 = \frac{I_C}{66} \times 17.469 \times 10^3 + 0.3 + I_C \left(\frac{1}{66} + 1\right) \times 3 \times 10^3$$

$$4.21 = I_C (0.265 + 3.045) \times 10^3$$

Therefore, $I_C = \frac{4.21}{3.31 \times 10^3}$

$$= 1.27 \text{ mA}$$

Since I_B is very small,

$$I_C \simeq I_E = 1.27 \text{ mA}$$

Therefore, $V_{CE} = V_{CC} - I_C R_C - I_E R_E$

$$= V_{CC} - I_C [R_C + R_E]$$
$$= 15 - 1.27 \times 10^{-3} \times (4 + 3) \times 10^3$$
$$= 6.11 \text{ V}$$

Therefore, the co-ordinates of the operating point are $I_C = 1.27$ mA and $V_{CE} = 6.11$ V

(*b*) **To find the stability factor S**

$$S = (1+\beta)\frac{1+\dfrac{R_B}{R_E}}{1+\beta+\dfrac{R_B}{R_E}}$$

$$= (1+66)\frac{1+\dfrac{17.469\times10^3}{3\times10^3}}{1+66+\dfrac{17.469\times10^3}{3\times10^3}}$$

$$= 67 \times \frac{6.823}{72.823}$$

$$= 6.277$$

2.5 FIELD EFFECT TRANSISTORS

Bipolar transistor has two principal disadvantages. First it has low input impedance. Secondly, it has considerable noise level.

The field effect transistor (FET) has, by virtue of its construction and biasing, large input impedance which may be more than 100 mega ohms. The FET is generally much less noisy than the bipolar transistor.

Types of Field Effect Transistors

A bipolar junction transistor (BJT) is a current controlled device *i.e.,* output characteristics of the device are controlled by base current and not by base voltage. However, in a field effect transistor (FET), the output characteristics are controlled by input voltage (*i.e.,* electric field) and not by input current. There are two basic types of field effect transistors :

(*i*) Junction Field Effect Transistor (JFET)

(*ii*) Metal Oxide Semiconductor Field Effect Transistor (MOSFET)

(i) Junction Field Effect Transistor (JFET)

A junction field effect transistor is a three terminal semiconductor device in which current conduction is by one type of carrier *i.e.,* electrons or holes.

Constructional Details

A JFET consists of a *p*-type or *n*-type silicon bar containing two pn junctions at the sides shown in Fig. 2.25. The bar forms the conducting channel for the charge carriers. If the bar is *n*-type, it is called *n*-channel JFET as shown in Fig. 2.25(*i*) and if the bar is of *p*-type, it is called *p*-channel JFET as shown in Fig. 2.25(*ii*). The two *pn* junction forming diodes are connected internally and a common terminal called gate is taken out. Other terminals are source and drain.

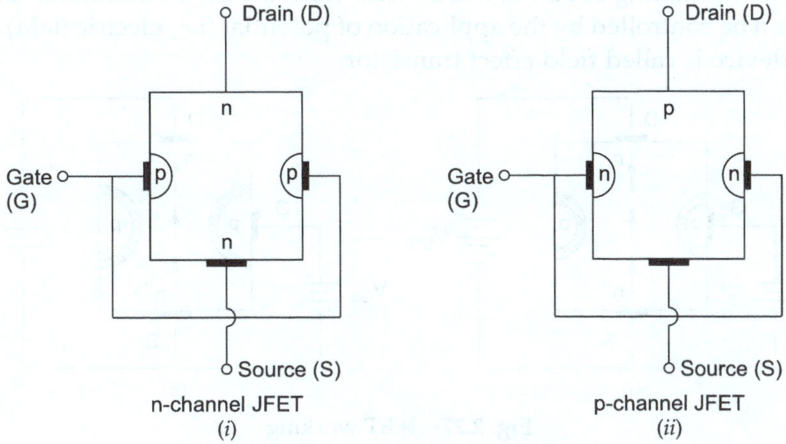

Fig. 2.25 JFET construction

JFET Polarities : Figure 2.26(*i*) shows *n* channel JFET polarities whereas Fig. 2.26(*ii*) shows the *p*-channel JFET polarities. Note that in each case, the voltage between the gate and source is such that the gate is reverse biased. This is the normal way or JFET connection. The drain and sources are interchangeable (for low frequency applications) *i.e.*, either end can be used as source and the other end as drain.

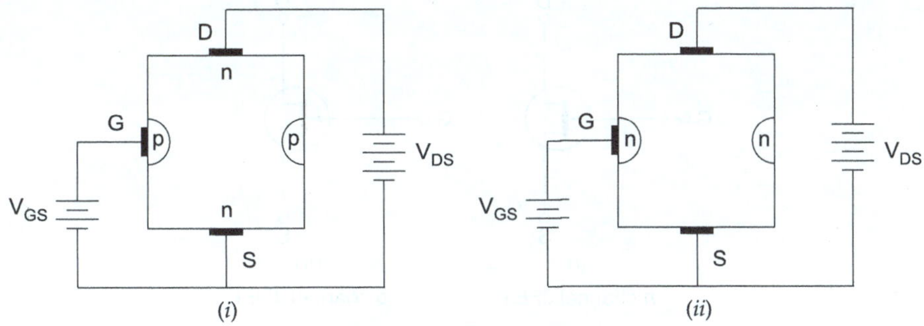

Fig. 2.26 JFET polarities

Working Principle of JFET

Figure 2.27 shows the circuit of *n*-channel JFET with normal polarities.

(*i*) When a voltage V_{DS} is applied between drain and source terminals and voltage on the gate is zero the two PN junctions at the sides of the bar establish depletion layers. The size of these layers determines the width of the channel and hence the current conduction through the bar.

(*ii*) When a reverse voltage V_{GS} is applied between the gate and source, the width of the depletion layers is increased. This reduces the width of the conducting channel, thereby increasing the resistance of *n*-type bar consequently, the current from source to drain is decreased. On the otherhand, if the reverse voltage on the gate is decreased, the width of the depletion layer also decreases. This increases the width of the conducting channel, and hence source to drain current.

Naming of FET

The reason behind the naming of FET is will be clear from the above discussion that current from source to drain can be controlled by the application of potential (*i.e.*, electric field) on the gate. For this reason, the device is called field effect transistor.

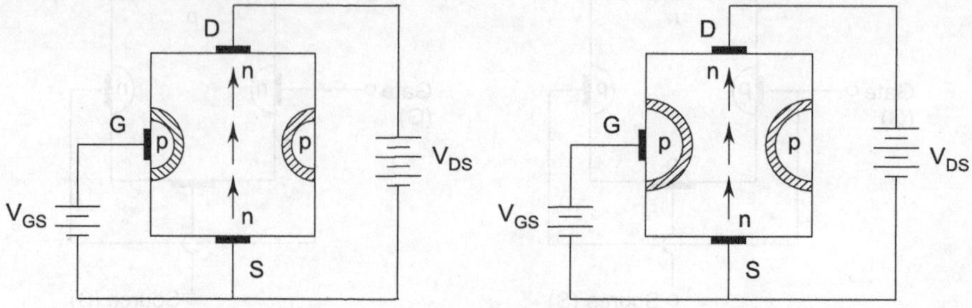

(*i*) (*ii*)

Fig. 2.27 JFET working

Schematic Symbol of JFET

Figure 2.28 shows the schematic symbol of JFET. The vertical line in the symbol may be thought as channel and source (S) and drain (D) connected to this line. If the channel is *n*-type, the arrow on the gate points towards the channel as shown in Fig. 2.28(*i*). For *p*-type channel, the arrow on the gate points from channel to gate Fig. 2.28(*ii*).

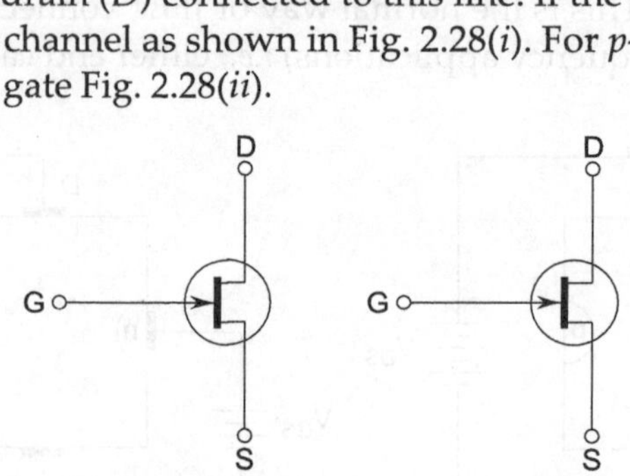

(*i*) *n* channel JFET (*ii*) *p* channel JFET

Fig. 2.28 JFET symbol

Difference between JFET and BJT

The JFET differs from an ordinary or bipolar transistor in the following ways :

(*i*) In a JFET, there is only one type of carrier, holes in *p*-type channel and electrons in *n*-type channel. For this reason, it is also called a unipolar transistor. However, in BJT both electrons and holes paly part in current conduction.

(*ii*) As the input circuit (*i.e.*, gate to source) of a JFET is reverse biased, therefore the device has high input impedance. However, the input of an Bipolar Junction Transistor is forward biased and hence has low input impedance.

(*iii*) As the gate is reverse biased, therefore it carries very small current. In JFET Gate carries extremely small current and input voltage controls the output current. For this reason, JFET is essentially a voltage driven device. However, BJT is a current controlled device, is input current controls output current.

(*iv*) A bipolar transistor uses a current into its base to control a large current between collector and emitter whereas a JFET uses voltage on the 'gate' (= base) terminal to control the current between drain (= collector) and source (= emitter). Thus a bipolar transistor gain is characterised by current gain whereas, the JFET gain is characterised by as a transconductance *i.e.*, the ratio of change in output current to the input voltage.

(*v*) In JFET, there are no junction as in BJT. The conduction is through as *n*-type or *p*-type semiconductor material. For this reason, noise level in JFET is very small.

Characteristics of JFET

Output characteristics: The curve between drain current (I_D) and drain-source voltage (V_{DS}) of a JFET at constant gate source voltage (V_{GS}) is known as output characteristics of JFET. Figure 2.29 shows the circuit for determining the output characteristics of JFET. Keeping V_{GS} fixed at some value, say IV, the drain source voltage is changed in steps. Corresponding to each value of V_{DS}, the drain current I_D is noted. A plot of these values gives the output characteristics of JFET at V_{GS} = IV. Repeating similar procedure, output characteristics can be drawn. Figure 2.30 shows output characteristic.

(*i*) At first, the drain current I_D rises rapidly with drain. Source voltage V_{DS} but then becomes constant. The drain-source voltage above which drain current becomes constant is known as pinch-off voltage. Thus in figure, OA is the pinch-off voltage.

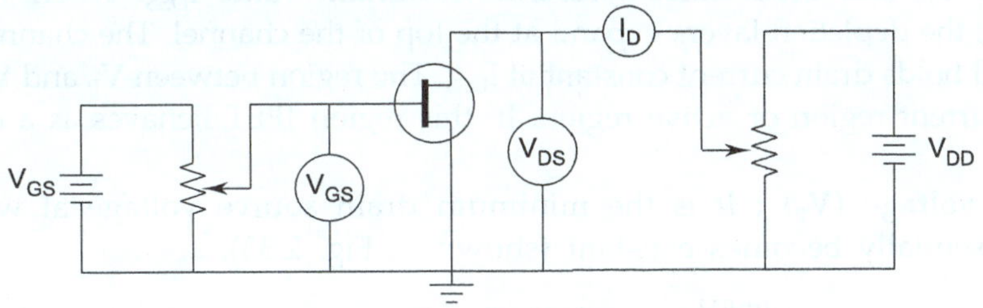

Fig. 2.29 JFET circuit

(*ii*) After pinch-off voltage, the channel width becomes so narrow that depletion layers almost touch each other. The drain current passes through the small passage between these layers. Therefore, increase in drain current is very small with V_{DS} above pinch off voltage consequently drain current remains constant.

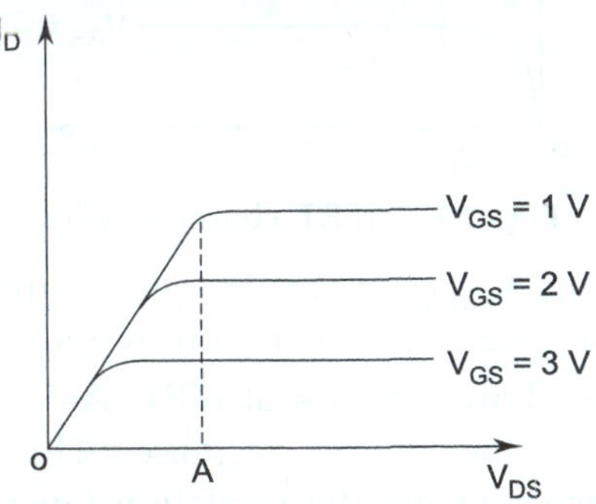

Fig. 2.30

FET Analysis: In the analysis of a JFET circuit, the important terms are :

(*i*) Shorted-gate drain current (I_{DSS})

(*ii*) Pinch-off voltage (V_P)

(*iii*) Gate source cut-off voltage ($V_{GS\ off}$)

(*i*) **Shorted-gate drain current (I_{DSS}) :** It is the drain current with source short-circuited to gate (*i.e.*, $V_{GS} = 0$) and drain voltage (V_{DS}) equal to pinch-off voltage. It is sometimes called zero-bias current.

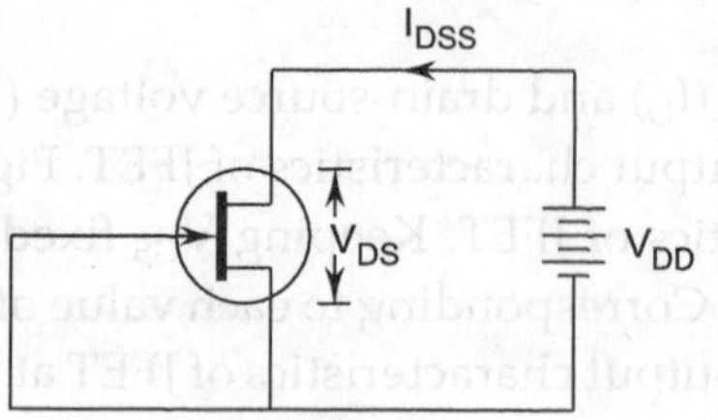

Fig. 2.31 JFET circuit with $V_{GS} = 0$

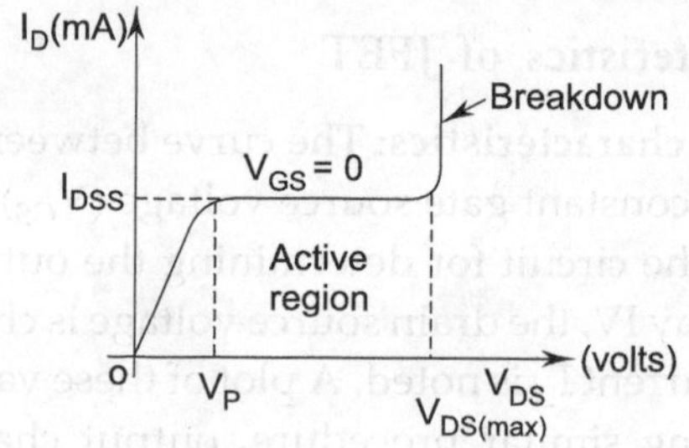

Fig. 2.32 JFET drain characteristics

Figure 2.31 shows the JFET circuit with $V_{GS} = 0$, *i.e.*, source shorted to gate. This is normally called shorted-gate condition. Figure 2.32 shows the graph between I_D and V_{DS} for the shorted gate condition. The drain current rises rapidly at first and then levels off at pinch-off voltage V_P, and drain current reaches maximum value I_{DSS}. When V_{DS} is increased beyond V_P, the depletion layers expand at the top of the channel. The channel acts a current emitter and holds drain current constant at I_{DSS}. The region between V_P and $V_{DS\ (max)}$ is called constant current region or active region. In this region JFET behaves as a constant-current device.

(*ii*) **Pinch-off voltage (V_P) :** It is the minimum drain source voltage at which the drain current essentially becomes constant (shown in Fig. 2.33).

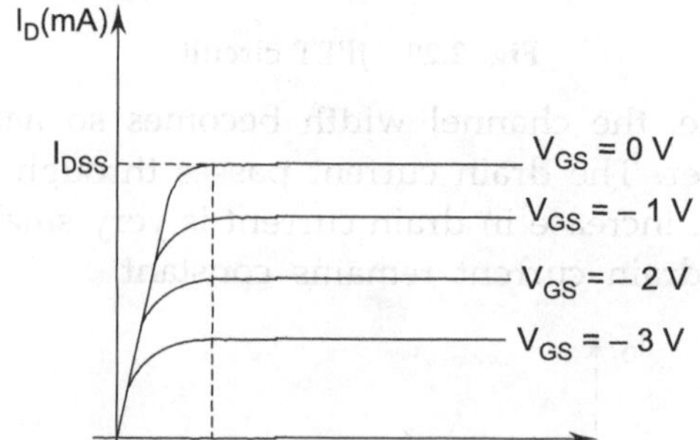

Fig. 2.33 JFET characteristics

(*iii*) **Gate-Source Cut-off voltage $V_{GS\ (off)}$:** It is the gate source voltage where the channel is completely cut-off and the drain current becomes zero.

Figure 2.34 shows the transfer characteristics of JFET. As the reverse gate source voltage is increased, the cross-sections area of the channel decreases. This in turn decreases the drain current. At some reverse gate source voltage, the depletion layers extend completely across the channels. In this condition, the channel is cut-off and drain current reduces to zero. The gate voltage at which the channel is cut-off is called $V_{GS\ (off)}$.

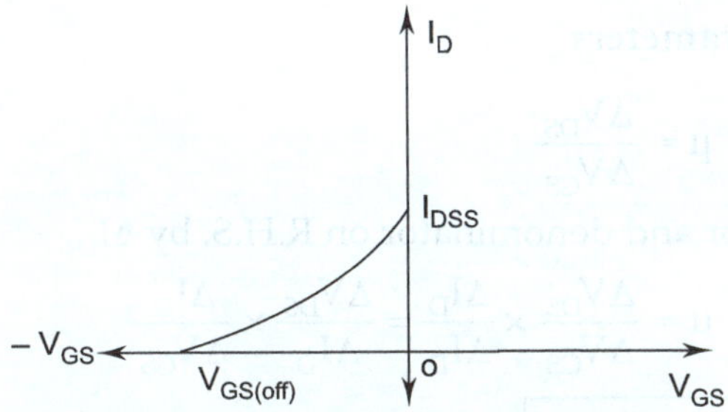

Fig. 2.34 JFET transfer characteristics

Drain current (I_D)

It is interesting to note that $V_{GS\,(off)}$ will always have the same magnitude value as V_P. The relationship between V_P and I_{DSS} is shown in Fig. 2.34. We note that gate-source cut-off voltage (*i.e.,*) $V_{GS\,(off)}$ on the transfer characteristics is equal to pinch off voltage V_P on the drain characteristics *i.e.,*

$$V_P = |\,V_{GS\,(off)}\,|.$$

For example, $V_{GS} = -4$ V, then $V_P = 4$ V.

A complex mathematical analysis yields the following expression for drain current.

$$I_D = I_{DSS}\left[1 - \frac{V_{GS}}{V_{GS\,(off)}}\right]^2$$

I_D = drain current at given V_{GS}

I_{DSS} = shorted gate-drain current

V_{GS} = gate-source voltage

$V_{GS\,(off)}$ = gate-source cut-off voltage.

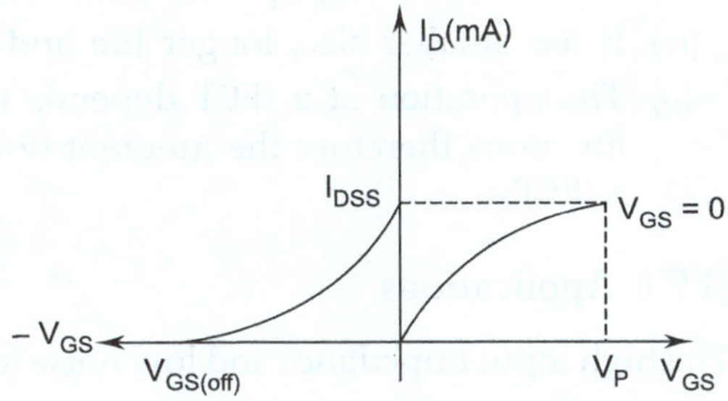

Fig. 2.35 JFET characteristics

Parameter of JFET

The main parameter or JFET are :

(*i*) **AC drain resistance :** It is the ratio of change in drain-source voltage (ΔV_{DS}) to the change in drain current (ΔI_D) at constant gate source voltage *i.e.,*

AC drain resistance, $r_d = \dfrac{\Delta V_{DS}}{\Delta I_D}$ at constant V_{GS}.

(*ii*) **Transconductance (g_{fs}) :** It is the ratio of change in drain current (ΔI_D) to change in gate-source voltage (ΔV_{GS}) at constant drain. Source voltage *i.e.,*

Transconductance, $g_{fs} = \dfrac{\Delta I_D}{\Delta V_{GS}}$ at constant V_{DS}.

(*iii*) **Amplification Factor (μ) :** It is the ratio of change in drain source voltage (ΔV_{DS}) for the change in gate-source voltage (ΔV_{GS}) at constant drain current *i.e.,*

Amplification factor, $\mu = \dfrac{\Delta V_{DS}}{\Delta V_{GS}}$ at constant I_D.

Relation among JFET Parameters

We know $$\mu = \frac{\Delta V_{DS}}{\Delta V_{GS}}$$

Multiplying the numerator and denominator on R.H.S. by ΔI_D,

$$\mu = \frac{\Delta V_{DS}}{\Delta V_{GS}} \times \frac{\Delta I_D}{\Delta I_D} = \frac{\Delta V_{DS}}{\Delta I_D} \times \frac{\Delta I_D}{\Delta V_{GS}}$$

$$\boxed{\mu = r_d \times g_{fs}.}$$

Advantages of JFET

Some advantages of JFET are :

(*i*) It has a very high input impedance (of the order of 100 MΩ). This permits high degree of isolation between the input and output circuits.

(*ii*) It has negative temperature co-efficient of resistance. This avoids the risk of thermal runaway.

(*iii*) It has a very high power gain. This eliminates the necessity of driver stages.

(*iv*) It has smaller size, longer life and high efficiency.

(*v*) The operation of a JFET depends upon bulk material current carriers that do not cross junctions therefore the inherent noise of tubes and those of transistor are not present in a JFET.

JFET Applications

The high input impedance and low noise level make JFET superior to BJT. Some of its applications are :

(*i*) **As a Buffer Amplifier :** A buffer amplifier is a stage of amplification that isolates the preceeding stage from the following stage. Because of the high input impedance and low output impedance, a JFET can act as an excellent buffer amplifier (Fig. 2.36). The high input impedance of a JFET means light loading of the preceeding stage. This permits almost the entire output from first stage to appear at the buffer inputs. Low output impedance of JFET can drive heavy loads or small load resistances. This ensures that all the output from the buffer reaches the input of the second stage.

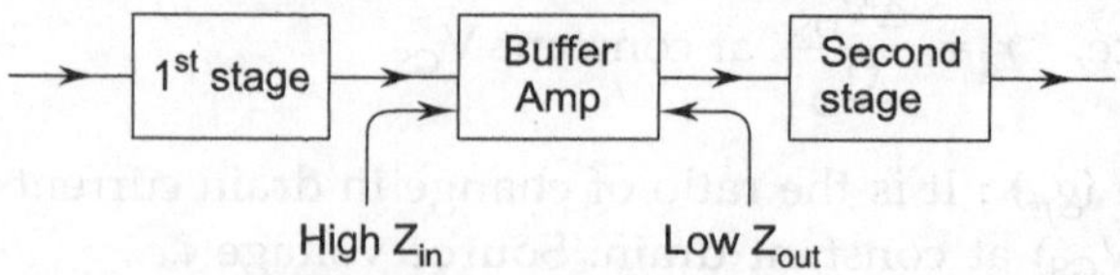

Fig. 2.36 JFET as buffer

(*ii*) **As RF Amplifier :** In communication circuits, we have to use JFET RF amplifier in a receivers instead of BJT for the reasons :

(*a*) Noise level of JFET is very low.

(*b*) The antenna of the receiver receives a very weak signal that has an extremely low amount or current. Since JFET is a voltage controlled device, it will respond well to low current signal provided by the antenna.

(*iii*) **Phase-shift Oscillators :** Oscillators will work with JFETs. The high input impedance of JFET is valuable in phase shift oscillators to minimise loading effect.

Example 2.6. *When V_{GS} of a JFET change from – 3.1 V to – 3 V, the drain current changes from 1 mA to 1.3 mA. What is the value of transconductance ?*

Solution. $\Delta V_{GS} = 3.1 - 3 = 0.1\ V$

$\Delta I_D = 1.3 - 1 = 0.3\ mA$

Transconductance, $$g_{fs} = \frac{\Delta I_D}{\Delta V_{GS}} = \frac{0.3\ mA}{0.1\ V} = \frac{3\ mA}{V} = 3000\ \mu\mho.$$

2.6 BIASING THE FET

For the proper functioning of a linear FET amplifier, it is necessary to maintain the operating point Q stable in the central portion of the pinch off region. The Q-point should be independent of device parameter variations and ambient temperature changes. This can be achieved by suitably selecting the gate to source voltage (V_{GS}) and drain current (I_D) which is referred to as biasing.

An FET may have a combination of self bias and fixed bias to provide stability of the quiescent drain current against device and temperature variations.

1. Self-Bias Circuit

Figure 2.37 shows the self-bias circuit for an N-channel FET when the drain voltage V_{DD} is applied, a drain current I_D flows even in the absence of gate voltage (V_G). The voltage drop across resistor R_S produced by the drain current is given by $V_S = I_D R_S$. This voltage drop reduces the gate to source reverse voltage required for FET operation. The resistor R_S, the feedback resistor prevents any variation in FET drain current.

The drain voltage, $V_D = V_{DD} - I_D R_D$

The drain to source voltage,

$$V_{DS} = V_D - V_S = (V_{DD} - I_D R_D) - I_D R_S$$
$$= V_{DD} - I_D (R_D + R_S)$$

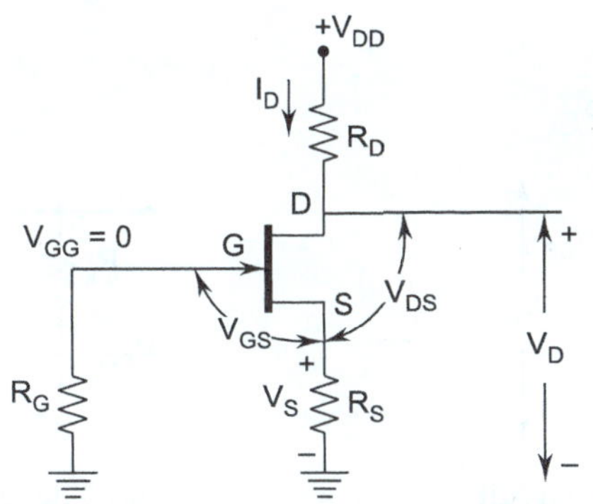

Fig. 2.37 Self bias circuit for JFET

The gate to source voltage,

$$V_{GS} = V_{GG} - V_S = 0 - I_D R_S = - I_D R_S$$

When drain current increases, the voltage drop across R_S increases. The increased voltage drop increases the reverse gate to source voltage, which decreases the effective width of the channel and hence reduces the drain current. Now the reduced drain current decreases, the gate to source voltage which, in turn, increases the effective width of channel thereby increasing the value of drain current.

Example 2.7. *Calculate the operating point of the self biased JFET having the supply voltage $V_{DD} = 15$ V, maximum value of drain current $I_{DSS} = 10$ mA and $V_{GS} = -2.5$ V at $I_D = 3$ mA. Also, determine the values of resistors R_D and R_S to obtain this bias condition.*

Solution. w.r.t., the value of drain current at Q-point,

$$I_{DQ} = \frac{I_{DSS}}{2} = \frac{10 \times 10^{-3}}{2} = 5 \text{ mA}$$

and the value of drain to source voltage at Q-point,

$$V_{DSQ} = \frac{V_{DD}}{2} = \frac{15}{2} = 7.5 \text{ V}$$

Therefore, the operating point is at $V_{DS} = 7.5$ V and $I_D = 5$ mA. Also, we known that the drain to source voltage,

$$V_{DS} = V_{DD} - I_D R_D$$

$$7.5 = 15 - (3 \times 10^{-3}) R_D$$

Therefore, $$R_D = \frac{15 - 7.5}{3 \times 10^{-3}} = 2.5 \text{ k}\Omega.$$

The source voltage or voltage across the source resistor R_S is,

$$V_S = -V_{GS} = +2.5 \text{ V. Also, } V_S = I_D R_S$$

i.e., $$2.5 = (3 \times 10^{-3}) R_S$$

Therefore, $$R_S = \frac{2.5}{3 \times 10^{-3}} = 833.33\ \Omega$$

2. Voltage Divider Bias Circuit

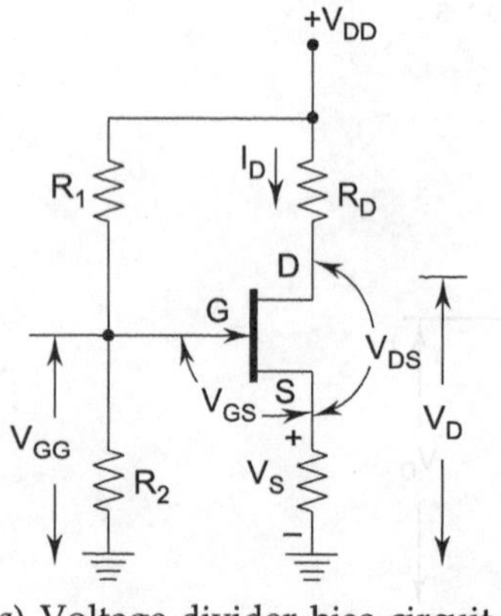

(*a*) Voltage divider bias circuit

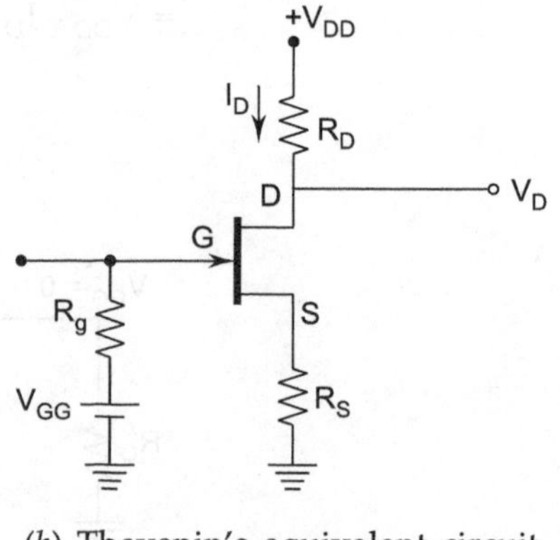

(*b*) Thevenin's equivalent circuit

Fig. 2.38 JFET biasing

Figure 2.38 (*a*) shows the voltage divider bias circuit and its Thevenin's equivalent is shown in Fig. 2.38(*b*). Resistors R_1 and R_2 connected on the gate side forms a voltage divider. The gate voltage,

$$V_{GG} = \left(\frac{R_2}{R_1 + R_2}\right)V_{DD} \text{ and } R_g = \frac{R_1 \cdot R_2}{R_1 + R_2}$$

The bias line satisfies the equation $V_{GS} = V_{GG} - I_D R_S$.

The drain to ground voltage, $V_D = V_{DD} - I_D R_D$. If the gate voltage V_{GG} is very large as compared to gate to source voltage V_{GS}, the drain current is approximately constant. In practice, the voltage divider bias is less effective with JFET than BJT. This is because in BJT, $V_{BE} \approx 0.7$ V (silicon) with only minor variations from one transistor to another. But in a JFET, the V_{GS} can vary several volts from one JFET to another.

Example 2.8. *A JFET amplifier with a voltage divider biasing circuit has the following parameters : $V_P = -3$ V, $I_{DSS} = 4$ mA, $R_S = 3.5$ kΩ, $R_D = 950$ Ω, $R_1 = 15$ MΩ, $R_2 = 10$ MΩ and $V_{DD} = 25$ V. Find the value of the drain current I_D at the operating point. Verify whether the FET will operate in the pinch off region.*

Solution. From Fig. 2.38(*a*), we obtain

$$V_{GG} = V_{DD}\frac{R_2}{R_1 + R_2} = 25.\frac{10 \times 10^6}{(15+10) \times 10^6} = 10 \text{ V}$$

w.r.t.,

$$I_D = I_{DSS}\left(1 - \frac{V_{GS}}{V_P}\right)^2$$

$$= I_{DSS}\left(1 - \frac{V_{GG} - I_D R_S}{V_P}\right)^2, \text{ where } V_{GS} = V_{GG} - I_D R_S$$

Expressing I_D and I_{DSS} in mA, we have

$$I_D = 4 \times \left(1 - \frac{10 - I_D \times 3.5}{-3}\right)^2$$

i.e., $5.39I_D^2 - 40.04I_D + 74.36 = 0$

Therefore, $I_D = 4.6$ mA or 2.97 mA

As $I_D = 4.6 \text{ mA} > 4 \text{ mA} = I_{DSS}$, this value is inappropriate.

So, $I_{DQ} = 2.97$ mA is selected.

Therefore, $V_{GSQ} = V_{GG} - I_{DQ} R_S = 10 - (2.97 \times 10^{-3} \times 3.5 \times 10^3) = -0.395$ V

and $V_{DSQ} = V_{DD} - I_{DQ}(R_D + R_S) = 25 - 2.97 - 10^{-3}(950 + 3.5 \times 10^3)$ m

$= 11.78$ V

Then $V_{DGQ} = V_{DSQ} - V_{GSQ} = 11.78 + 0.395 = 12.1785$ V

which is greater than $|V_P| = 2$ V. Hence, the FET is in the pinch-off region.

3. Fixed Bias Circuit

Figure 2.39 shows the fixed bias circuit for the *n*-channel FET. This is the simplest, biasing arrangement. To make gate-source junction reverse biased, a separate supply V_{CC} is connected such that gate is more negative than the source.

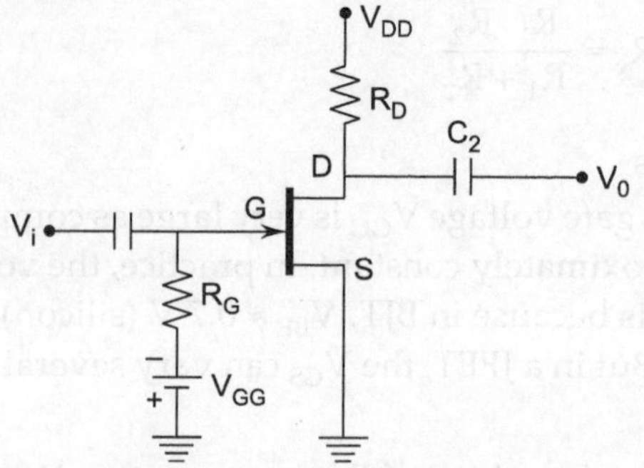

(a) Fixed bias circuit for JFET

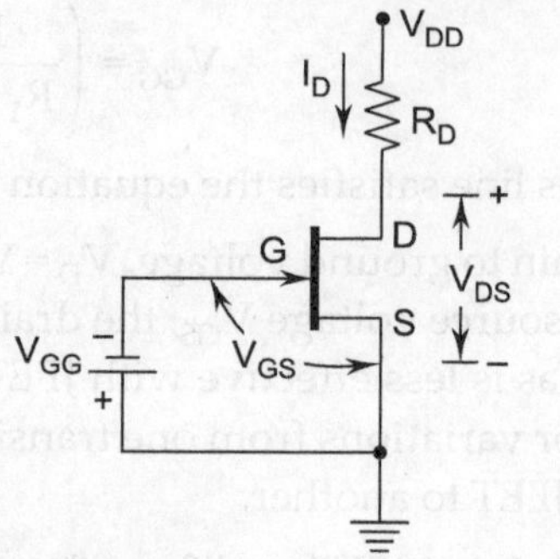

(b) Simplified fixed bias circuit

Fig. 2.39 JFET fixed biasing

For the DC analysis coupling capacitors are open circuits. The current through R_G is I_G which is zero. This permits R_G to replace by short circuit equivalent, simplifying the fixed bias circuit as shown in Fig. 2.39(*b*).

We know for DC analysis,

$$I_G = 0 \text{ A}$$

and applying KVL to the input circuit, we get

$$V_{GS} + V_{GG} = 0$$

$$\therefore \quad V_{GS} = -V_{GG}$$

Since V_{GG} is a fixed DC supply, the voltage V_{GS} is fixed in magnitude, and hence the name fixed bias circuit.

For fixed bias circuit, the drain current, I_D can be calculated using the following equation

$$I_D = I_{DSS}\left(1 - \frac{V_{GS}}{V_P}\right)^2$$

The drain to source voltage of output circuit can be determined by applying KVL.

$$V_{DS} + I_D R_D - V_{DD} = 0$$

$$\therefore \quad V_{DS} = V_{DD} - I_D R_D$$

The Q point of the JFET amplifier with fixed bias circuit is given by

$$I_{DQ} = I_{DSS}\left(1 - \frac{V_{GS}}{V_P}\right)^2$$

$$V_{DQ} = V_{DD} - I_{DQ} R_D$$

Example 2.9. *For the circuit shown in Fig. 2.40, calculate (a) V_{GSQ}, (b) I_{DQ}, (c) V_{DSQ}, (d) V_D.*

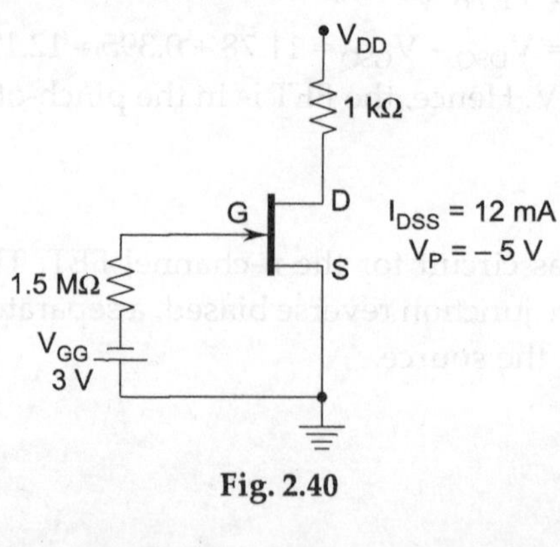

Fig. 2.40

Solution.

(*a*) $V_{GSQ} = -V_{GG} = -3\text{ V}$

(*b*) $$I_{DQ} = I_{DSS}\left(1 - \frac{V_{GS}}{V_P}\right)^2 = 12 \times 10^{-3}\left(1 - \frac{-3}{-5}\right)^2$$
$$= 12 \times 10^{-3}\,(1 - 0.6)^2$$
$$= 12 \times 10^{-3}\,(0.16)$$
$$= 1.92\text{ mA}$$

(*c*) $$V_{DSQ} = V_{DD} - I_{DQ}R_D = 10 - 1.92 \times 10^{-3}\,(1 \times 10^3)$$
$$= 10 - 1.92$$
$$= 8.08\text{ V}$$

(*d*) $$V_D = V_{DS} + V_S = 8.08 + 0 = 8.08\text{ V}.$$

2.7 POWER AMPLIFIERS

Voltage amplifiers are used for voltage amplification but they are not capable of supplying a large power to the loads such as loudspeakers. Whenever the load demands a large power, we have to use specially designed amplifiers called "power amplifiers". The input signal to the large signal amplifier is a high voltage low current signal obtained from a chain of voltage amplifiers. Power is equal to the product of voltage and current. Hence high voltage, low current signal at the input corresponds to a low power signal. Power amplifier will increase the current sourcing and sinking capability, so at its output we get a high voltage, high current signal that means a high power signal. Thus, the power amplifier is basically a current amplifier.

Applications of Power Amplifier

Applications of power amplifiers are :

1. Radio receivers
2. Public address (P.A.) systems
3. CD/Casette players.
4. TV receivers.

Figure 2.41 shows the position of a power amplifier in a public address (P.A.) system. Primary aim here is to supply a large power, the voltage is not important.

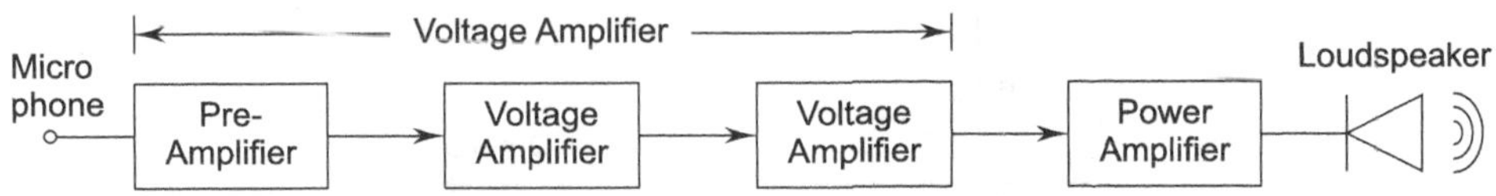

Fig. 2.41 Block diagram of a P.A. system

Classification of Power Amplifiers

Depending on the position of Q point or operating point on the load line, the amplifiers are classifed in following categories :

- Class-A amplifier – At the center of load line
- Class-B amplifier – In the cut-off region
- Class-AB amplifier – Just above the cut-off
- Class-C amplifier – Below the cut-off.

Class A Amplifier

An amplifier is referred to as a class A amplifier if the transistor used for amplification conducts for the full cycle duration of the input AC signal. The Q-point is adjusted exactly at the center of the load line. Due to this the output signal is obtained for the full cycle of the AC input *i.e.,* for 360°. The transistor is biased such that the operating point (Q-point) is at the center of the load line. The input signal is amplified, without any distortions. Typically the efficiency (η) of a class A power-amplifier lies between 25 to 50%. Here the transistor conducts for the complete cycle of AC input *i.e.,* for 360°.

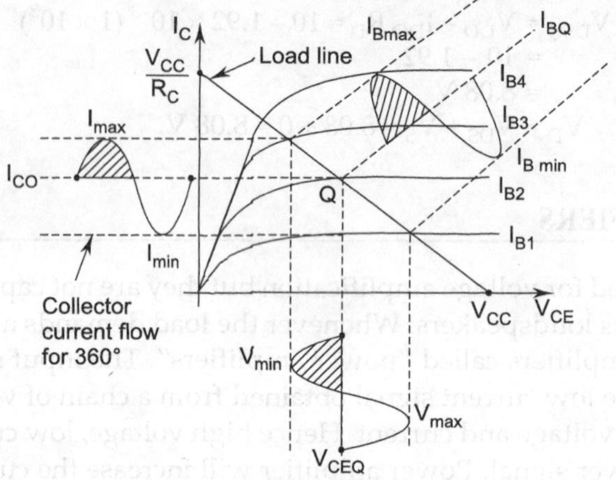

Fig. 2.42 Class A operation

Class B Amplifier

An amplifier is referred to as class B amplifier if the output signal is obtained only for one half cycle period of the input AC signal. The transistor conducts only in one half cycle of the input and the collector current therefore flows for only 180°. For this the Q-point is adjusted to be at cut-off *i.e.,* on the X-axis. Thus in the absence of AC input signal the transistor remains in the off state. Collector current flows for 180° of the input signal. Only one half cycle is obtained at the output. Therefore, the output voltage waveform is distorted. The maximum efficiency can be 78.5%.

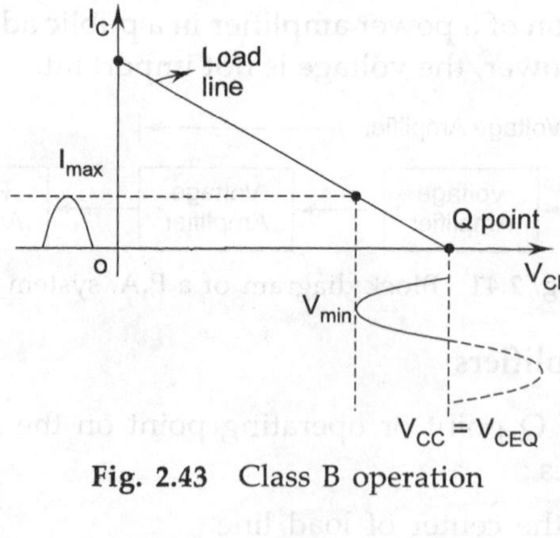

Fig. 2.43 Class B operation

Class C Amplifier

An amplifier is referred to as a class C power amplifier if its output is obtained for less that a half cycle period of the input AC signal. Thus the transistor will conduct for a duration which is less than the period of a half cycle of the AC input signal. For this the operating point is adjusted to be below the X-axis. Thus, the transistor is biased below the cut-off. Hence the transistor can remain in the active region for less than a half cycle period. Thus, collector current flows for less than 180°. Due to this the output signal is heavily distorted. The efficiency of class C is very high. Typically above 95%.

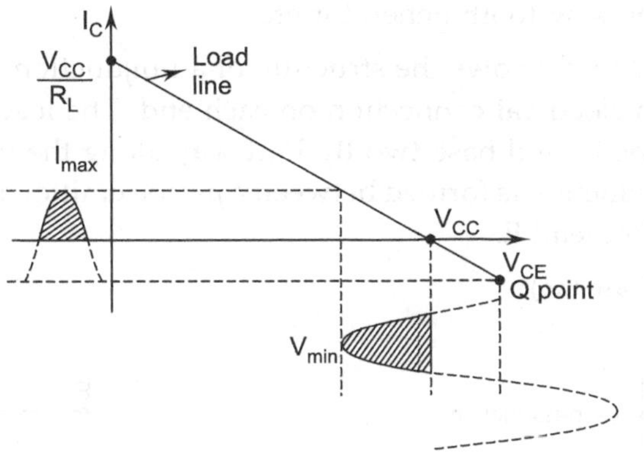

Fig. 2.44 Class C operation

Class AB Amplifiers

The amplifier is said to be class AB amplifier, if the Q point and the input signal are selected such that the output signal is obtained for more than 180° but less than 360°, for a full input cycle. The Q point is above X-axis but below the mid-point or a load line. The output signal is distorted in class AB operation. The efficiency is more than class A but less than class B operation.

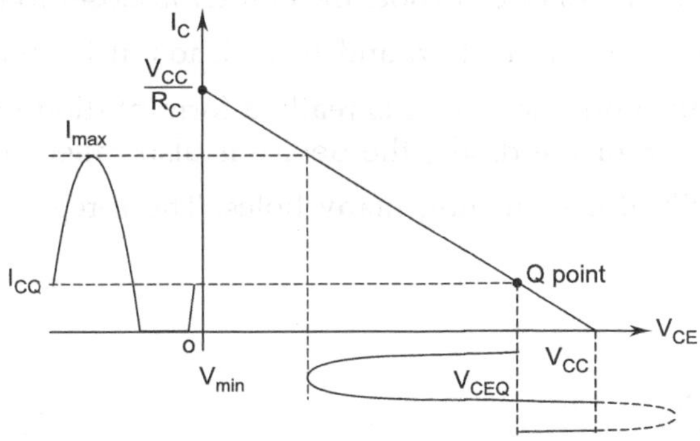

Fig. 2.45 Class AB operation.

Comparison of Power Amplifier

S.No.	Parameter	Class A	Class B	Class C	Class AB
1.	Operating cycle	360°	180°	Less than 180°	180° to 360°
2.	Position of the Q point	At the center	On the X-axis	Below the X-axis	Above X-axis but below the mid-point
3.	Efficiency	Lowest 25% to 50%	Higher (78.5%)	Very high (95%)	Between 50 to 78.5 V
4.	Distortion	No distortion	More than class A	More than A, B and AB	Low

2.8 UNIJUNCTION TRANSISTOR (UJT)

A UJT is a three terminal semiconductor switching device. This device has a unique characteristic that when it is triggered, the emitter current increases regeneratively until it is limited by emitter power supply. Due to this characteristic the UJT can be employed in a variety of application *e.g.*, switching, pulse generator, saw tooth generator etc.

Construction : Figure 2.46(*a*) shows the structure of a unijunction transistor. It consists of an *n*-type silicon bar with an electrical connection on each end. The leads to these connections are called base leads base. One B_1 and base two B_2. Part way along the bar between the two bases, nearer to B_2 than B_1, a *p-n* junction is formed between a *p*-type emitter and the bar. The lead to this junction is called the emitter lead E.

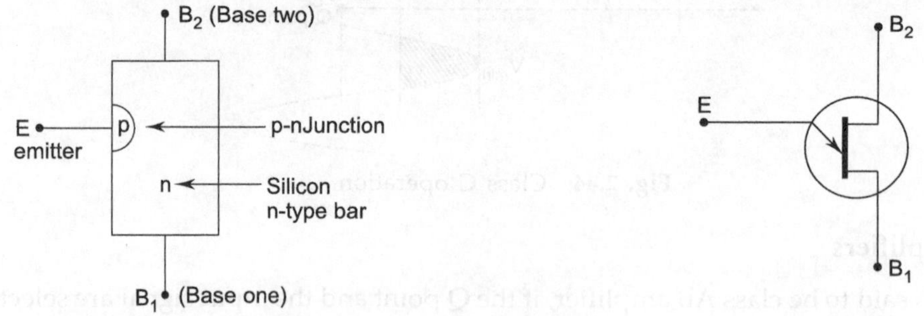

(*a*) UJT construction (*b*) Symbol of UJT

Fig. 2.46

Figure 2.46(*b*) shows the symbol of UJT (note the emitter is closer to B_2 than B_1).

(*i*) Since the device has one *p-n* junction and three leads, it is commonly called UJT.

(*ii*) With only one *p-n* junction, the device is really a form of diode because the two base are taken from one section of the diode, the device is also called double-based diode.

(*iii*) The emitter is heavily doped having many holes. The *n*-region is lightly doped.

Working

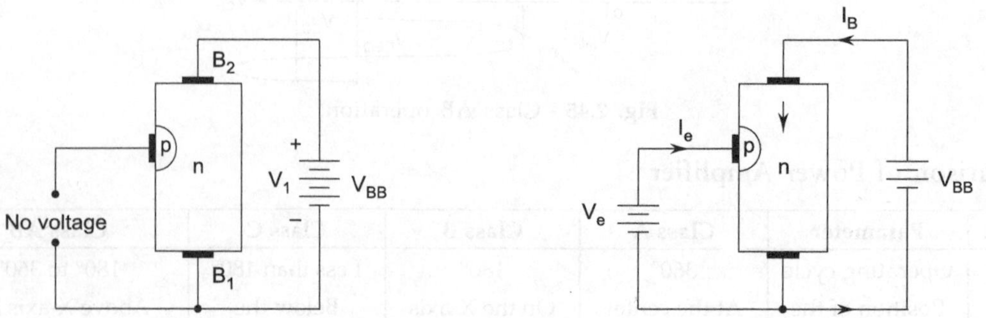

(*a*) (*b*)

Fig. 2.47 UJT working

Figure 2.47 shows the basic circuit operation of a unijunction transistor. The device has normally B_2 positive w.r.t. B_1.

(*i*) If voltage V_{BB} is applied between B_2 and B_1 with emitter open [Fig. 2.47(*a*)] a voltage gradient is established along the *n*-type bar. Since the emitter is located nearer to B_2. More than half of V_{BB} appears between the emitter and B_1. The voltage V_1 between emitter and B_1 establishes a reverse bias on the *p-n* junction and the emitter current is cut-off. A small leakage current flows from B_2 to emitter due to minority carriers.

(*ii*) If a positive voltage is applied at the emitter [Fig. 2.47(*b*)], the *pn* junction will remain reverse biased so long as the input voltage is less than V_1. If the input voltage to the emitter exceeds V_1, the *pn* junction becomes forward biased. Under these conditions, holes are injected from *p*-type material into the *n*-type bar. These holes are repelled by positive B_2 terminal and they are attracted towards B_1 terminal of the bar. This accumulation of holes in the emitter to B_1 region results in the decrease of resistance in this section of the bar. The result is that internal voltage drop from emitter to B_1 is decreased and hence the emitter current I_E increase. As more holes are injected, a condition of saturation will eventually be reached. At this point, the emitter current is limited by emitter power supply only. The device is now in the ON state.

(*iii*) If a negative pulse is applied to the emitter, the *pn* junction is reverse biased and the emitter current is cut-off. The device is then said to be in the OFF state.

Equivalent Circuit of a UJT

Figure 2.48 shows the equivalent circuit of a UJT. The resistance of the silicon bar is called the interbase resistance R_{BB}. The inter-base resistance is represented by two resistors in series.

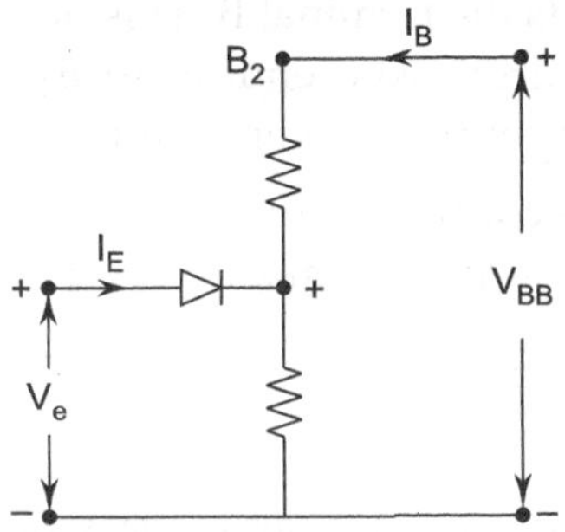

Fig. 2.48 UJT Equivalent circuit

(*a*) R_{B_2} is the resistance of silicon bar between B_2 and the point at which the emitter junction lies.

(*b*) R_{B_1} is the resistance of the bar between B_1 and emitter junction. This resistance is shown variable because its value depends upon the bias voltage across the *pn* junction.

The *pn* junction is represented in the emitter by a diods D.

(*i*) With no voltage applied to the UJT, the interbase resistance is given by :

$$R_{BB} = R_{B_1} + R_{B_2}$$

The value of R_{BB} generally lies between 4 kΩ and 10 kΩ.

(*ii*) If a voltage V_{BB} is applied between the bases with emitter open, the voltage will divide up across R_{B_1} and R_{B_2}.

Voltage across R_{B_1}, $V_1 = \frac{R_{B_1}}{R_{B_1} + R_{B_2}} V_{BB}$

or $\frac{V_1}{V_{BB}} = \frac{R_{B_1}}{R_{B_1} + R_{B_2}}$

The ratio $\frac{V_1}{V_{BB}}$ is called intrinsic stand-off ratio is represented by Greek letter η.

Obviously, $\eta = \frac{R_{B_1}}{R_{B_1} + R_{B_2}}$

The value of η usually lies between 0.51 and 0.82

Voltage across $R_{B_1} = \eta V_{BB}$.

The voltage ηV_{BB} appearing across R_{B_1} reverse bias the diode. Therefore, $I_E = 0$.

(*iii*) If a rising positive voltage is applied to the emitter the diode will become forward biased when input voltage exceeds ηV_{BB} by V_D, the forward voltage drop across the silicon diode *i.e.*,

when $V_P = \eta V_{BB} + V_D$

V_P = 'Peak point voltage'

V_D = Forward voltage drop across silicon diode ($\approx$ 0.7 V)

When the diode D starts conducting, holes an injected from *p*-type material to the *n*-type bar. These holes are swept down towards the terminal B. This decreases the resistance between emitter and B_1 and hence the internal drop from emitter to B_1. The emitter current now increase regeneratively until it is limited by emitter power supply.

The above discussion leads to the conclusion that when input positive voltage to emitter is less than peak-point voltage V_P, the *pn* junction remains reverse biased, and the emitter current is practically zero.

Characteristics of UJT

Figure 2.49 shows the curve between emitter voltage (V_E) and emitter current (I_E) of a UJT at a given voltage V_{BB} between the bases. This is known as the emitter characteristics of UJT.

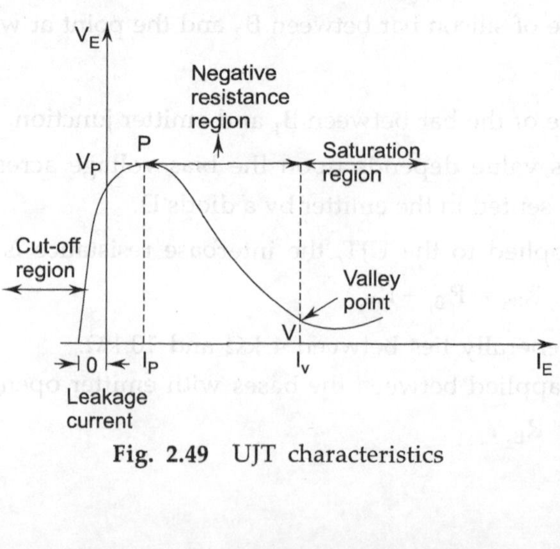

Fig. 2.49 UJT characteristics

(*i*) Initially, in the cuts-off region, as V_E increases from zero, slight leakage current flows from terminal B_2 to the emitter. This current is due to minority carriers in reverse biased diode.

(*ii*) Above a certain value of V_E, forward I_E begins to flow, increasing until the peak voltage V_P and current I_P are reached at point P.

(*iii*) After the peak point P, an attempt to increase V_E is followed by a sudden increase in emitter current I_E with a corresponding decrease in V_E. This is a negative resistance portion of the curve because with increase in I_E, V_E decreases. The device, therefore, has a negative resistance region which is stable enough to be used with a great deal of reliability in many areas, *e.g.*, trigger circuits, saw tooth generators, biasing circuits.

(*iv*) The negative portion of the curve lasts until the valley point V is reached with valley-point voltage V_V and valley-point current I_V. After the valley point, the device is driven to the saturation.

Advantages of UJT

(*i*) It is a low cost device

(*ii*) It has excellent characteristics

(*iii*) It is a low-power absorbing device under normal operating conditions.

Applications of UJT

Unijunction transistors are used extensively in oscillator, pulse and voltage sensing circuits. Some of the important applications of UJT are :

(*i*) **Overvoltage detector :** Figure 2.50 shows a simple DC over voltage indicator. A warning pilot-lamp L is connected between the emitter and B_1 circuit. So long as the input voltage is less than the peak-point voltage (V_P) of the UJT, the device remains switched off. However, when the input voltage exceeds V_P, the UJT is switched on and the capacitor discharges through the low resistance path between terminals E and B_1. The current flowing in the pilot lamp L lights it, thereby indicating the overvoltage in the circuit.

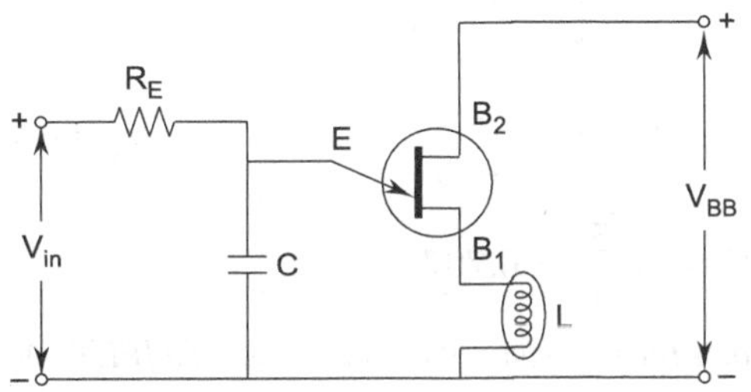

Fig. 2.50 Overvoltage detector

(*ii*) **UJT relaxation oscillator :** Figure 2.51 shows UJT relaxation oscillator where the discharging of a capacitor through UJT can develop a saw-tooth output as shown.

When battery V_{BB} is turned on, the capacitor C charges through resistor R_1. During the charging period, the voltage across the capacitor C rises in an exponential manner until it reaches the peak-point voltage. At this instant of time, the UJT switches to its low resistance conducting mode and the capacitor is discharged between E and B_1. As the capacitor voltage flys back to zero, the emitter ceases to conduct and the UJT is switched off. The next cycle then begins, allowing the capacitor C to charge again. The frequency of the output saw-tooth wave can be varied by changing the value of R_1 since this controls the time constant R_1C of the capacitor charging circuit.

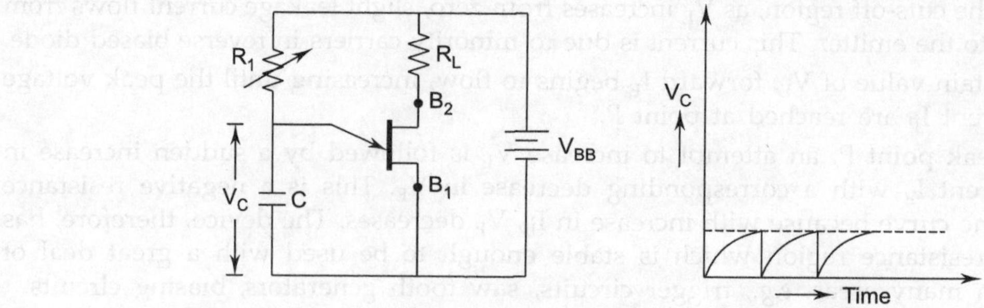

Fig. 2.51 UJT Relaxation oscillator

Example 2.10. *When a reverse gate voltage of 15 V is applied to a JFET the gate current is 10^{-3} μA. Find the resistance between gate and source.*

Solution. $V_{GS} = 15\text{ V}; I_G = 10^{-8}\ \mu\text{A} = 10^{-9}\text{ A}.$

Gate to source resistance $= \dfrac{V_{GS}}{I_G} = \dfrac{15\text{ V}}{10^{-9}\text{ A}} = 15 \times 10^9\ \Omega = 15000\text{ M}\Omega.$

Example 2.11. *A JFET has the following parameters. $I_{DS} = 32$ mA ; $V_{GS\,(off)} = -8$ V ; $V_{GS} = -4.5$ V. Find the value of drain current.*

Solution.
$$I_D = I_{DSS}\left[1 - \frac{V_{GS}}{V_{GS\,(off)}}\right]^2 = 32\left[1 - \frac{(-4.5)}{-8}\right]^2 \text{ mA}$$
$$= 6.12\text{ mA}.$$

Example 2.12. *A JFET has a drain current of 5 mA. If $I_{DSS} = 10$ mA and $V_{GS\,(off)} = -6$ V, find the value of V_{GS} and V_P.*

Solution.
$$I_D = I_{DSS}\left[1 - \frac{V_{GS}}{V_{GS\,(off)}}\right]^2$$
$$5 = 10\left[1 + \frac{V_{GS}}{6}\right]^2$$
$$1 + \frac{V_{GS}}{6} = \sqrt{\frac{5}{10}} = 0.707.$$

$$V_{GS} = -1.76\text{ V}$$
$$V_P = -V_{GS\,(off)} = 6\text{ V}$$

Example 2.13. *Figure 2.52 shows the transistor characteristics of a JFET. Write the equation for drain current.*

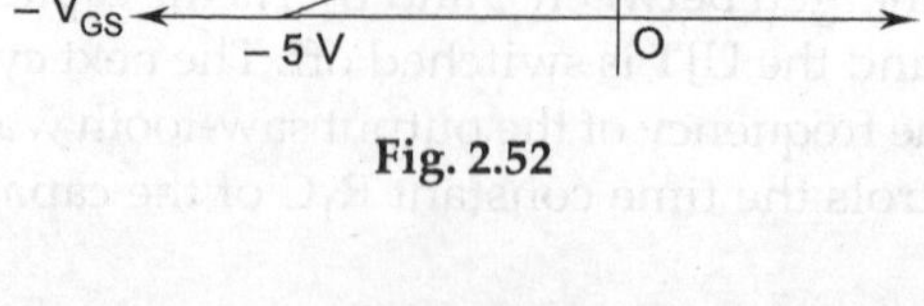

Fig. 2.52

Solution. $I_{DSS} = 12$ mA

$V_{GS\,(off)} = -5$ V

$$I_D = I_{DSS}\left[1 - \frac{V_{GS}}{V_{GS\,(off)}}\right]^2$$

$$I_D = 12\left[1 + \frac{V_{GS}}{5}\right]^2 \text{ mA}$$

Example 2.14. *The intrinsic stand-off ratio for a UJT is determined to be 0.6. If the interbase resistance is 10 kΩ, what are the values of R_{B_1} and R_{B_2} ?*

Solution. $R_{BB} = 10\text{ k}\Omega, \eta = 0.6$

$$R_{BB} = R_{B_1} + R_{B_2}$$

$$10 = R_{B_1} + R_{B_2}$$

Also
$$\eta = \frac{R_{B_1}}{R_{B_1} + R_{B_2}}.$$

$$0.6 = \frac{R_{B_1}}{10} \qquad (\because R_{B_1} + R_{B_2} = 10\text{ k}\Omega)$$

$\therefore$
$$R_{B_1} = 10 \times 0.6 = 6\text{ k}\Omega$$

and
$$R_{B_2} = 10 - 6 = 4\text{ k}\Omega$$

Example 2.15. *A UJT has 10 V between the bases. If the intrinsic stand off ratio is 0.65. Find the value of stand-off voltage. What will be the peak-point voltage, if the forward voltage drop in the pn junction is 0.7 V ?*

Solution. $V_{BB} = 10\text{ V}\,;\ \eta = 0.65\,;\ V_D = 0.7$ V

Stand-off voltage $= \eta V_{BB} = 0.65 \times 10 = 6.5$ V

Peak-point voltage, $V_P = \eta V_{BB} + V_D = 6.5\text{ V} + 0.7\text{ V} = 7.2$ V.

2.9 SILICON CONTROLLED RECTIFIER (SCR)

SCR is an unidirectional device and like diode, it allows current to flow in one direction only. It can be used as a switch. The switching of SCR is controlled by the additional input called gate and biasing conditions. It can be used as a rectifier to convert AC signals to DC signals.

Construction

The SCR is a four layer *p-n-p-n* device, where *p* and *n* layers are alternately arranged. The outer layers are heavily doped. There are three *p-n* junctions called J_1, J_2 and J_3. The outer P layer is called anode and outer *n* layer is called cathode, middle P layer is called gate. The terminals are taken out from their layers, it is shown in Fig. 2.53. Anode must be positive with respect to cathode so that SCR will be forward biased. To turn on an SCR, a current is to be passed through the gate terminal. Hence it is a current operated device.

To turn on an SCR, the anode must be kept at positive potential with respect to cathode, and gate must be also kept positive w.r.t. cathode.

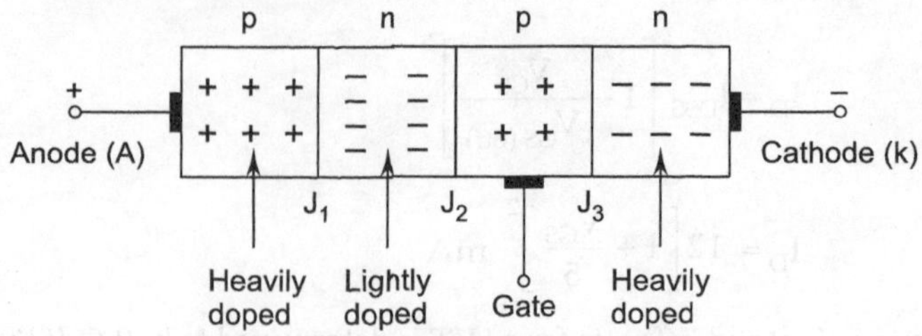

Fig. 2.53 SCR construction

Working of SCR

1. **When gate is open :** Assume that anode is kept positive with respect to cathode and the gate is open. Junctions J_1 and J_3 are forward biased and junction J_2 is reverse biased. In J_2 a small leakage current flows which is negligibly small. Now the SCR is said to be in off condition. This is called forward blocking state of SCR. The voltage applied to anode and cathode is called forward voltage.

   With gate open, if cathode is made positive with respect to anode, the junctions J_1, J_3 become reverse biased and J_2 forward biased. Still the current flowing is leakage current, which can be neglected as it is very small. The SCR is said to be in reverse blocking state and voltage applied is called reverse voltage.

   In forward blocking state, if the forward voltage is increased, the current remain almost zero upto a limit. At a particular value, the reverse biased junction J_2 breaks down and SCR conducts heavily. This voltage is called forward breakover voltage V_{BO} of SCR. In this condition, SCR is said to be ON.

2. **When gate is closed :** Consider that the voltage is applied between gate and cathode when the SCR is in forward blocking state. The gate is made positive with respect to the cathode. The electrons from *n*-type cathode ; cross the junction J_3 to reach positive terminal of battery. While holes from *p*-type move towards the negative terminal of battery. This constitutes the gate current. This current increases the anode current as some of the electrons cross junction J_2.

   As anode current increases, more electron cross J_2 and it further increases. Due to this regenerative action, Junction J_2 breaks and SCR conducts heavily. Once the SCR conducts, the gate loses its control. The only way to turn off an SCR is to bring the anode to cathode voltage to zero.

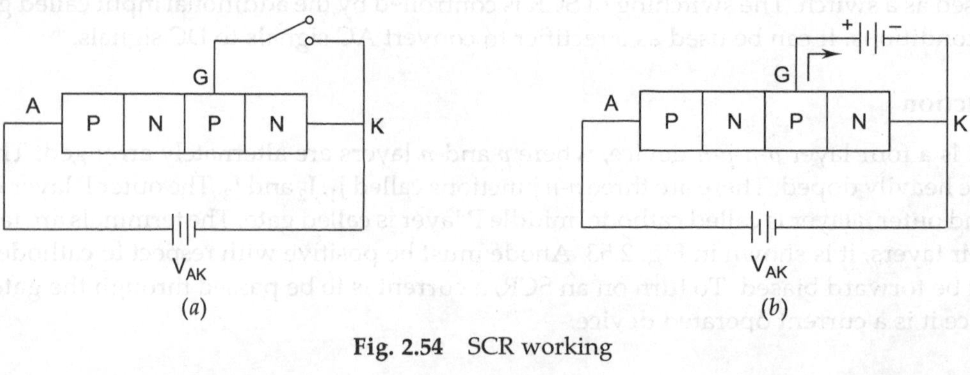

Fig. 2.54 SCR working

V-I Characteristics of SCR

1. **Forward Characteristics :** When $I_G = 0$, it shows a forward blocking region. When forward voltage increases upto V_{BO}, the SCR turn ON and high current-flows.

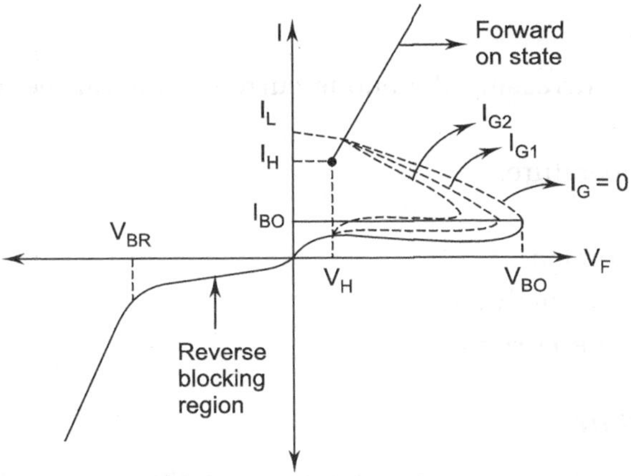

Fig. 2.55 SCR characteristics

It also shows that, if gate bias is used, then gate current increases, less voltage is required to turn ON the SCR. If the forward current falls below the holding current I_H, then depletion region begins to develop around J_2 and device goes into the forward blocking region. When SCR is turned ON from OFF state, the resulting forward current is called latching current I_L. Latching current is slightly higher than the holding current.

2. **Reverse Characteristics :** If the anode to cathode voltage is reversed, then the device enters into the reverse blocking region. The current is negligibly small. If the reverse voltage is increased, at a particular value breakdown occurs and a large current flows. This is called reverse breakdown and the voltage at which this happens is called reverse breakdown voltage.

Important Terms in SCR

1. **Forward Breakover Voltage (V_{BO}) :** This is the maximum forward voltage that can be applied between anode and cathode, without initiating forward conduction. This voltage is defined for a zero gate current. In short this is the maximum forward voltage across SCR in its off state.
2. **Forward Leakage Current :** The small current flowing in the forward direction in the off state of device. This current is generated due to the minority current carriers and therefore is dependent on the operating temperature.
3. **Holding Current (I_H) :** It represents the minimum current that can flow through SCR and still 'hold' it in the on state. The corresponding voltage is termed as holding voltage V_B. If the forward anode current is reduced below holding current, SCR will be turned off. The holding current is defined for zero gate current ($I_G = 0$).
4. **ON State Voltage :** Voltage across SCR in its on state is very low as compared to the off state voltage. The on state voltage is equal to drop across one junction (*i.e.*, of the order of 1 to 1.5 volts).

5. **Latching Current (I_L)** : It is the minimum anode current that must flow through SCR to latch it into the on state. The latching current is higher than the holding current. The latching current is important when SCR is being turned on.

SCR Turn on Methods

SCR can be turned on by increasing the anode current. This can be achieved with one of the following ways :

- By raising the temperature.
- By focussing light.
- By applying high voltage.
- By applying voltage at high *dv/dt*.
- By increasing the gate current.

By raising the temperature

If the temperature of SCR increases, there is an increase in the number of electron-hole pairs. This will increase the leakage current. Then SCR may be turned on due to increased temperature, but it may cause thermal runaway.

By focussing light

If light of adequate frequency and intensity is allowed to strike SCR junction, then the photons will strike the electrons and increase the 'number of electron hole pairs. SCR may turn on due to the increased leakage current. This principle is used to turn on the light activated SCR (LASCR).

By applying high voltage

SCR will turn on if the forward anode to cathode voltage is greater than the forward breakdown voltage V_{BO}, because then a sufficient leakage current will flow to turn on SCR.

By applying voltage at high $\frac{dv}{dt}$

We should avoid this type of turn on. It may even damage SCR.

By increasing the gate current

With increase in the gate current, the breakdown voltage will decrease.

Drawbacks of SCR

1. It can conduct only in one direction. So it can control power only during one half cycle of AC.
2. It can turn on accidentally due to high $\frac{dv}{dt}$ of the source voltage.
3. Gate current cannot be negative.
4. SCR cannot used at high frequencies. The maximum frequency of its operation is 400 Hz.

Advantages of SCR

1. It can handle large voltages, currents and power
2. Easy to turn on.
3. Triggering circuits are simple
4. It can be protected with the help of fuse
5. We can control the power delivered to the load.

2.10 BIDIRECTIONAL TRIODE SCRS (TRIACS)

A TRIAC is a bidirectional device is a device which allows current to flow through it in either directions.

Construction of a TRIAC

In the SCR family, after the SCR. TRIAC is the most widely used device for power control. It is a three terminal device, with the terminals named as main terminals 1, 2 (MT_1 and MT_2) and gate, out of which gate is the control terminal. TRIAC is a bidirectional device, *i.e.*, the current can flow in both the directions through it.

The basic structure of triac is shown in Fig. 2.56(*a*). Its operation is equivalent to two SCRs connected in anti-parallel. The two main terminals are designated MT_2 and MT_1 (main terminal-MT). The gate is near MT_1.

TRIAC can be turned on by applying either a positive or negative voltage to the gate with respect to terminal MT_1, whereas the SCR can be triggered only by a positive gate signal. The other difference is that the SCR conducts only in the forward direction whereas the triac conducts in both the directions.

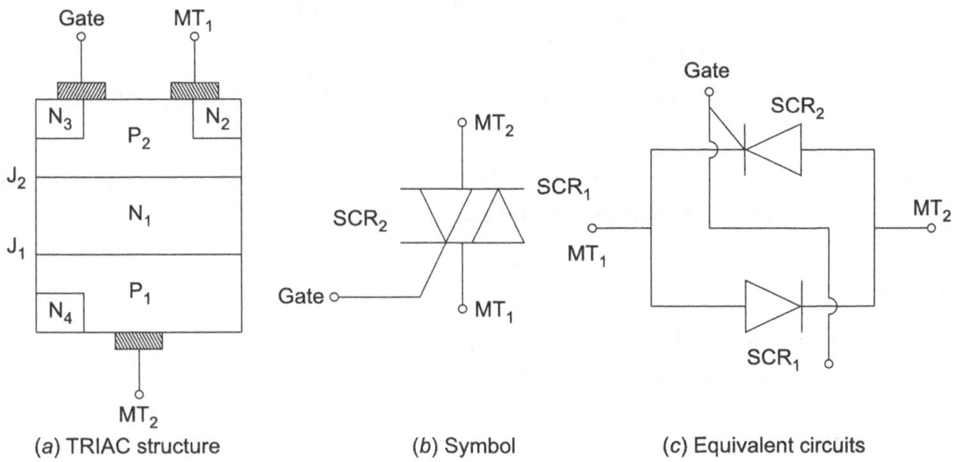

Fig. 2.56 TRIAC

Characteristics of TRIAC

The characteristics of triac is same as that of two back to back connected SCRs. In triac the gate current is also same. *i.e.*, with increase in the gate current the breakdown voltage decreases.

Its characteristics can be divided into three regions of operation :

1. Blocking state or off state
2. Transition or unstable state.
3. Conduction state or on state.

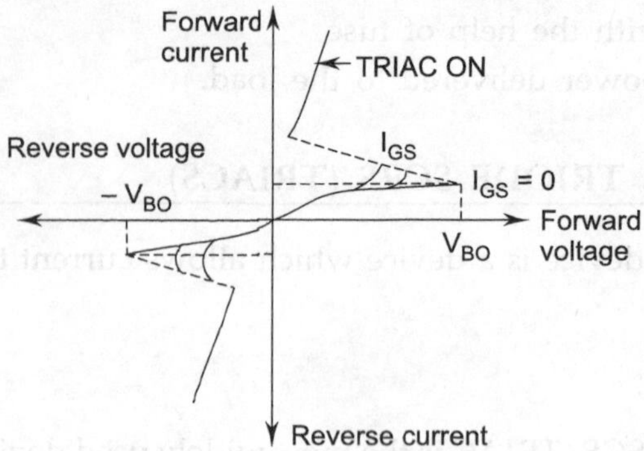

Fig. 2.57 TRIAC characteristics

Different States of TRIAC Operation

The three states of triac operations are :

1. Forward blocking state
2. Reverse blocking state
3. Conduction or ON state

Forward blocking state : (MT_2 positive with respect to MT_1) :

When forward voltage is less than V_{BO}, applied with gate terminal open, the TRIAC can successfully block the forward voltage without getting turned ON.

Reverse blocking state : (MT_1 positive with respect to MT_2) :

When reverse voltage is less than V_{BO} with gate open circuited, the TRIAC will block the reverse voltage without getting turned ON.

Conduction or ON state: The gate current can be positive or negative. The forward and reverse breakover voltage reduces with increase in the gate current.

Modes of Operation

I^+, I^-, III^+ and III^- are the four modes of operation of a TRIAC, where I or III represents the quadrant of operation and (+) and (–) indicates the direction of gate current.

Advantages of TRIAC

1. It is a bi-directional device. So we can control the power delivered to load in both the half cycles of AC supply.
2. It is equivalent to two SCRs connected back to back. We can turn it on by using a positive as well as negative gate current.

3. It is more economical than SCRs, since inside the same package, we get two SCRs connected back to back.
4. It is not necessary to use a protection diode across the TRIAC.

Disadvantages of TRIAC

1. We cannot use it as a controlled rectifier.
2. Low $\frac{di}{dt}$ rating, and low $\frac{dv}{dt}$ rating than SCR.
3. Suitable for resistive loads only.
4. Power rating is lower than that of SCR.

Applications of TRIAC

1. As a static switch
2. Fan speed regulator
3. Flasher circuit
4. Temperature controller
5. In AC voltage stabilizer.

2.11 DIAC

Diac is one of the most important devices used to trigger the TRIAC.

Construction of DIAC

The DIAC is basically a two terminal device. It is sometimes calld as a triac without gate terminal. It is used as a triggering device for the triac circuits. The arrangement of semicondcutor layers is shown in Fig. 2.58(a).

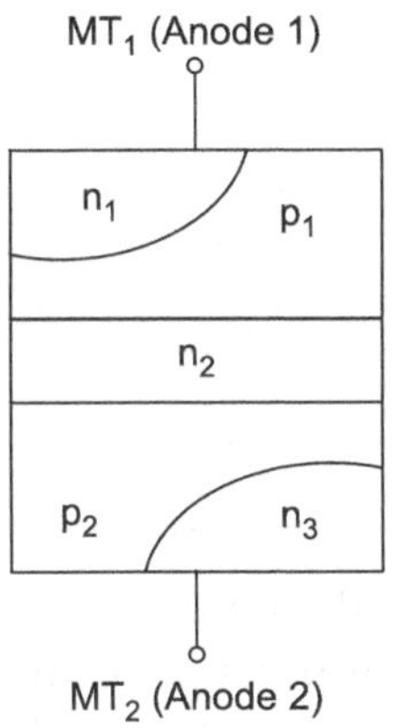

(a) Structure

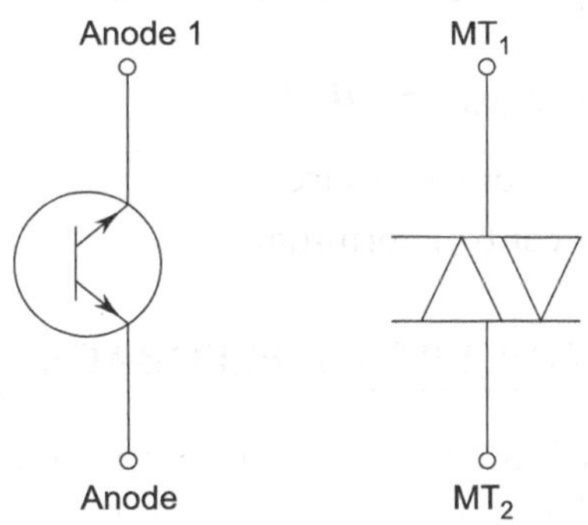

(b) DIAC Symbols

Fig. 2.58 DIAC

The two terminals are MT_1 and MT_2. (Main Terminal = MT). When MT_1 is positive with respect to MT_2, the conduction takes place through the layer p_1-n_2-p_2-n_3 which is equivalent to SCR. When MT_2 is positive with respect to MT_1 conduction takes place through another SCR. Hence we can assume it as two SCRs connected back to back. Symbols for diac are shown in Fig. 2.58(b).

V-I Characteristics of a DIAC

The device can be turned on by either positive or negative half cycle of an a.c. voltage. For the positive cycle of an ac voltage, if the applied voltage is less than the forward breakover voltage, a very small current flows through the device called 'leakage current'. It is due to the drift of electron and holes at the depletion region and not sufficient to cause conduction. Hence the diac remain in non-conducting state called 'blocking state'.

As soon as the applied voltage reaches the breakover voltage, the device starts conducting. The current starts increasing and voltage starts decreasing. This region is the conduction state. The characteristics is much similar to that of TRIAC.

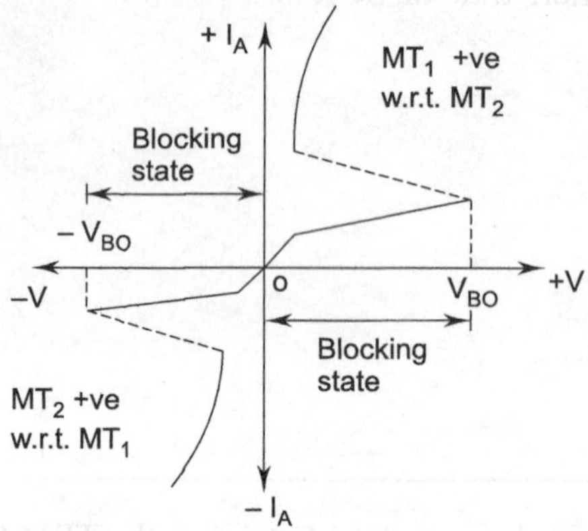

Fig. 2.59 DIAC characteristics

DIAC Applications

1. Lamp dimmer
2. Fan speed regulator
3. Temperature controller
4. Triggering device for TRIAC.

Disadvantages of DIAC

1. Low power device
2. No control terminal.

2.12 CONCEPT OF FEEDBACK

In feedback process, a part of output is sampled and feedback to the input. It plays an important role in all electronic circuits. At the input, we have two signals : input signal, and part of output which is feedback when both are in phase, the feedback is called positive feedback. When they are out of phase, the feedback is called negative feedback.

Negative Feedback

Let us see negative feedback in a process control system. Figure 2.60 shows the block diagram of process control system with negative feedback. Consider the process is a heating process and we have to maintain the temperature at a desired level. If the case is such that the temperature is below the desired level ; in this case feedback signal is less than desired signal and hence the difference signal is positive. It will increase the temperature by continuing heating. If the case is opposite such that the temperature is greater than the desired level, the difference signal is negative. It stops heating process. In this way, the heating process is controlled with help of negative feedback.

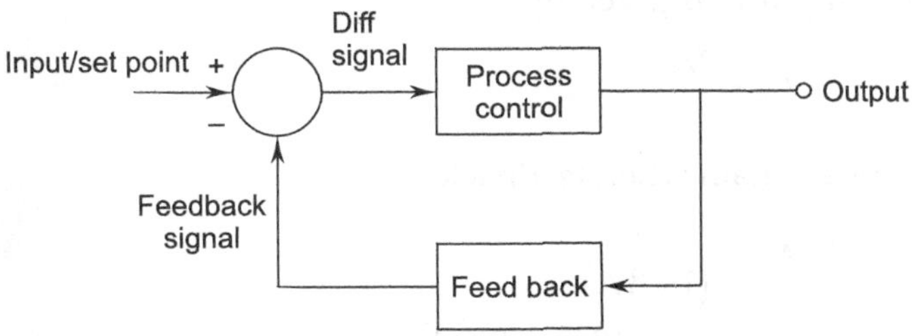

Fig. 2.60 Process control system

Negative Feedback in Amplifier

Figure 2.61 shows the block diagram of an amplifier with negative feedback. Here we assume that the amplifier is of inverting type which will provide a phase shift of 180° between its input and output signals. The input signal V_S is assumed to be sinusoidal.

Operation : The feedback factor of this network is β (part of output returned to input of inverting amplifier through a feedback network). Hence the relation between output and input of the feedback network is given by

$$V_F = \beta V_0 \qquad (V_F = \text{feedback signal})$$

$$\therefore \quad \text{Feedback factor} \quad \beta = \frac{V_F}{V_0}$$

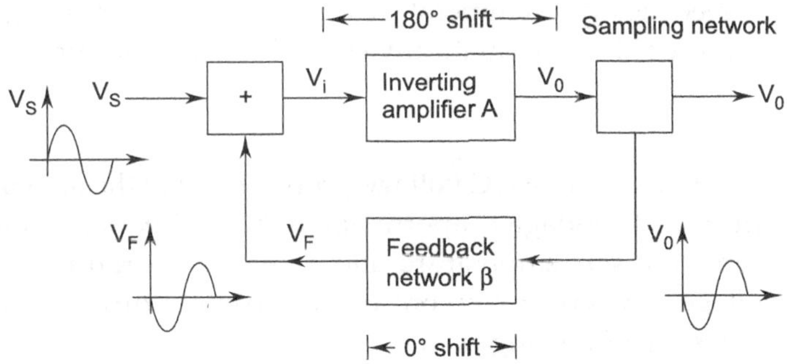

Fig. 2.61 Negative feedback

The value of the feedback factor (β) lies between 0 and 1. If β = 0, then the value of feedback voltage $V_F = 0$, so there is no feedback. If β = 1, then $V_F = V_0$ and all the output voltage is feedback. The feedback network is assumed to be purely resistive. Therefore, the phase shift introduced by it is 0°, hence V_0 and V_F are in phase.

The inverting amplifier introduces a 180° phase shift between V_0 and V_S. In this way the phase shift between V_S and V_0 is 180° and that between V_0 and V_F is 0°. So, the input signal V_S and feedback signal V_F are 180° out of phase. The mixer and sampling network do not introduce any phase shift. The job of sampling network is to sample either the output voltage or output current. In the mixing network the source signal V_S and the feedback signal are mixed together and its output is proportional to the difference between them. The actual input voltage V_i is given by

$$V_i = V_S - V_F$$

As the amplifier input is the difference between V_S and V_F, this feedback is called "negative feedback".

Gain of the feedback amplifier is given by

$$A_F = \frac{V_0}{V_S}.$$

The expression for voltage gain when feedback is

$$A_F = \frac{A}{(1+\beta A)}$$

where A_F is closed loop gain or gain with feedback and A is gain without feedback.

Applications of Feedback

Temperature Controller:

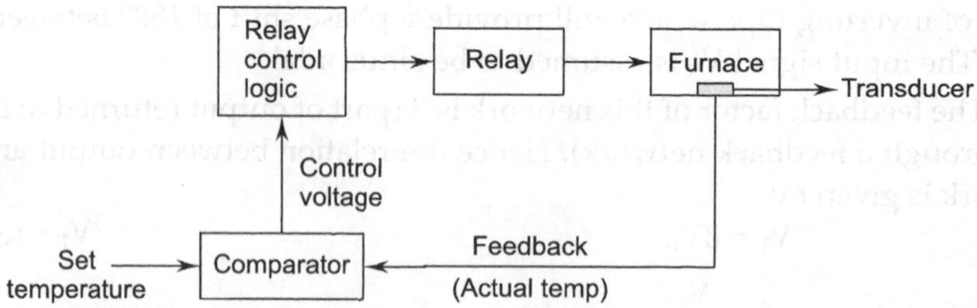

Fig. 2.62 Temperature controller

The block diagram of feedback type temperature controller is shown in Fig. 2.62. The principle of negative feedback can be used to automatically control the temperature to the desired value.

Working

The temperature transducer produces a DC voltage proportional to the actual temperature of the furnace. Set temperature is a DC voltage proportional to the desired temperature of the furnace. The comparator compares the set temperature and actual temperature to produce a control voltage. This control voltage drives the relay control logic to either turn ON or OFF the relay. The relay contacts divides the state of heating filament.

When we turn on the circuit, the actual temperature is lower than the set temperature. So the control voltage is low. The relay is OFF and the heating filament is ON so furnace temperature starts increasing. As soon as actual temperature crosses the set temperature, the control voltage becomes high. The relay turns ON and the heating filament is turned OFF. So furnace temperature starts decreasing. Thus, the temperature is kept constant at the set temperature level.

Speed Control System

Another application of negative feedback is in speed control.

Armature Voltage Control

There are different methods of controlling the speed of DC shunt motor. One of the method is armature voltage control method.

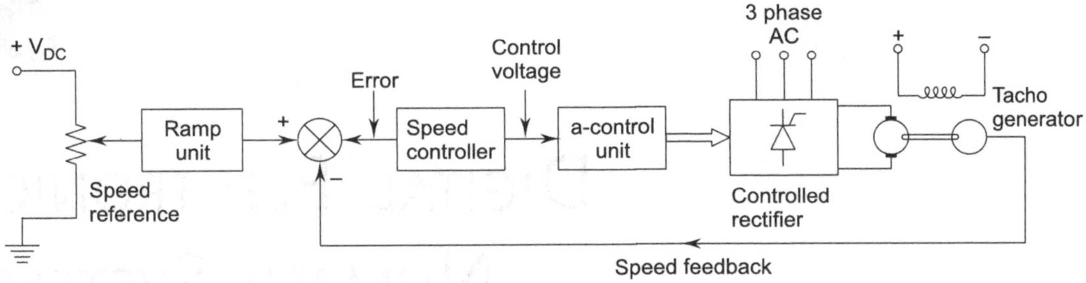

Fig. 2.63 Armature voltage control method

In this method, the field winding is excited by a constant DC voltage and the armature voltage is varied in order to vary the motor speed. This method is applicable only for speed variation below the rated speed of the motor. The reference voltage is passed through a unit called 'ramp' unit to provide the gradual acceleration of the motor. The output of ramp unit acts a speed reference. The reference voltage is compared with speed feedback signal obtained from the DC tacho generator. The difference between the reference speed and feedback speed is known as the "speed error". The speed controller will act on this speed error to produce a control voltage V_C which changes the firing angle α in such a way that armature voltage is automatically increased or decreased. This will automatically reduce the speed error.

Field Current Control

Here the armature voltage is kept constant and field current is varied. The output of the thyristor controlled rectifier drive the field winding of the motor. This method is used for ranging the speed of the DC shunt motor above the rated speed.

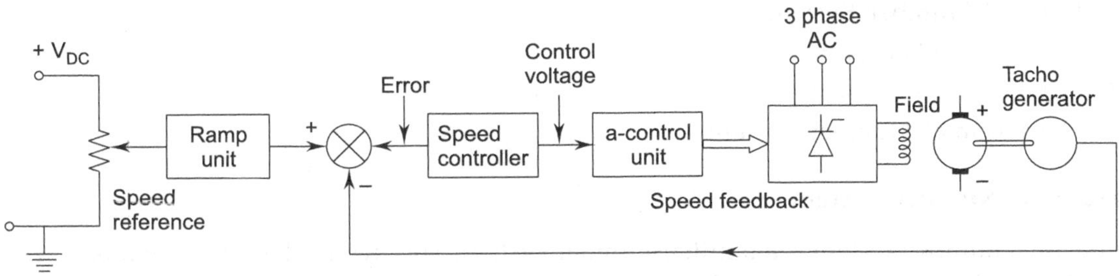

Fig. 2.64 Field current control method

3

DIGITAL ELECTRONICS NUMBER SYSTEMS

3.1 INTRODUCTION

The digital circuit operates with binary numbers. The output of the circuit is either low (zero) or high (one) in a positive logic system. Generally, 0 represents zero volts and 1 represents five volts. If the situation is reverse, it is known as negative logic system. Digital circuits, find application in computers, data processing, telephony, military systems, radar navigation, military systems, medical instruments and consumer products.

Number Systems

The number systems are used in the field of digital electronics and computers. There are several number systems but the following are the important ones in the field of digital electronics :

1. Decimal Number System
2. Binary Number System
3. Octal Number System
4. Hexadecimal Number System

1. Decimal Number System

The decimal number system is a familiar number system. This system has 10 numerals (0, 1, 2, 3, 4, 5, 6, 7, 8, 9). The decimal system is also called the base-10 system because it has 10 digits. Figure 3.1 shows the decimal position values as powers of 10.

MSB ····	10^4	10^3	10^2	10^1	10^0	•	10^{-1}	10^{-2}	10^{-3}	10^{-4}	10^{-5}	···· LSB
						↑ Decimal point						

Fig. 3.1 Decimal position values as power of 10.

For example, the decimal number 25.43 is equal to

$(2 \times 10^1) + (5 \times 10^0) \cdot (4 \times 10^{-1}) + (3 \times 10^{-2})$

$20 + 5 \cdot 0.4 + 0.03 = 25.43$

In this, number 25 is called integer part and the number 43 is called fractional part. Both are separated by a decimal point.

2. Binary Number System

The binary number system is a base 2 number system. This number system has two digits (bits) 0 and 1. The weights in a binary number are based in powers of two. Figure 3.2 shows the binary position values as powers of 2.

MSB ·····	2^4	2^3	2^2	2^1	2^0	•	2^{-1}	2^{-2}	2^{-3}	2^{-4}	2^{-5}	····· LSB

Fig. 3.2 Binary position values as power of 2.

For example, the binary number 1 1 1 0. 1 1 1 can be represented as

$$N = (1 \times 2^3) + (1 \times 2^2) + (1 \times 2^1) + (0 \times 2^0) \cdot (1 \times 2^{-1}) + (1 \times 2^{-2}) + (1 \times 2^{-3})$$
$$= (8 + 4 + 2 + 0) \cdot (0 + 0.5 + 0.25 + 0.125)$$
$$= 14.875$$

3. Octal Number System

The octal number system is a base 8 number system. This number system has 8 digits 0, 1, 2, 3, 4, 5, 6 and 7. The weights in a octal number are based in powers of eight. Figure 3.3 shows octal position values as power of 8.

MSB ·····	8^4	8^3	8^2	8^1	8^0	•	8^{-1}	8^{-2}	8^{-3}	8^{-4}	8^{-5}	····· LSB

Fig. 3.3 Octal position values as power of 8.

For example, the octal number 557.34 can be represented as

$$N = (5 \times 8^2) + (5 \times 8^1) + (7 \times 8^0) \cdot (3 \times 8^{-1}) + (4 \times 8^{-2})$$
$$= (320 + 40 + 7) \cdot (0.375 + 0.0625)$$
$$= 367.4375$$

4. Hexadecimal Number System

The hexadecimal number system is a base 16 number system. This number system has 16 digits 0, 1, 2, 3, 4, 5, 6, 7, 8, 9, A, B, C, D, E and F. The weights in a hexadecimal number are based in powers of sixteen. Figure 3.4 shows hexadecimal position values as power of 16.

MSB ·····	16^4	16^3	16^2	16^1	16^0	•	16^{-1}	16^{-2}	16^{-3}	16^{-4}	16^{-5}	····· LSB

Fig. 3.4 Hexadecimal position values as power of 16.

For example, the hexadecimal number AB.5F can be represented as

$$N = (A \times 16^1) + (B \times 16^0) \cdot (5 \times 16^{-1}) + (F \times 16^{-2})$$

$$= (10 \times 16) + (11 \times 1) \cdot (5 \times 16^{-1}) + (15 \times 16^{-2})$$
$$= (160 + 11) \cdot (0.3125 + 0.0586)$$
$$= 171.3711$$

Table 3.1 shows the binary, octal and hexadecimal numbers corresponding decimal numbers.

Table 3.1. Number systems

Decimal	Binary	Octal	Hexadecimal
0	0000	0	0
1	0001	1	1
2	0010	2	2
3	0011	3	3
4	0100	4	4
5	0101	5	5
6	0110	6	6
7	0111	7	7
8	1000	10	8
9	1001	11	9
10	1010	12	A
11	1011	13	B
12	1100	14	C
13	1101	15	D
14	1110	16	E
15	1111	17	F

3.2 NUMBER BASE CONVERSIONS

Man uses decimal number system while computer uses binary number system. Therefore, it is necessary to convert decimal number into its equivalent binary number while feeding the number into the computer. However, dealing with a large quantity of binary numbers of many bits is inconvenient for human beings. Therefore, octal and hexadecimal numbers are used as a shorthand means of expressing large binary numbers.

3.3 CONVERSIONS BETWEEN BINARY, OCTAL AND HEXADECIMAL NUMBERS

Binary to Octal Conversion

The base for octal number is 8 and the base for binary number is 2. We can write the base for octal number as 2^3. Therefore, by grouping 3 digits of binary numbers and then converting each group digit to its octal equivalent, we can convert binary numbers to its octal equivalent.

Example. $(1\,1\,0\,1\,1\,1\,0\,1\,0)_2 = (?)_8$

Solution.

$$\underbrace{110}_{6}\,\underbrace{111}_{7}\,\underbrace{010}_{2}$$

$$(1\,1\,0\,1\,1\,1\,0\,1\,0)_2 = (672)_8$$

Example. $(1\,1\,1\,1\,0\,1\,0)_2 = (?)_8$

Solution.

$$\underbrace{001}_{1}\underbrace{111}_{7}\underbrace{010}_{2}$$

$$(1\,1\,1\,1\,0\,1\,0)_2 = (172)_8$$

Octal to Binary Conversion

The octal to binary conversion is a reversal of the process explained in the previous section. To convert the octal to binary number, each digit of the octal number is individually converted to its binary equivalent.

Example. $(752)_8 = (?)_2$

Solution.

$$\begin{array}{ccc} 7 & 5 & 2 \\ \downarrow & \downarrow & \downarrow \\ 111 & 101 & 010 \end{array}$$

$$(752)_8 = (1\,1\,1\,1\,0\,1\,0\,1\,0)_2$$

Example. $(314.62)_8 = (?)_2$

Solution.

$$\begin{array}{cccccc} 3 & 1 & 4 & \cdot & 6 & 2 \\ \downarrow & \downarrow & \downarrow & & \downarrow & \downarrow \\ 011 & 001 & 100 & & 110 & 010 \end{array}$$

$$(314.62)_8 = (0\,1\,1\,0\,0\,1\,1\,0\,0 \cdot 1\,1\;0\,0\,1\,0)_2$$

Binary to Hexadecimal Conversion

The base for hexadecimal number is 16 and the base for binary number is 2. We can write the base for hexadecimal number as 2^4. Therefore, by grouping 4 digits of binary numbers and then converting each group digit to its hexadecimal equivalent, we can convert binary number to its hexadecimal equivalent.

Example. $(1\,1\,0\,1\,1\,0\,0\,1\,0\,1\,0\,1)_2 = (?)_{16}$

Solution.

$$\underbrace{1101}_{D}\;\underbrace{1001}_{9}\;\underbrace{0101}_{5}$$

$$(1\,1\,0\,1\,1\,0\,0\,1\,0\,1\,0\,1)_2 = (D95)_{16}$$

Example. $(1\,0\,1\,0\,1\,1\,1\,1\,0\,1\,0)_2 = (?)_{16}$

Solution.

$$\underbrace{0101}_{5}\;\underbrace{0111}_{7}\;\underbrace{1010}_{A}$$

$$(1\,0\,1\,0\,1\,1\,1\,1\,0\,1\,0)_2 = (57A)_{16}$$

Hexadecimal to Binary Conversion

The hexadecimal to binary conversion is a reversal of the process explained in the previous section. To convert the hexadecimal to binary number, each digit of the hexadecimal number is individually converted to its binary equivalent.

Example. $(A59)_{16} = (?)_2$

Solution.

```
A    5    9
↓    ↓    ↓
1010 0101 1001
```

$$(A59)_{16} = (101001011001)_2$$

Example. $(F1B \cdot C5)_{16} = (?)_2$

Solution.

```
F    1    B    ·   C    5
↓    ↓    ↓        ↓    ↓
1111 0001 1011     1100 0101
```

$$(F1B \cdot C5)_{16} = (111100011011 \cdot 11000101)_2$$

Octal to Hexadecimal Conversions

Two steps are involved in octal to hexadecimal conversion. They are :

1. Convert octal number to its binary equivalent
2. Convert binary number to its hexadecimal equivalent.

Example. $(724)_8 = (?)_{16}$

Solution.

Step 1. Octal to binary

```
 7   2   4
 ↓   ↓   ↓
111 010 100
```

Step 2. Binary to hexadecimal

$$\underbrace{0001}_{1}\ \underbrace{1101}_{D}\ \underbrace{0100}_{4}$$

$$(724)_8 = (1D4)_{16}$$

Hexadecimal to Octal Conversion

Two steps are involved in hexadecimal to octal conversion. They are :

1. Convert hexadecimal number to its binary equivalent.
2. Convert binary number to its octal equivalent.

Example. *$(C59)_{16} = (?)_8$*

Solution.

Step 1. Hexadecimal to binary

$$\begin{array}{ccc} C & 5 & 9 \\ \downarrow & \downarrow & \downarrow \\ 1100 & 0101 & 1001 \end{array}$$

Step 2. Binary to octal

$$\underbrace{110}_{6}\ \underbrace{001}_{1}\ \underbrace{011}_{3}\ \underbrace{001}_{1}$$

$$(C59)_{16} = (6131)_8$$

Converting any Radix to Decimal Number

(*a*) Binary to Decimal Conversion

Example. *$(1\,1\,0\,1 \cdot 0\,1)_2 = (?)_{10}$*

Solution.

$$\begin{aligned} N &= (1 \times 2^3) + (1 \times 2^2) + (0 \times 2^1) + (1 \times 2^0) \cdot (0 \times 2^{-1}) + (1 \times 2^{-2}) \\ &= 8 + 4 + 0 + 1 \cdot 0 + 0.25 \\ &= 13.25 \end{aligned}$$

(*b*) Octal to Decimal Conversion

Example. *$(257 \cdot 14)_8 = (?)_{10}$*

Solution.

$$\begin{aligned} N &= (2 \times 8^2) + (5 \times 8^1) + (7 \times 8^0) \cdot (1 \times 8^{-1}) + (4 \times 8^{-2}) \\ &= 128 + 40 + 7 \cdot 0.125 + 0.0625 \\ &= 175.1875. \end{aligned}$$

(*c*) Hexadecimal to Decimal conversion

Example. *$(B6C \cdot 5A)_{16} = (?)_{10}$*

Solution.

$$\begin{aligned} N &= (B \times 16^2) + (6 \times 16^1) + (C \times 16^0) \cdot (5 \times 16^{-1}) + (A \times 16^{-2}) \\ &= (11 \times 16^2) + (6 \times 16^1) + (12 \times 16^0) \cdot (5 \times 16^{-1}) + (10 \times 16^{-2}) \\ &= 2816 + 96 + 12 \cdot 0.3125 + 0.039 \\ &= 2924.3515 \end{aligned}$$

(*d*) Number with radix 4 to Decimal Conversion

Example. *$(132.03)_4 = (?)_{10}$*

Solution.

$$\begin{aligned} N &= (1 \times 4^2) + (3 \times 4^1) + (2 \times 4^0) \cdot (0 \times 4^{-1}) + (3 \times 4^{-2}) \\ &= 16 + 12 + 2 \cdot 0 + 0.1875 \\ &= 30.1875 \end{aligned}$$

(*e*) Number with radix 6 to Decimal Conversion

Example. *$(5541.23)_6 = (?)_{10}$*

Solution.

$$\begin{aligned} N &= (5 \times 6^3) + (5 \times 6^2) + (4 \times 6^1) + (1 \times 6^0) \cdot (2 \times 6^{-1}) + (3 \times 6^{-2}) \\ &= 1080 + 180 + 24 + 1 \cdot 0.333 + 0.083 \\ &= 1285.416 \end{aligned}$$

Conversion of Decimal Number to any Radix Number

We have to carry out the conversion of decimal number to any radix number in two steps.

Step 1. Conversion of integer part by successive division method.

Step 2. Conversion of fractional part by successive multiplication method.

Successive Division for Integer Part Conversion

In this method we repeatedly divide the integer part of the decimal number by *r* (the new radix) until quotient is zero. The remainder of each division becomes the numeral in the new radix. The remainders are taken in the reverse order to form a new radix number as illustrated in following examples.

(*a*) Decimal to binary conversion

Example. $(45)_{10} = (?)_2$

Solution. Here $r = 2$

```
2 | 45
2 | 22 – 1
2 | 11 – 0
2 | 5 – 1
2 | 2 – 1
    1 – 0
```

$(45)_{10} = (1\,0\,1\,1\,0\,1)_2$

(*b*) Decimal to octal conversion

Example. $(2053)_{10} = (?)_8$

Solution. Here $r = 8$

```
8 | 2053
8 | 256 – 5
8 | 32 – 0
    4 – 0
```

$(2053)_{10} = (4005)_8$

(*c*) Decimal to hexadecimal conversion

Example. $(3508)_{10} = (?)_{16}$

Solution. Here $r = 16$

```
16 | 3508
16 | 219 – 4
     13 – 11(B)
     (D)
```

$(3508)_{10} = (DB4)_{16}$

(*d*) Decimal to radix 4 number

Example. $(40)_{10} = (?)_4$

Solution. Here $r = 4$

```
4 | 40
4 | 10 – 0
    2 – 2
```

$(40)_{10} = (220)_4$

Successive Multiplication for Fractional Part Conversion

In this method, the number to be converted is multiplied by the radix of the new number, producing a product that has an integer part and a fractional part. The integer part (carry) of the product becomes a numeral in the new radix number. The fractional part is again multiplied by the radix and this process is repeated until fractional part reaches 0 or until the new radix number is carried out to sufficient digits. The integer part (carry) of each product is read downward to represent the new radix number as illustrated in following examples.

(*a*) Decimal to binary conversion

Example. $(0.85)_{10} = (?)_2$

Solution.

$0.85 \times 2 = 1.7 = 0.7$ with a carry of 1
$0.7 \times 2 = 1.4 = 0.4$ with a carry of 1
$0.4 \times 2 = 0.8 = 0.8$ with a carry of 0
$0.8 \times 2 = 1.6 = 0.6$ with a carry of 1
$0.6 \times 2 = 1.2 = 0.2$ with a carry of 1
$0.2 \times 2 = 0.4 = 0.4$ with a carry of 0 ↓

$(0.85)_{10} = (0 \cdot 1\,1\,0\,1\,1\,0)_2$

Example. $(0.625)_{10} = (?)_2$

Solution.

$0.625 \times 2 = 1.25 = 0.25$ with a carry of 1
$0.25 \times 2 = 0.50 = 0.50$ with a carry of 0
$0.50 \times 2 = 1.00 = 0.00$ with a carry of 1 ↓

$(0.625)_{10} = (0 \cdot 1\,0\,1)_2$

(*b*) Decimal to octal conversion

Example. $(0.55)_{10} = (?)_8$

Solution.

$0.55 \times 8 = 4.4 = 0.4$ with a carry of 4
$0.4 \times 8 = 3.2 = 0.2$ with a carry of 3
$0.2 \times 8 = 1.6 = 0.6$ with a carry of 1
$0.6 \times 8 = 4.8 = 0.8$ with a carry of 4
$0.8 \times 8 = 6.4 = 0.4$ with a carry of 6 ↓

$(0.55)_{10} = (0.43146)_8$

Example. $(0.640625)_{10} = (?)_8$

Solution. $0.640625 \times 8 = 5.125 = 0.125$ with a carry of 5
$0.125 \times 8 = 1.0 = 0.0$ with a carry of 1 ↓

$(0.640625)_{10} = (0.51)_8$

(*c*) Decimal to hexadecimal conversion

Example. $(0.62)_{10} = (?)_{16}$

Solution.

$0.62 \times 16 = 9.92 = 0.92$ with a carry of 9
$0.92 \times 16 = 14.72 = 0.72$ with a carry of 14 (E)
$0.72 \times 16 = 11.52 = 0.52$ with a carry of 11 (B)
$0.52 \times 16 = 8.32 = 0.32$ with a carry of 8
$0.32 \times 16 = 5.12 = 0.12$ with a carry of 5
$0.12 \times 16 = 1.92 = 0.92$ with a carry of 1 ↓

$(0.62)_{10} = (0.9EB851)_{16}$

Example. $(0.1289062)_{10} = (?)_{16}$

Solution. $0.1289062 \times 16 = 2.0625 = 0.0625$ with a carry of 2

$0.0625 \times 16 = 1.00 = 0.0$ with a carry of 1

$(0.1289062)_{10} = (0.21)_{16}$

(*d*) Decimal to radix 4 number conversion

Example. $(0.4125)_{10} = (?)_4$

Solution. $0.4125 \times 4 = 1.65 = 0.65$ with a carry of 1

$0.65 \times 4 = 2.6 = 0.6$ with a carry of 2

$0.6 \times 4 = 2.4 = 0.4$ with a carry of 2

$0.4 \times 4 = 1.6 = 0.6$ with a carry of 1

$(0.4125)_{10} = (0.1221)_4$

Example. $(0.375)_{10} = (?)_4$

Solution. $0.375 \times 4 = 1.5 = 0.5$ with a carry of 1

$0.5 \times 4 = 2.0 = 0.0$ with a carry of 2

$(0.375)_{10} = (0.12)_4$

Complements

For simplifying the subtraction operation in digital computers, complements are used. There are two types of complements in every number system. For binary numbers (base = 2), 1's complement and 2's complement and for decimal numbers (base = 10), 9's complement and 10's complement.

1's complement

The 1's complement of a binary number can be obtained by changing all 1's to 0's and all 0's to 1's.

Example. *1's complement of* $(1\,0\,0\,1)_2 = (?)_2$

Solution.

number → 1 0 0 1

1's complement → 0 1 1 0

2's complement

The 2's complement of a binary number can be obtained by adding 1 to its 1's complement.

Example. *2's complement of* $(1\,0\,1\,0)_2 = (?)_2$

Solution. number → 1 0 1 0

1's complement → 0 1 0 1

+1

0 1 1 0

1's Complement Subtraction

Subtraction of binary numbers using the 1's complement method allows subtraction only by addition. For subtraction of two numbers, we have two cases.

(*a*) Subtraction of smaller number from larger number and

(*b*) Subtraction of larger number from smaller number

(*a*) *Subtraction of smaller number from larger number:*

To subtract a smaller number from a larger number, the 1's complement method is as follows :

1. Determine the 1's complement of the smaller number
2. Add this to the larger number
3. Remove the carry and add it to the result. This carry is called 'end around carry'.

Example. *Subtract* $(0\ 1\ 0\ 0)_2$ *from* $(1\ 1\ 0\ 0)_2$ *using the 1's complement method.*

Solution.

Step 1. 1's complement of 0 1 0 0 *is* 1 0 1 1

Step 2. 1 1 0 0 larger number

+ 1 0 1 1 1's complement of 0 1 0 0

0 1 1 1

Step 3. + 1 Add end-around carry

1 0 0 0 final answer

(*b*) *Subtraction of larger number from smaller number:*

To subtract a larger number from a smaller number, the 1's complement method is as follows :

1. Determine the 1's complement of the larger number
2. Add this to the smaller number
3. Answer is in 1's complement form. To get the answer in correct form, take the 1's complement and assign negative sign to the answer.

Example. *Subtract* $(1\ 1\ 1\ 1)_2$ *from* $(0\ 0\ 1\ 1)_2$ *using the 1's complement method.*

Solution.

Step 1. 1's complement of 1 1 1 1 is 0 0 0 0

Step 2. 0 0 1 1 smaller number

+ 0 0 0 0 1's complement of 1 1 1 1

0 0 1 1 Answer in 1's complement form

Step 3. – 1 1 0 0 Answer in correct form

2's Complement Subtraction

Like 1's complement subtraction, in 2's complement subtraction, the subtraction is accomplished by only addition. For 2's complement subtraction also, we have two cases

(*a*) Subtraction of smaller number from larger number

(*b*) Subtraction of larger number from smaller number

(*a*) *Subtraction of smaller number from larger number:*

To subtract a smaller number from larger number, the 2's complement method is as follows :

1. Determine the 2's complement of the smaller number
2. Add the 2's complement to the larger number
3. Discard the carry.

Example. *Subtract $(0\ 0\ 0\ 1)_2$ from $(1\ 0\ 0\ 1)_2$ using the 2's complement method.*

Solution.

Step 1. 2's complement of a 0 0 0 1 is 1 1 1 1

Step 2. 1 0 0 1 Larger number
1 1 1 1 2's complement of 0 0 0 1

Step 3. 1 0 0 0 discard carry

∴ 1 0 0 0 final answer

(*b*) *Subtraction of larger number from smaller number:*

To subtract larger number from smaller number, the 2's complement method is as follows :

1. Determine the 2's complement of the larger number
2. Add the 2's complement to the smaller number
3. Answer is in the 2's complement form. To get the answer in the correct form take the 2's complement and assign negative sign to the answer.

Example. *Subtract $(1\ 1\ 1\ 0)_2$ from $(0\ 1\ 0\ 0)_2$ using 2's complement method*

Solution.

Step 1. 2's complement of 1 1 1 0 is 0 0 1 0

Step 2. 0 1 0 0 smaller number
+ 0 0 1 0 2's complement of 1 1 1 0

0 1 1 0 Answer in 2's complement form

Step 3. 1 0 0 1 1's complement of answer
+ 1

– 1 0 1 0 2's complement of answer

9's Complement

The 9's complement of a decimal number can be found by subtracting each digit in the number from 9. The 9's complement of decimal digits 0 to 9 is shown in Table 3.2.

Table 3.2. 9's complement

Decimal digit	9's complement
0	9
1	8
2	7
3	6
4	5
5	4
6	3
7	2
8	1
9	0

Example. *9's complement of 21 = (?)*

Solution.

$$\begin{array}{r} 99 \\ 21 \\ \hline 78 \end{array} \rightarrow \text{9's complement of 21}$$

10's Complement

The 10's complement of a decimal number can be obtained by adding 1 to its 9's complement.

Example. *10's complement of 47 = (?)*

Solution.

$$\begin{array}{rl} 99 & \\ -\ 47 & \\ \hline 52 & \text{9's complement of 47} \\ +\ \ 1 & \\ \hline 53 & \text{10's complement of 47} \end{array}$$

9's Complement Subtraction

We know that for the subtraction of two numbers there are two cases :

(*a*) Subtraction of smaller number from larger number

(*b*) Subtraction of larger number from smaller number

(*a*) *Subtraction of smaller number from larger number:*

To subtract a smaller number from a larger number, the 9's complement method is as follows

1. Determine the 9's complement of the smaller number
2. Add this to the larger number
3. Remove the carry and add it to the result. This carry is called 'End-around carry'.

Example. *Subtract 25 from 38 using the 9's complement method.*

Solution.

$$\begin{array}{lrl} & 99 & \\ \textbf{Step 1.} & -\ 25 & \text{Smaller Number} \\ \hline & 74 & \text{9's complement of 25} \end{array}$$

$$\begin{array}{lrl} \textbf{Step 2.} & 38 & \text{Larger Number} \\ & +\ 74 & \text{9's complement of 25} \\ \hline & 112 & \\ \textbf{Step 3.} & \rightarrow 1 & \text{Add end-around carry} \\ \hline & 13 & \text{final answer.} \end{array}$$

(*b*) *Subtraction of larger number from smaller number:*

To subtract a larger number from a smaller number, the 9's complement method is as follows :

1. Determine the 9's complement of the larger number
2. Add this to the smaller number
3. Answer is in 9's complement form. To get the answer in correct form, take the 9's complement and assign negative sign to the answer.

Example. *Subtract 85 from 51 using the 9's complement method.*

Solution.

Step 1.
```
   99
 – 85 Larger Number
   14 9's complement of 85
```

Step 2.
```
   51 Smaller Number
 + 14 9's complement of 85
   65 Answer in 9's complement form
```

Step 3.
```
   99
   65
 – 34 Answer in correct form
```

10's Complement Subtraction

For the subtraction of two numbers, there are two cases :

(*a*) Subtraction of smaller number from larger number

(*b*) Subtraction of larger number from smaller number.

(*a*) *Subtraction of smaller number from larger number:*

To subtract a smaller number from a larger number, the 10's complement method is as follows :

1. Determine the 10's complement of the smaller number
2. Add this to the larger number
3. Discard the carry

Example. *Subtract 25 from 35 using 10's complement method.*

Solution :

Step 1.
```
   99
 – 25 Smaller number
   74 9's complement of 25
 +  1
   75 10's complement of 25
```

Step 2.
```
   35 Larger number
 + 75
```

Step 3.
```
 1 10 Final answer
```

(*b*) *Subtraction of larger number from smaller number:*

To subtract a larger number from a smaller number, the 10's complement method is as follows:

1. Determine the 10's complement of the larger number
2. Add this to the smaller number

3. Answer is in 10's complement form. To get the answer in correct form, take the 10's complement and assign negative sign to the answer.

Example. *Subtract 42 from 11 using 10's complement method.*

Solution.

Step 1.

```
   99
 − 42   Larger number
 ────
   57   9's complement of 42
 +  1
 ────
   58   10's complement of 42
```

Step 2.

```
   11
 + 58
 ────
   69   Answer in 10's complement
```

Step 3.

```
   99
 − 69
 ────
   30
 +  1
 ────
 − 31   Final answer
```

3.4 LOGIC GATES

Introduction

Logic gates are the basic elements of a digital system. This gate operates on a number of binary inputs in order to perform a particular logical operation. The most common logic gates are AND, OR, NOT, NAND, EX-OR and EX-NOR gates. The NAND and NOR gates are called as the universal gates. Except for the exclusive-NOR gate, they are available in monolithic integrated circuit form.

Logic gates have two or more inputs and only one output except for the NOT gate, which has only one input. Each gate has a distinct logic symbol and its operation can be described by means of an algebraic function. The operation of a logic gate can be easily understood with the help of "truth table". A truth table is a table that shows all the input-output possibilities of a logic circuit.

OR Gate

The OR gate performs logical addition, commonly known as OR function. An OR gate has two or more inputs and one output. The logic symbol for a 2-input OR gate is shown in Fig. 3.5(*a*) where OR gate with a 3 input is shown in Fig. 3.5(*b*).

(*a*) Logic symbol of two input OR gate

(*b*) Logic symbol of three input OR gate

Fig. 3.5

Operation

An OR gate produces a HIGH (1) on the output when any of the inputs is HIGH (1). The output is LOW (0) when all of the inputs are LOW (0). The Table 3.3 shows the truth table of 2 input OR gate which describes its logical operation.

Table 3.3. Truth table of 2 input OR gate

Inputs		Output
A	B	Y = A + B
0	0	0
0	1	1
1	0	1
1	1	1

Figure 3.6 shows one way to build a 2-input OR gate in which A and B represent the inputs and Y the output. The resistance R_L is the load resistance.

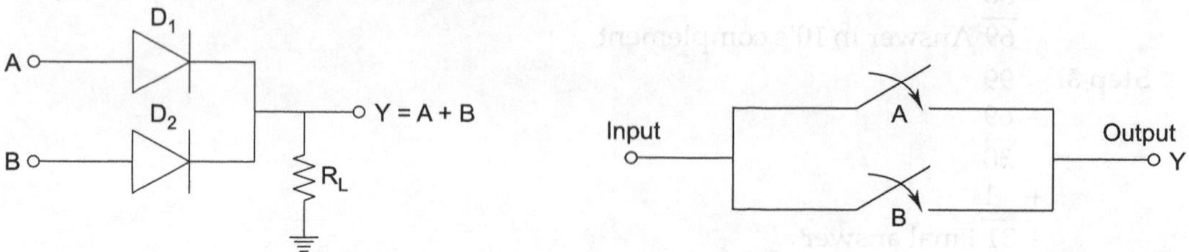

(*a*) 2 input OR gate using diodes (*b*) Electrical equivalent circuit of 2-input OR gate

Fig. 3.6

With 2 inputs, there are four possible input cases :

Case 1. A = 0 and B = 0

When both input voltages are low, the diodes D_1 and D_2 will not conduct, and hence the output Y = 0

Case 2. A = 0 and B = 1

When the B input is high, it makes the diode D_2 to conduct, then $V_0 \approx 5$ V and so Y = 1.

Case 3. A = 1 and B = 0

When the A input is high, it makes the diode D_1 to conduct, then $V_0 \approx 5$ V and so Y = 1.

Case 4. A = 1 and B = 1

When both input voltages are high, the diodes D_1 and D_2 conducts and hence Y = 1.

The electrical equivalent circuit of a 2 input OR gate is shown in figure where switches A and B are connected in parallel. If either A or B is closed or if both are closed, then the output is high.

AND Gate

The AND gate performs logical multiplication, commonly known as AND function. An AND gate has two or more inputs and one output. The logic symbol for a 2-input AND gate is shown in Fig. 3.7(*a*), where AND gate with a 4 input is shown in Fig. 3.7(*b*).

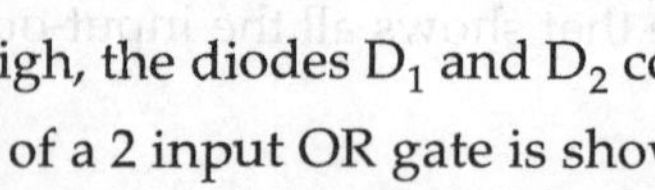

Fig. 3.7 (*a*) Logic symbol of two input AND gate

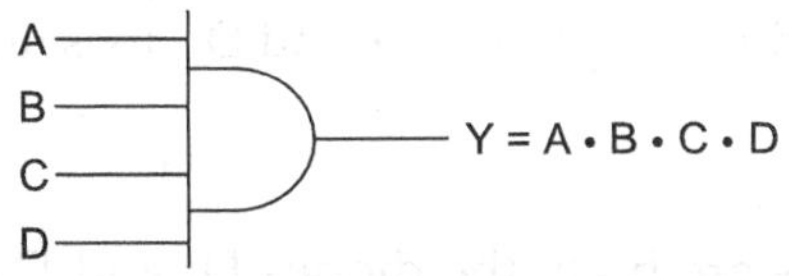

Fig. 3.7 (*b*) Logic symbol of 4-input AND gate

Operation

An AND gate produces a HIGH (1) on the output when all the inputs are HIGH (1). The output is LOW (0) when any of the inputs is LOW (0). The logical operation of the two input AND gate is described in the truth Table 3.4.

Table 3.4. Truth table of a 2 input AND gate

Inputs		Output
A	**B**	**Y = A · B**
0	0	0
0	1	0
1	0	0
1	1	1

Figure 3.8 shows the 2 input AND gate using diodes in which A and B represent the inputs and Y the output. The resistance R_L is the load resistance.

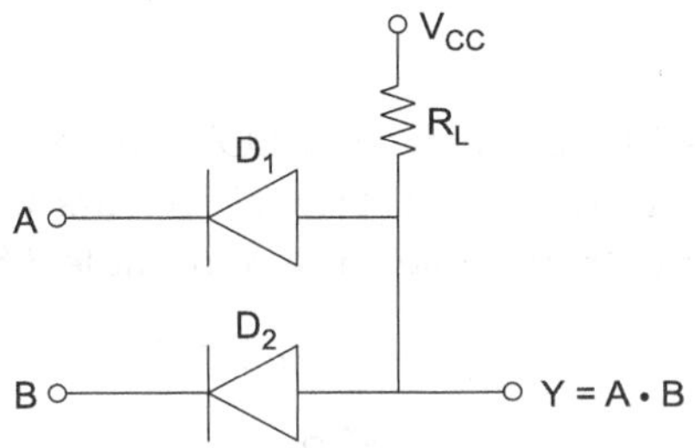

Fig. 3.8 2 input AND gate using diodes

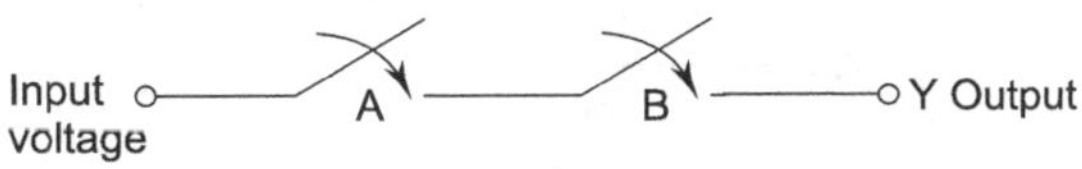

Fig. 3.9. Electrical equivalent circuit of a 2 input AND gate

With 2 inputs, there are four possible input cases.

Case 1. A = 0 and B = 0

When both input voltages are low, the diodes D_1 and D_2 conducts, and hence the output Y = 0.

Case 2. A = 0 and B = 1

When the B input is high, the diode D_1 conducts and D_2 does not conduct and hence the output Y = 0.

Case 3. A = 1 and B = 0

When the A input is high, the diode D_2 conducts and D_1 does not conduct, and hence the output Y = 0.

Case 4. A = 1 and B = 1

When both the input voltages are high, the diodes D_1 and D_2 do not conduct and hence the output Y = 1.

The electrical equivalent circuit of an 2-input AND gate is shown in Fig. 3.9 where two switches A and B are connected in series. If both A and B are closed, then the output is high.

NOT Gate (Inverter)

The NOT gate performs the basic logical function called inversion or complementation. An NOT gate has one input and one output. The logic symbol of an NOT gate is shown in Fig. 3.10.

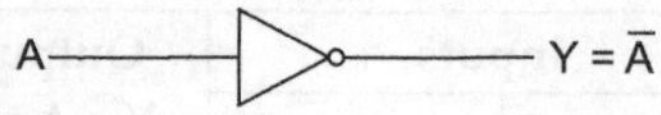

Fig. 3.10. Logic symbol of a NOT gate

Table 3.5. Truth table of an NOT gate

Input	Output
A	**Y = Ā**
0	1
1	0

Operation

When a HIGH (1) level is applied to an inverter, a LOW (0) level appears at its output. When a LOW (0) level is applied to an inverter, a HIGH (1) level appears at its output. The logical operation of an NOT gate is described by the truth table shown in Table 3.5.

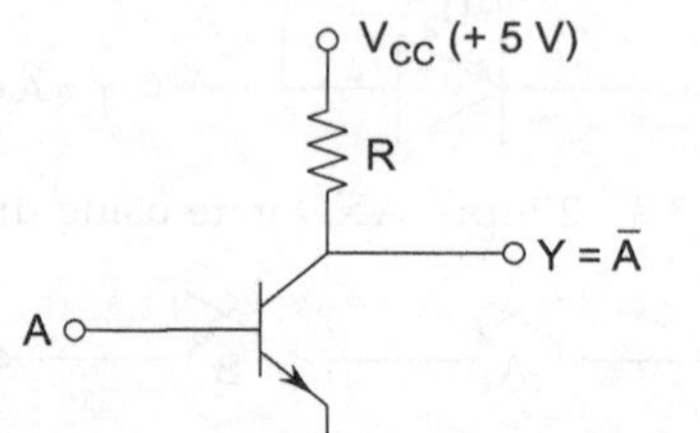

Fig. 3.11 NOT gate using transistor

Figure 3.11 shows the NOT gate using transistor in which A represent the input and Y the output. With 1 input, there are two possible input cases.

Case 1. A = 0

When the input is low, the transistor is in the OFF state and hence the output Y = 1.

Case 2. A = 0

When the input is HIGH, the transistor is in the ON state and hence the output Y = 0.

NOR Gate

NOR is a contraction of NOT-OR gates. An NOR gate has two or more inputs and only one output. The logic symbol for the NOR gate is shown in Fig. 3.12. The small circle or bubble represents the operation of inversion.

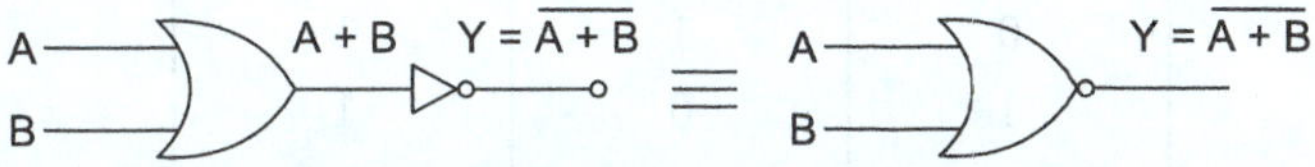

Fig. 3.12 Logic symbol of NOR gate

Operation

An NOR gate produces a HIGH (1) on the output when all the inputs are LOW (0). If any one or both the inputs are HIGH (1), then the output is LOW (0). The Table 3.6 shows the truth table of 2-input NOR gate which describes its logical operation.

Table 3.6. Truth table of a two input NOR gate

Inputs		Output
A	**B**	**Y = $\overline{A + B}$**
0	0	1
0	1	0
1	0	0
1	1	0

The NOR gate is equivalent to an AND gate with a bubble at its inputs. This is shown in Fig. 3.13.

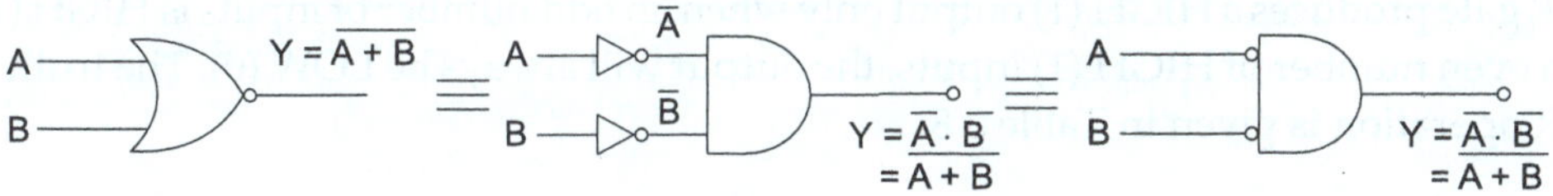

Fig. 3.13 NOR gate ≡ Bubbled AND gate

NAND Gate

NAND is a contraction of the NOT-AND gates. An NAND gate has two or more inputs and only one output. The logic symbol for a NAND gate is shown in Fig. 3.14. The small circle or bubble represents the operation of inversion.

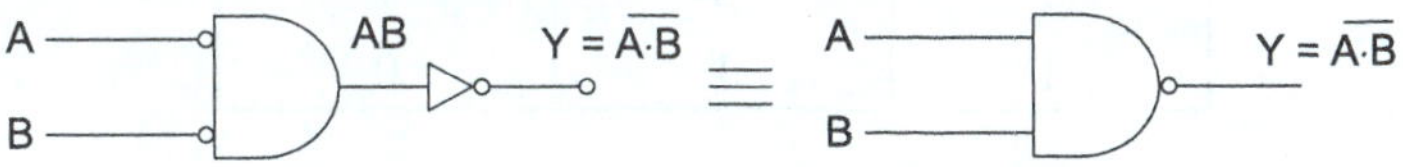

Fig. 3.14 Logic symbol of NAND gate

Operation

An NAND gate produces LOW (0) output, when all the inputs are HIGH (1). If any one or both the inputs are LOW (0), then the output is HIGH (1). The Table 3.7 shows the truth table for the NAND gate which describes its logical operation.

Table 3.7. Truth table of a 2-input NAND gate

Inputs		Output
A	B	$Y = \overline{A \cdot B}$
0	0	1
0	1	1
1	0	1
1	1	0

The NAND gate is equivalent to an OR gate with a bubble at its inputs which is shown in Fig. 3.15.

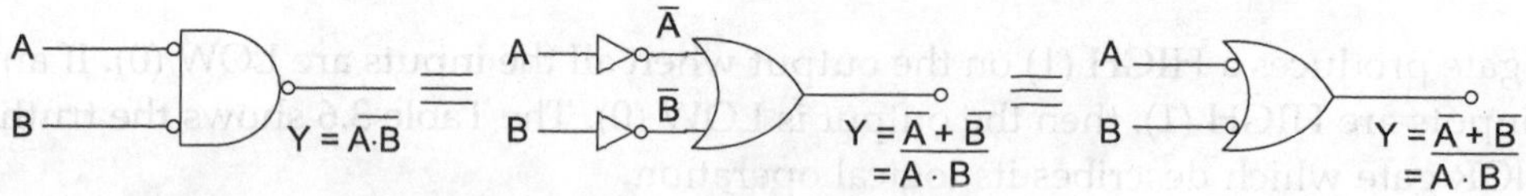

Fig. 3.15 NAND gate ≡ Bubbled OR gate

Exclusive-OR (Ex-OR) Gate

An Exclusive-OR gate is a gate with two or more inputs and one output. The Logic Symbol for a Ex-OR gate is shown in Fig. 3.16.

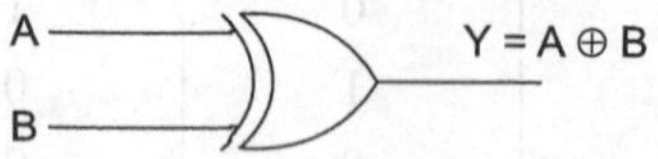

Fig. 3.16 Logic symbol of a Ex-OR gate

Operation

The Ex-OR gate produces a HIGH (1) output only when an odd number of inputs is HIGH (1). When there is an even number of HIGH (1) inputs, the output will always be LOW (0). The truth table for the Ex-OR operation is given in Table 3.8.

Table 3.8. Truth table of a 2-input Ex-OR gate

Inputs		Output
A	B	$Y = A \oplus B$
0	0	0
0	1	1
1	0	1
1	1	0

From the truth table of a 2-input Ex-OR gate, the Ex-OR function can be written as $Y = \overline{A}B + A\overline{B} = A \oplus B$. Using this expression, we can implement the 2-input Ex-OR gate using basic gates like AND, OR and NOT gates using Fig. 3.17.

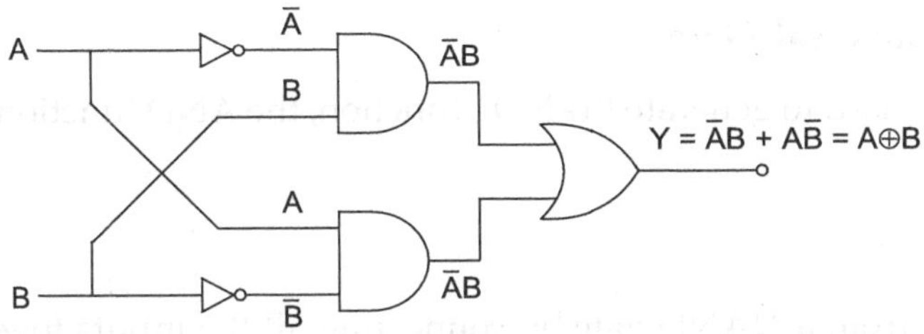

Fig. 3.17 Ex-OR gate using AND-OR-NOT gates

Exclusive-NOR (Ex-NOR) Gate

Ex-NOR gate is an Ex-OR gate, followed by an inverter. An exclusive-NOR gate has two or more inputs and one output. The Logic Symbol for a Ex-NOR gate is shown in Fig. 3.18.

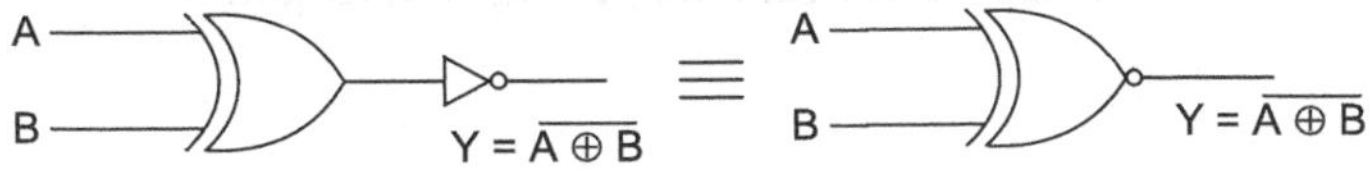

Fig. 3.18 Logic symbol of 2 input Ex-NOR gate

Operation

The output of an Ex-NOR gate is HIGH (1) if both are 0 or 1 ; otherwise, its output is LOW (0). It is also called a coincidence circuit. The truth table for the Ex-NOR operation is given in Table 3.9.

Table 3.9. Truth table of a 2-input Ex-NOR gate

Inputs		Output
A	**B**	$\mathbf{Y = \overline{A \oplus B}}$
0	0	1
0	1	0
1	0	0
1	1	1

From the truth table of a 2 input Ex-NOR gate, the Ex-NOR function can be written as $Y = \overline{A}\overline{B} + AB = \overline{A \oplus B} = A \odot B$. Using this expression, we can implement the 2 input Ex-NOR gate using basic gates like AND, OR and NOT gates using Fig. 3.19.

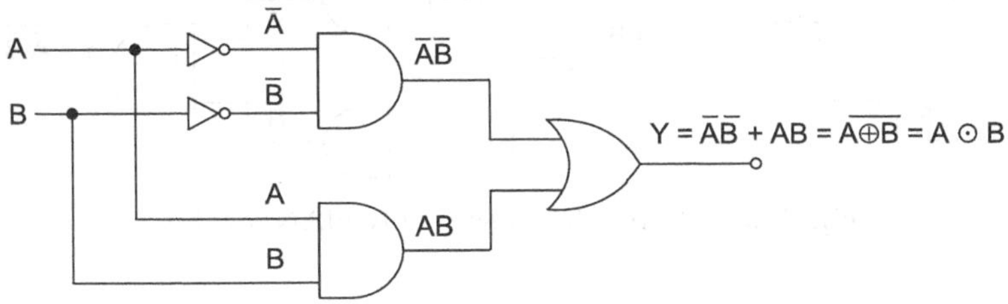

Fig. 3.19 Ex-NOR gate using AND-OR-NOT gates

Universal Gates

NAND and NOR gates are called universal gates because both can be used to implement any gate like AND, OR and NOT gates or any combination of these basic gates.

NAND Gate as a Universal Gate

The NAND gate can be used to generate the NOT function, the AND function, the OR function and the NOR function.

NOT function

An NOT can be made from a NAND gate by connecting all the inputs together and creating, in effect, a single common input as shown in Fig. 3.20, for a two input gate.

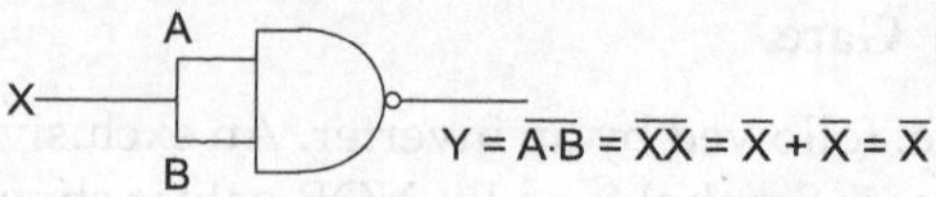

Fig. 3.20 NOT function, using NAND gate

AND function

An AND function can be made from a NAND gate by simply inverting output of NAND gate. This is shown in Fig. 3.21 for a two input AND gate.

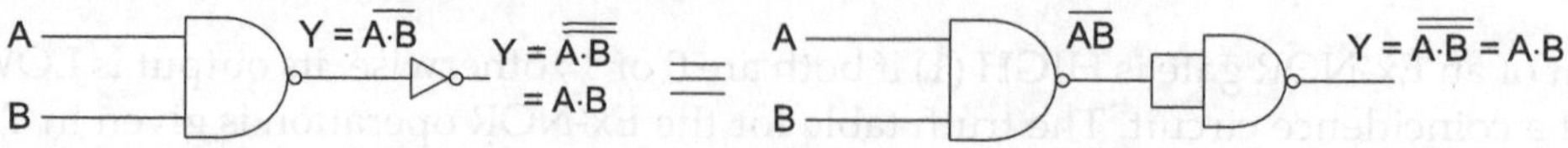

Fig. 3.21 AND function using NAND gates

OR function

An OR function can be generated using only NAND gates. We know that Boolean expression for OR gate is

$$Y = A + B$$

$$= \overline{\overline{A}} + \overline{\overline{B}}$$

$$= \overline{\overline{A} \cdot \overline{B}} \qquad \text{(By Demorgan's theorem 1)}$$

The implementation of above equation is shown in Fig. 3.22.

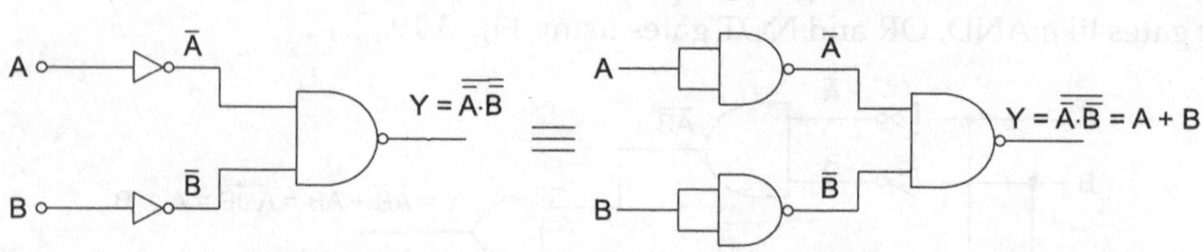

Fig. 3.22 OR function using only NAND gates

NOR function

An NOR function can be generated using only NAND gates. We know that Boolean expression for NOR gate is

$$Y = \overline{A + B}$$

$$= \bar{A} \cdot \bar{B} \qquad \text{(By Applying Demorgan's theorem 2)}$$

$$= \overline{\overline{\bar{A} \cdot \bar{B}}}$$

The implementation of above equation is shown in Fig. 3.23.

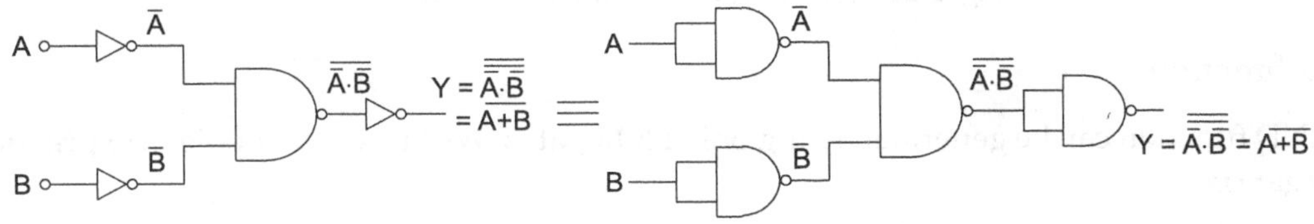

Fig. 3.23 NOR function using only NAND gates

NOR Gate as a Universal Gate

The NOR gate can be used to generate the NOT function, OR function, AND function and the NAND function, since it is also a universal gate.

NOT function

An NOT can be made from a NOR gate by connecting all the inputs together and creating, in effect, a single common input as shown in Fig. 3.24, for a two input gate.

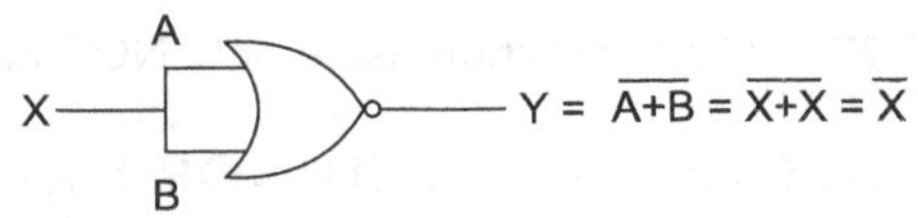

Fig. 3.24 NOT function using NOR gate

OR function

An OR function can be made from NOR gate by simply inverting output of NOR gate. This is shown in Fig. 3.25, for a two input OR gate.

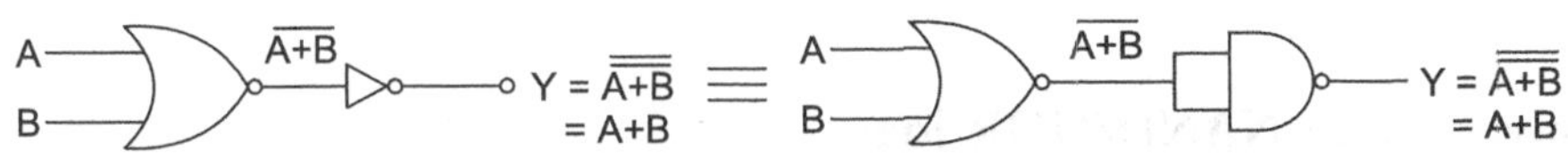

Fig. 3.25 OR function using NOR gates

AND function

An AND function can be generated using only NOR gates. We know that Boolean expression for AND gate is

$$Y = A \cdot B$$

$$= \bar{\bar{A}} \cdot \bar{\bar{B}}$$

$$= \overline{\bar{A} + \bar{B}} \qquad \text{(By applying Demorgan's theorem 2)}$$

Figure 3.26 shows the implementation of above equation.

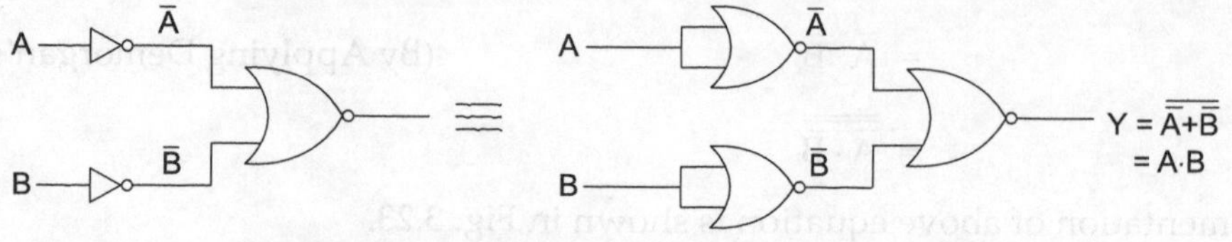

Fig. 3.26 AND function using only NOR gates

NAND function

An NAND function can be generated using only NOR gates. We know that Boolean expression for NAND gate is

$$Y = \overline{A \cdot B}$$

$$= \overline{A} + \overline{B}$$

$$= \overline{\overline{\overline{A} + \overline{B}}} \quad \text{(By applying Demorgan's theorem 1)}$$

Figure 3.27 shows the implementation of above equation.

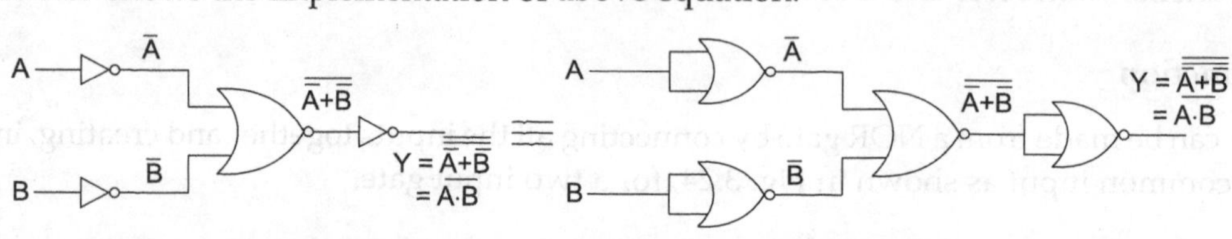

Fig. 3.27 NAND function using only NOR gates

Conversion of AND | OR | NOT Logic to NAND/NOR Logic

Digital circuits are frequently constructed with NAND or NOR gates rather than with AND, OR and NOT gates. NAND and NOR gates are easier to fabricate with electronic components and are the basic gates used in all IC digital logic families. Because of the prominence of NAND and NOR gates in the design of digital circuits, rules and procedures have been developed for the conversion from Boolean functions given in terms of AND, OR and NOT into equivalent NAND and NOR logic diagrams.

Steps for converting to NAND/NOR logic

1. Draw AND/OR logic.
2. If NAND hardware has been chosen, add bubbles on the output of each AND gate; and bubbles on input side to all OR gates.
3. If NOR hardware has been chosen, add bubbles on output of each OR gate and bubble on input of each AND gate.
4. Add or subtract an inverter on each line that received a bubble in step 2 or 3.
5. Replace bubbled OR by NAND and bubbled AND by NOR.
6. Eliminate Double inversions.

NAND Circuit

Example. *Implement the Boolean expression* $Y = \overline{A} + B \cdot C + \overline{D + E}$ *using NAND-NAND Logic.*

Solution.

Step 1. Draw : AND/OR Logic [Fig. 3.28 (*a*)]

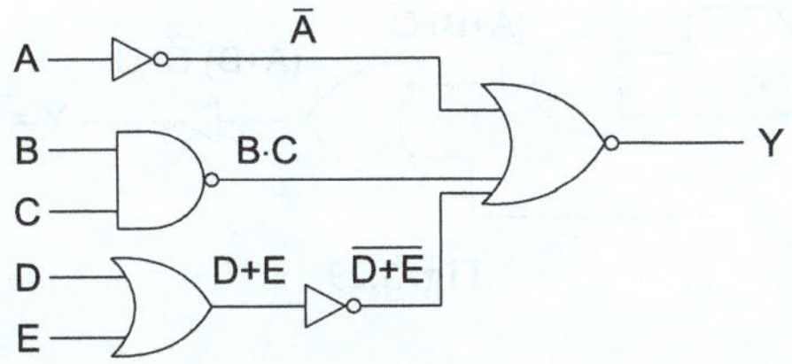

Fig. 3.28 (*a*)

Step 2. Add bubbles on the output of each AND gate and bubbles on input side to all OR gates [Fig. 3.28 (*b*)]

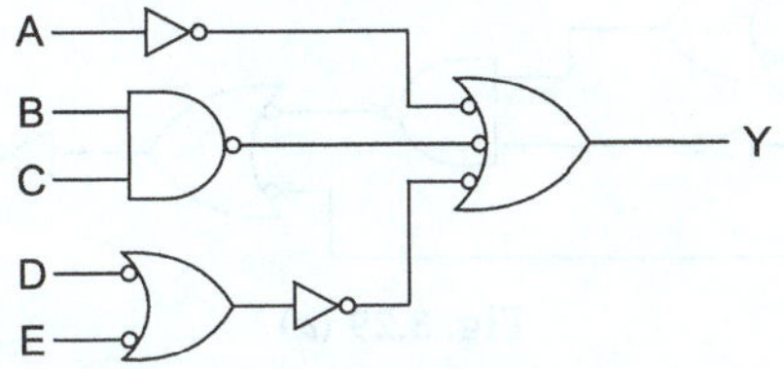

Fig. 3.28 (*b*)

Step 3. Add or subtract an inverter [Fig. 3.28 (*c*)]

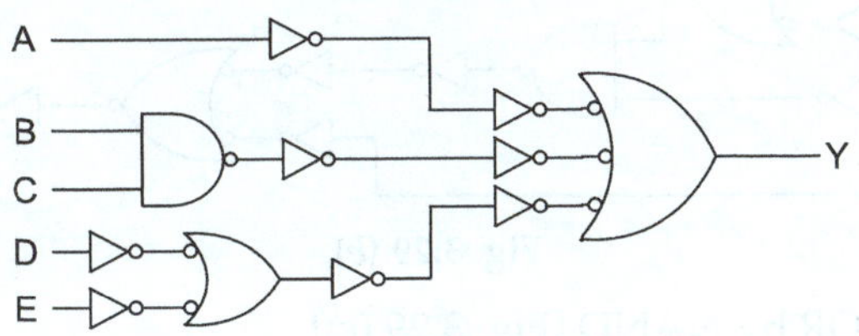

Fig. 3.28 (*c*)

Step 4. Replace bubbled OR by NAND [Fig. 3.28 (*d*)]

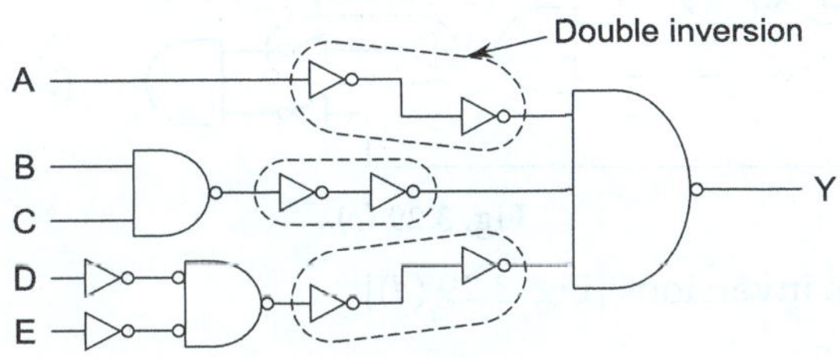

Fig. 3.28 (*d*)

Step 5. Eliminate double inversions [Fig. 3.28 (*e*)]

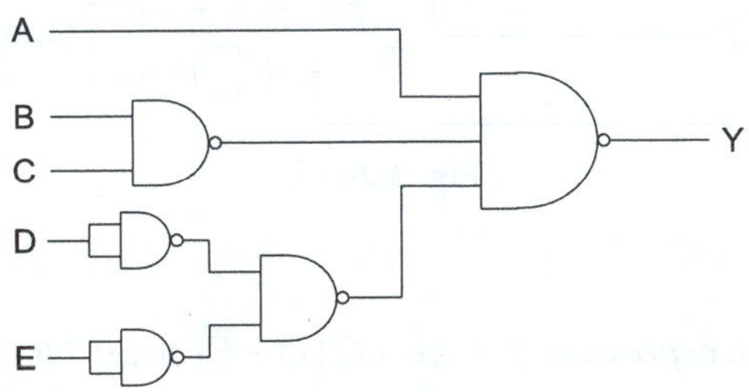

Fig. 3.28 (*e*)

Example. *Draw the following circuit using only NAND-NAND logic*

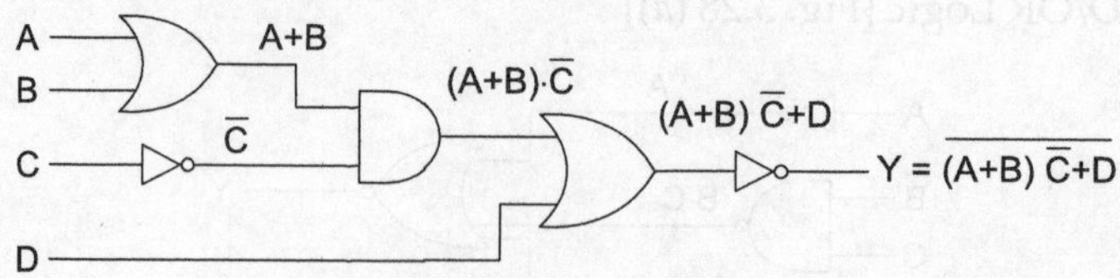

Fig. 3.29

Solution.

Step 1. Add bubbles on the output of each AND gate, and bubbles on input side to all OR gates [Fig. 3.29 (*a*)]

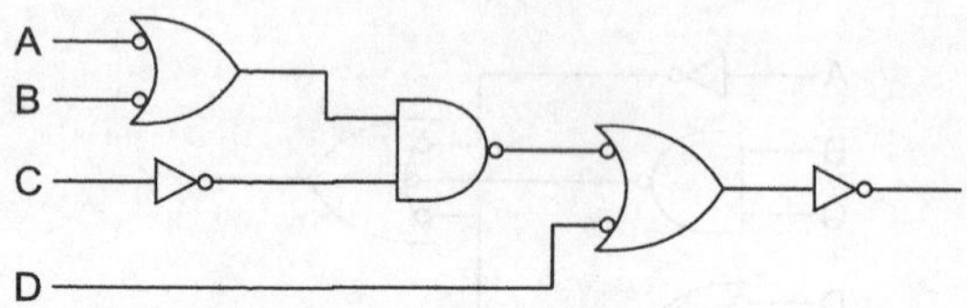

Fig. 3.29 (*a*)

Step 2. Add or subtract an inverter [Fig. 3.29 (*b*)]

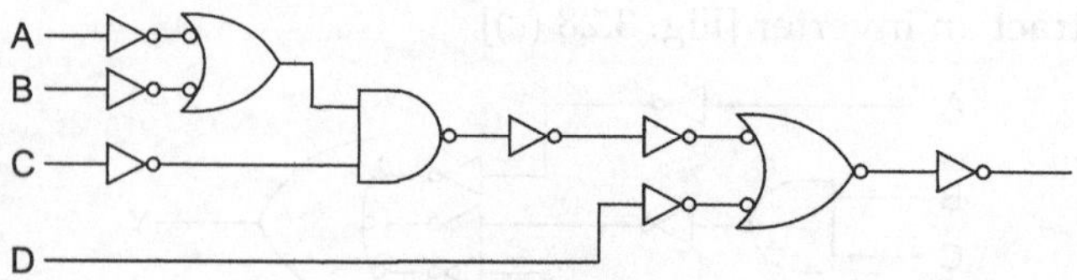

Fig. 3.29 (*b*)

Step 3. Replace bubbled OR by NAND [Fig. 3.29 (*c*)]

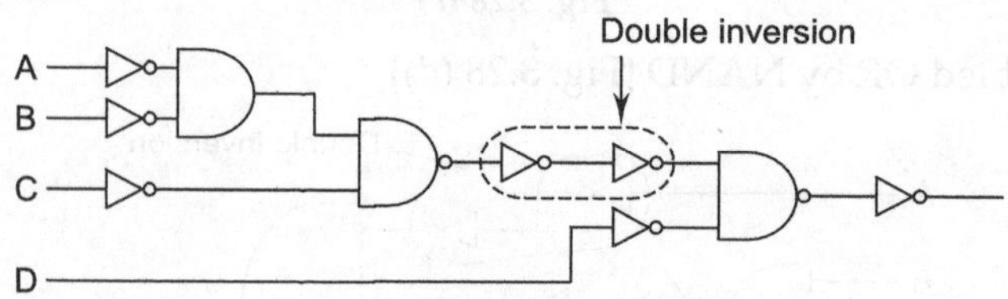

Fig. 3.29 (*c*)

Step 4. Eliminate double inversions [Fig. 3.29 (*d*)]

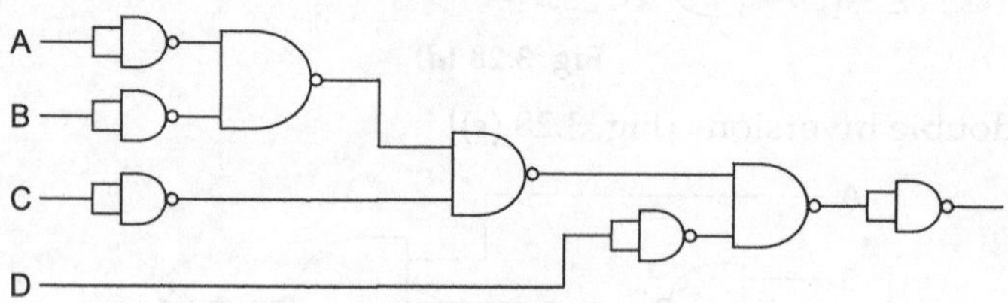

Fig. 3.29 (*d*)

NOR Circuit

Example. *Implement the Boolean expression* $Y = \overline{A}\overline{B} + C\left(D + \overline{E}\right)$ *using NOR-NOR logic.*

Solution.

Step 1. Draw AND/OR logic [Fig. 3.30 (*a*)]

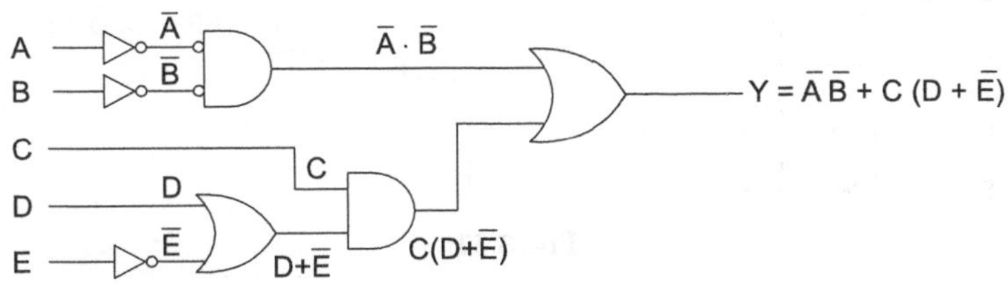

Fig. 3.30 (*a*)

Step 2. Add bubbles on the output of each OR gate and bubble on input of each AND gate [Fig. 3.30 (*b*)]

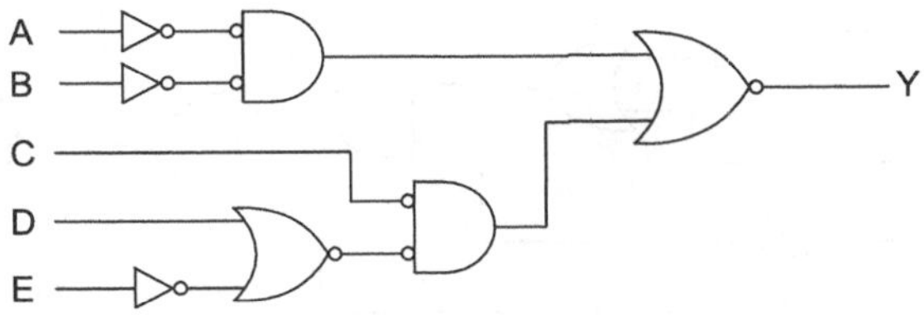

Fig. 3.30 (*b*)

Step 3. Add or subtract an inverter [Fig. 3.30 (*c*)]

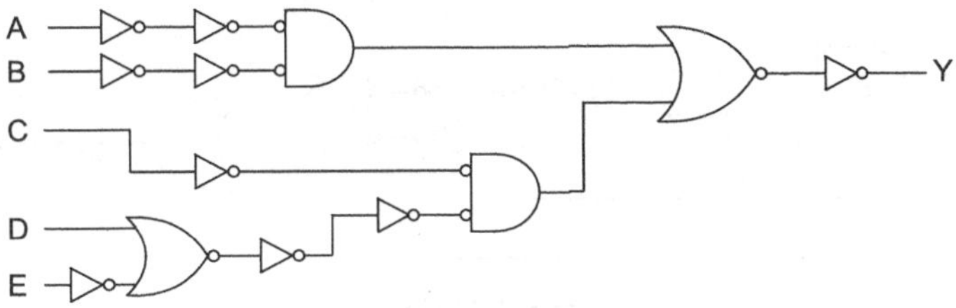

Fig. 3.30 (*c*)

Step 4. Replace bubbled AND by NOR [Fig. 3.30 (*d*)]

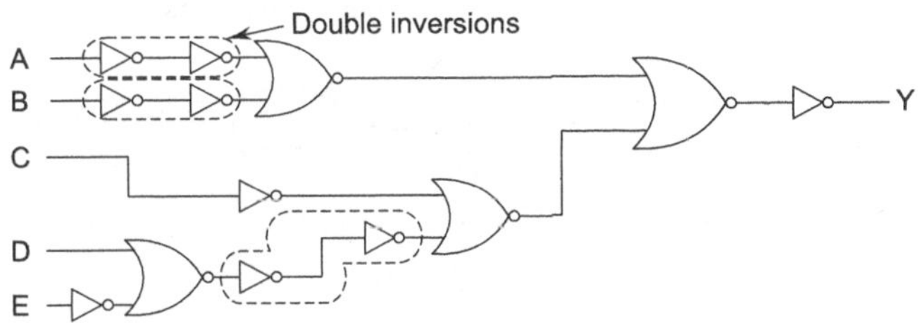

Fig. 3.30 (*d*)

Step 5. Eliminate double inversions [Fig. 3.30 (*e*)]

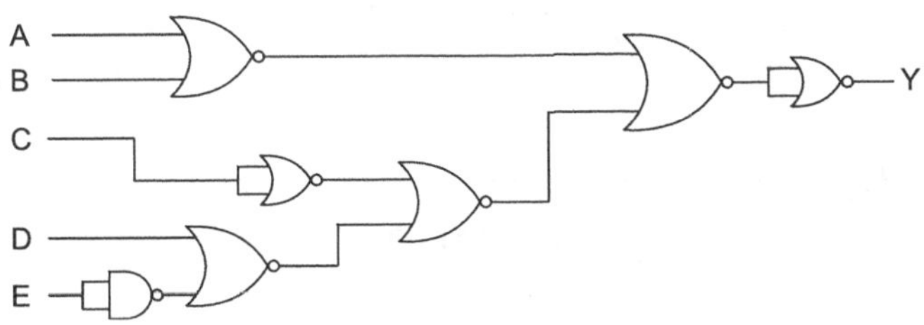

Fig. 3.30 (*e*)

Example. *Draw the following circuit using only NOR-NOR logic*

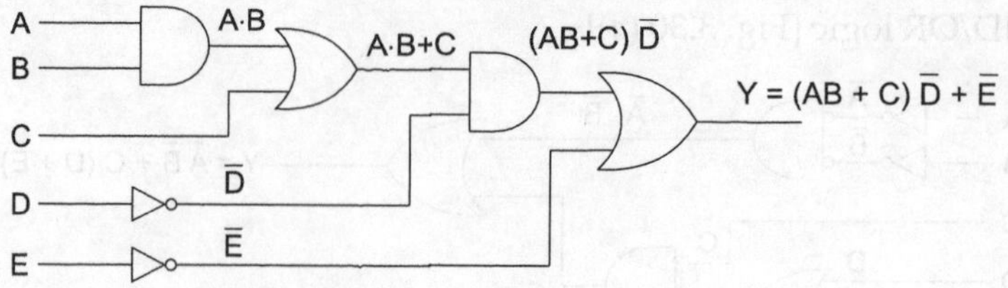

Fig. 3.31

Solution.

Step 1. Add bubbles on the output of each OR gate and bubble on input of each AND gate [Fig. 3.31 (*a*)]

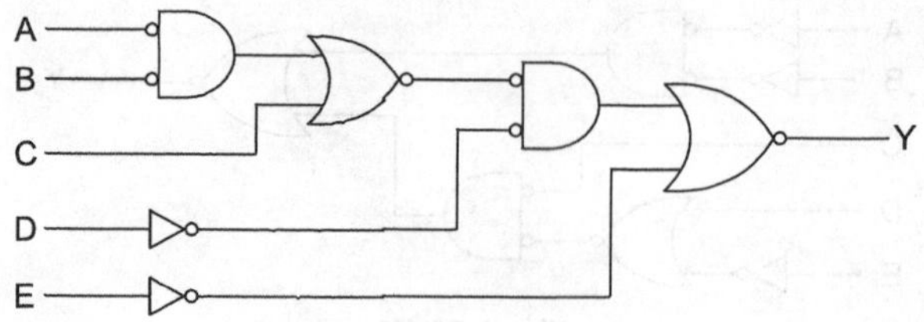

Fig. 3.31 (*a*)

Step 2. Add or subtract an inverter [Fig. 3.31 (*b*)]

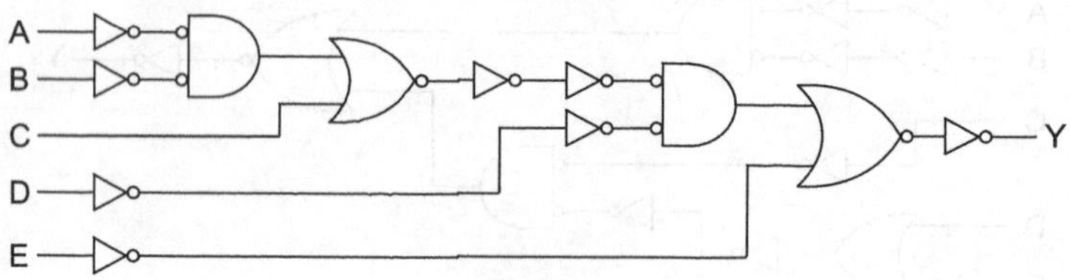

Fig. 3.31 (*b*)

Step 3. Replace bubbled AND by NOR [Fig. 3.31 (*c*)]

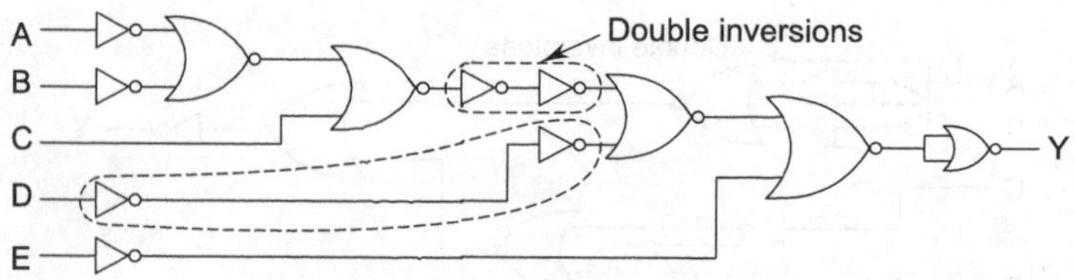

Fig. 3.31 (*c*)

Step 4. Eliminate double inversions [Fig. 3.31 (*d*)]

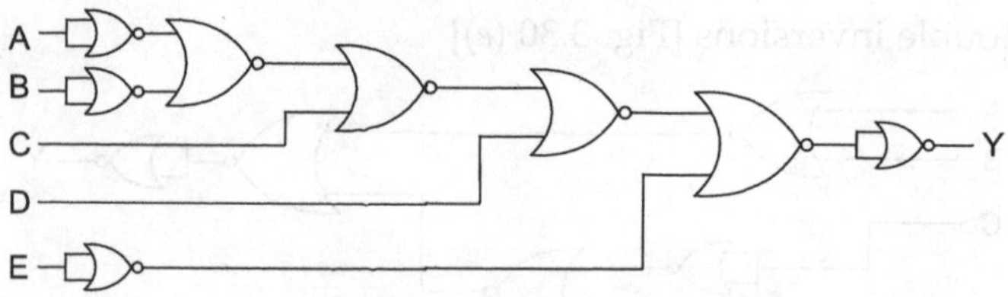

Fig. 3.31 (*d*)

Example. *Implement Boolean expression for Ex-NOR gate using NAND gates.*

Solution. The Boolean expression for Ex-NOR gate is $Y = \bar{A}\bar{B} + AB$

Step 1. Draw the circuit [Fig. 3.32 (*a*)]

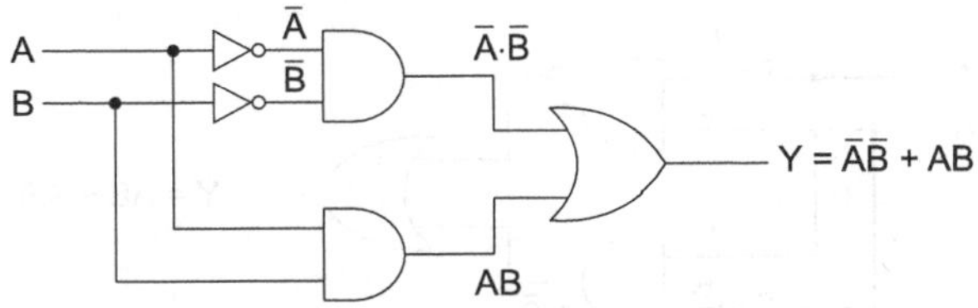

Fig. 3.32 (*a*)

Step 2. Add bubbles on the output of each AND gate and bubbles on input side to all OR gates [Fig. 3.32 (*b*)]

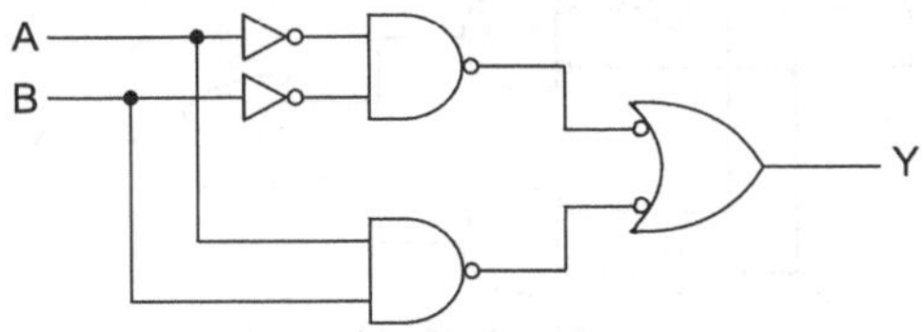

Fig. 3.32 (*b*)

Step 3. Add or subtract an inverter [Fig. 3.32 (*c*)]

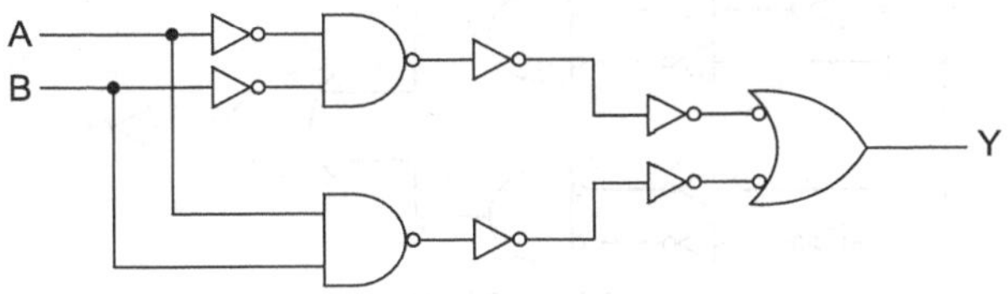

Fig. 3.32 (*c*)

Step 4. Replace bubbled OR by NAND [Fig. 3.32 (*d*)]

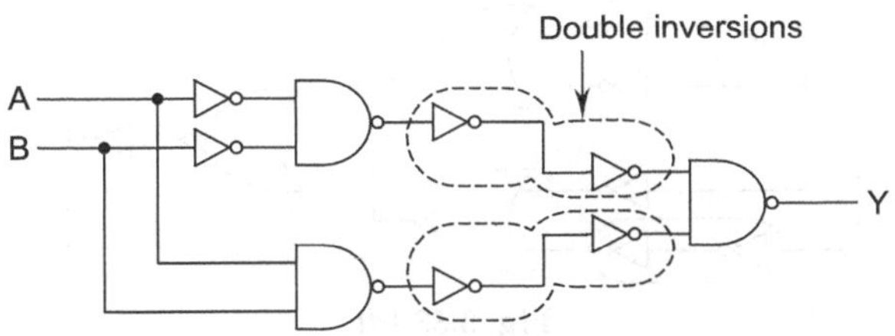

Fig. 3.32 (*d*)

Step 5. Eliminate double inversions [Fig. 3.32 (*e*)]

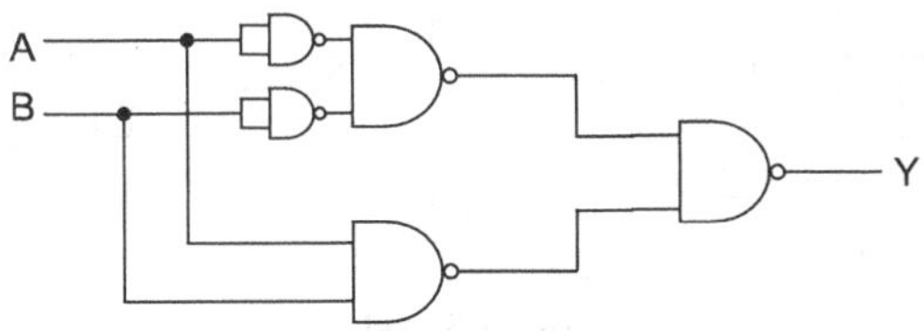

Fig. 3.32 (*e*)

Example. *Implement Boolean expression for Ex-OR gate using NOR gates.*

Solution. The Boolean expression for Ex-OR gate is $Y = \bar{A}B + A\bar{B}$

Step 1. Draw the circuit [Fig. 3.33 (*a*)]

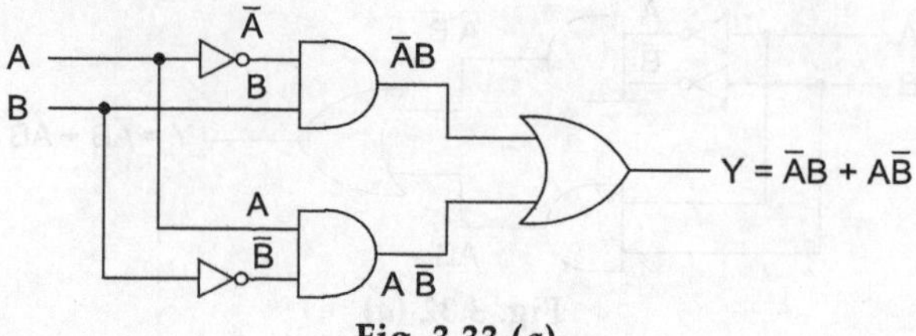

Fig. 3.33 (*a*)

Step 2. Add bubbles on output of each OR gate and bubble on input of each AND gate [Fig. 3.33 (*b*)]

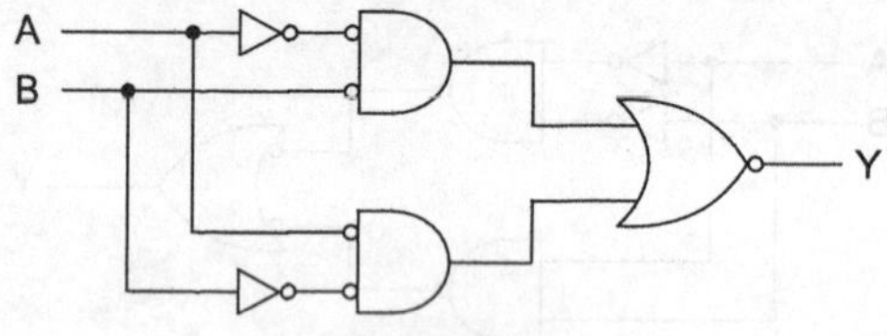

Fig. 3.33 (*b*)

Step 3. Add or subtract an inverter [Fig. 3.33 (*c*)]

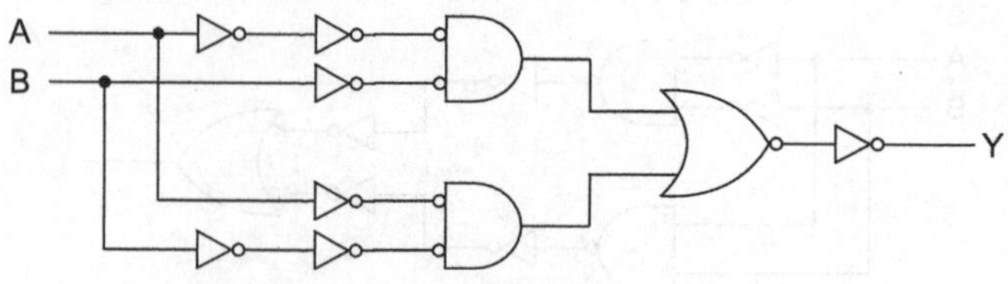

Fig. 3.33 (*c*)

Step 4. Replace bubbled AND by NOR [Fig. 3.33 (*d*)]

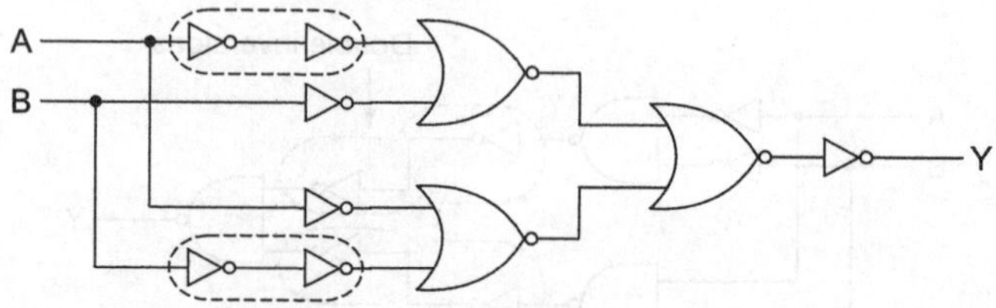

Fig. 3.33 (*d*)

Step 5. Eliminate double inversions [Fig. 3.33 (*e*)]

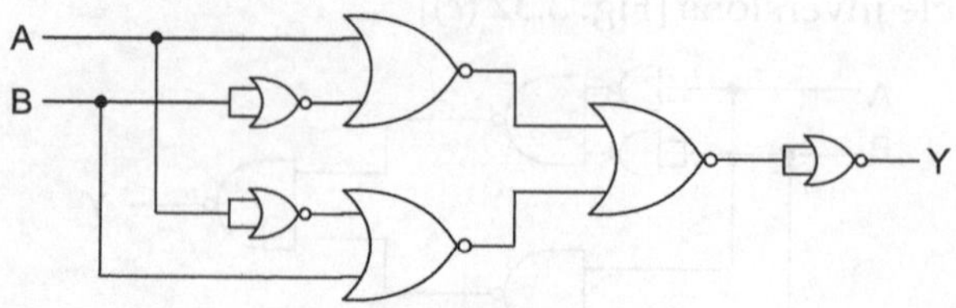

Fig. 3.33 (*e*)

Boolean Algebra

Boolean Algebra is used to simplify/rearrange Boolean equation to make simple logic circuit. In this section, we will see, logic operations, laws and theorems used in Boolean Algebra.

Boolean Logic Operations

A Boolean function is an algebraic expression formed using binary constants, binary variables and basic logical operation symbols. Basic logical operation include the AND function (logical multiplication), the OR function (logical addition) and the NOT function (logical complementation). A Boolean function can be converted into a logic diagram composed of the AND, OR and NOT gates.

Logical AND Operation

We know that, the logical AND operation of two Boolean variables A and B, given as $Y = A \cdot B$. The common symbol, for this operation is the multiplication sign ($\cdot$). Usually, the dot denoting the AND function is omitted and $A \cdot B$ is written as AB. The logical AND operation can be described by the Table 3.10.

Table 3.10. Logical AND Operation

Inputs		Output
A	**B**	**$Y = A \cdot B$**
0	0	0
0	1	0
1	0	0
1	1	1

Logical OR Operation

We know that, the logical OR operation of two Boolean variables A and B, given as $Y = A + B$. The common symbol for this operation is the Addition Sign (+). The logical OR operation can be described by the Table 3.11.

Table 3.11. Logical OR Operation

Inputs		Output
A	**B**	**$Y = A + B$**
0	0	0
0	1	1
1	0	1
1	1	1

Logical Complementation (Inversion)

We know that the logical inverse operation converts the logical 1 to the logical 0 and *vice-versa*. The symbol used for this operation is a bar over the function or the variable. The complement of a variable A will be denoted by $\bar{A}$. So, if A = 1, $\bar{A}$ = 0 and if A = 0, $\bar{A}$ = 1. Sometimes, a prime symbol (′) is used to denote the complement.

Basic laws of Boolean Algebra

Logical operations can be expressed and minimized mathematically using the rules, laws, and theorems of Boolean Algebra. It is a convenient and systematic method of expressing and analyzing the operation of digital circuits and systems.

Boolean Addition

Addition by the Boolean method involves variables having values of either a binary 1 [high level] or a 0 [low level]. The basic rules of Boolean addition are given below :

$$0 + 0 = 0$$

$$0 + 1 = 1$$

$$1 + 0 = 1$$

$$1 + 1 = 1$$

Boolean Addition is same as the logical OR operation.

Boolean Multiplication

Multiplication by the Boolean method also involves variables having values of either a binary 1 [high level] or a 0 [low level]. The basic rules of the Boolean multiplication method are as follows:

$$0 \cdot 0 = 0$$

$$0 \cdot 1 = 0$$

$$1 \cdot 0 = 0$$

$$1 \cdot 1 = 1$$

Boolean multiplication is same as the logical AND operation.

Laws of Boolean Algebra

Three of the basic laws of Boolean algebra are the same as in ordinary algebra : (1) the commutative laws, (2) associative laws, and (3) the distributive law.

1. **Commutative Laws :**

   Boolean addition is commutative, given by

   $$A + B = B + A$$

   This states that the order in which the variables are ORed makes no difference in the output. Boolean algebra is also commutative over multiplication, given by

   $$A \cdot B = B \cdot A.$$

   This states that the order in which the variables are ANDed makes no difference in the output.

2. **Associative Laws :**

   The associative property of addition is given by

   $$A + (B + C) = (A + B) + C$$

This states that in the ORing of several variables, the result is the same regardless of the grouping of the variables.

The associative law of multiplication is given by

$$A \cdot (B \cdot C) = (A \cdot B) \cdot C$$

This states that it makes no difference in what order the variable are grouped when ANDing several variables.

3. **Distributive Laws :**

The Boolean addition is distributive over Boolean multiplication, given by

$$A + BC = (A + B)(A + C)$$

This law states that the AND operation (multiplication) of several variables and then the OR operation (addition) of the result with a single variable is equivalent to the OR operation of the single variable with each of the several variables and then the AND operation of the sums.

The Boolean multiplication is also distributive over Boolean addition given by

$$A \cdot (B + C) = A \cdot B + A \cdot C$$

This law states that the OR operation of several variables and then the AND operation of the result with a single variable is equivalent to the AND operation of the single variable with each of the several variables and then the OR operation of the products.

The Table 3.12 summarizes the several basic rules that are useful in manipulating and simplifying Boolean algebra expressions.

Table 3.12

Rule No.	Rules	Law
1.	$A + 0 = A$	—
2.	$A + 1 = 1$	—
3.	$A \cdot 0 = 0$	—
4.	$A \cdot 1 = A$	—
5.	$A + A = A$	Idempotent
6.	$A + \bar{A} = 1$	Complementary
7.	$A \cdot A = A$	Idempotent
8.	$A \cdot \bar{A} = 0$	Complementary
9.	$\bar{\bar{A}} = A$	Involution
10.	$A + AB = A$	—
11.	$A + \bar{A}B = A + B$	—
12.	$(A + B)(A + C) = A + BC$	—

Absorption laws

(*i*) $A + AB = A$

Proof : $A + AB = A \cdot 1 + AB$

$= A(1 + B)$

$= A \cdot 1$

$= A$

(*ii*) $A \cdot (A + B) = A$

Proof : $A(A + B) = A \cdot A + AB$

$= A + AB$

$= A(1 + B)$

$= A \cdot 1$

$= A$

(*iii*) $A + \bar{A}B = A + B$

Proof : $A + \bar{A}B = (A + \bar{A})(A + B)$ $[\because A + BC = (A + B)(A + C)]$

$= 1 \cdot (A + B)$

$= A + B$

Proof of Rule 12 :

It is given by $(A + B)(A + C) = A + BC$

$$\begin{aligned}(A+B)(A+C) &= A \cdot A + AB + AC + BC \\ &= A + AB + AC + BC \\ &= A(1+B) + AC + BC \\ &= A \cdot 1 + AC + BC \\ &= A + AC + BC \\ &= A(1+C) + BC \\ &= A \cdot 1 + BC \\ &= A + BC\end{aligned}$$

Duality Theorem

The duality theorem say that, starting with a Boolean relation, we can derive another Boolean relation by

1. Changing each OR sign to an AND sign
2. Changing each AND sign to an OR sign
3. Complementing any 0 or 1 appearing in the expression.

For example, the Rule 1 say that

$A + 0 = A$

The dual relation is $A \cdot 1 = A$

Consensus Theorem

(*i*) $AB + \bar{A}C + BC = AB + \bar{A}C$

Proof :

$$\begin{aligned}AB + \bar{A}C + BC &= AB + \bar{A}C + BC \cdot 1 \\ &= AB + \bar{A}C + BC(A + \bar{A}) && (\because A + \bar{A} = 1) \\ &= AB + \bar{A}C + ABC + \bar{A}BC \\ &= AB(1+C) + \bar{A}C(1+B) && (\because 1 + B = 1 = 1 + C) \\ &= AB + \bar{A}C\end{aligned}$$

(*ii*) $(A+B)(\bar{A}+C)(B+C) = (A+B)(\bar{A}+C)$

Proof :

$$\begin{aligned}(A+B)(\bar{A}+C)(B+C) &= (A+B)(\bar{A}+C)(B+C+0) \\ &= (A+B)(\bar{A}+C)(B+C+A\bar{A})\end{aligned}$$

$$= (A+B)(\bar{A}+C)(B+C+A)(B+C+\bar{A})$$

$$[\because \quad A+BC=(A+B)(A+C)]$$

$$= (A+B)(A+B+C)(\bar{A}+C)(\bar{A}+C+B)$$

$$= (A+B)(\bar{A}+C) \qquad [\because \quad A(A+B)=A]$$

De-Morgan's Theorems

De-Morgan suggested two theorems that form an important part of Boolean Algebra.

1. $\overline{A \cdot B} = \bar{A}+\bar{B}$

   This theorem states that the complement of a product is equal to the sum of the complements. This is illustrated by the gate equivalency in Fig. 3.34 and by truth Table 3.13.

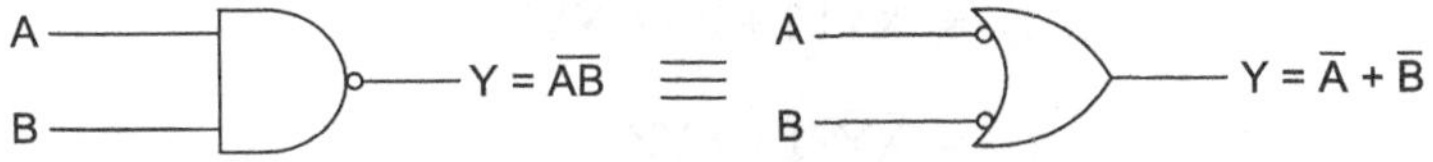

Fig. 3.34 Gate equivalency of first theorem

Truth table :

Table 3.13

A	B	$\overline{AB}$	$\bar{A}+\bar{B}$
0	0	1	1
0	1	1	1
1	0	1	1
1	1	0	0

2. $\overline{A+B} = \bar{A} \cdot \bar{B}$

   This theorem states that the complement of a sum is equal to the product of the complements. This is illustrated by the gate equivalency in Fig. 3.35 and by truth Table 3.14.

Fig. 3.35 Gate equivalency of second theorem

Truth table :

Table 3.14

A	B	$\overline{A+B}$	$\bar{A} \cdot \bar{B}$
0	0	1	1
0	1	0	0
1	0	0	0
1	1	0	0

Simplification of Boolean expressions using algebraic method.

3.5 EXAMPLES

1.
$$
\begin{aligned}
A+AB+AC &= A(1+B+C) \\
&= A && [\because \; A+1=1]
\end{aligned}
$$

2.
$$
\begin{aligned}
A+\bar{A}B+AC &= A+B+AC && \left[\because \; A+\bar{A}B=A+B\right] \\
&= A(1+C)+B \\
&= A\cdot 1+B && [\because \; A+1=A] \\
&= A+B
\end{aligned}
$$

3.
$$
\begin{aligned}
ABCD+AB\bar{C}D &= ABD\left(C+\bar{C}\right) \\
&= ABD && \left[\because \; A+\bar{A}=1\right]
\end{aligned}
$$

4.
$$
\begin{aligned}
\bar{X}\bar{Y}+\bar{X}\bar{Y}Z+\bar{X}\bar{Y}+\bar{X}YZ &= \bar{X}\bar{Y}+\bar{X}\bar{Y}Z+\bar{X}YZ && [\because \; A+A=A] \\
&= \bar{X}\bar{Y}\left[1+Z\right]+\bar{X}YZ \\
&= \bar{X}\bar{Y}\cdot 1+\bar{X}YZ && [\because \; A+1=1] \\
&= \bar{X}\bar{Y}+\bar{X}YZ \\
&= \bar{X}\left[\bar{Y}+YZ\right] \\
&= \bar{X}\left[\bar{Y}+Z\right] && \left[\because \; A+\bar{A}B=A+B\right]
\end{aligned}
$$

5.
$$
\begin{aligned}
A+A\bar{B}+\bar{A}B &= A\left[1+\bar{B}\right]+\bar{A}B \\
&= A\cdot 1+\bar{A}B && [\because \; A+1=1] \\
&= A+\bar{A}B \\
&= A+B && \left[\because \; A+\bar{A}B=A+B\right]
\end{aligned}
$$

6.
$$
\begin{aligned}
\bar{A}B+AB+\bar{A}\bar{B} &= B\left[A+\bar{A}\right]+\bar{A}\bar{B} \\
&= B\cdot 1+\bar{A}\bar{B} && \left[\because \; A+\bar{A}=1\right] \\
&= B+\bar{A}\bar{B} \\
&= B+\bar{A} && \left[\because \; A+\bar{A}B=A+B\right]
\end{aligned}
$$

7.
$$
\begin{aligned}
\overline{\bar{A}B+C\bar{D}} &= \overline{\bar{A}B}\cdot\overline{C\bar{D}} && \left[\because \; \overline{A+B}=\bar{A}\cdot\bar{B}\right] \\
&= \bar{\bar{A}}+\bar{B}\cdot\bar{C}+\bar{\bar{D}} && \left[\because \; \overline{A\cdot B}=\bar{A}+\bar{B}\right] \\
&= \left(A+\bar{B}\right)\cdot\left(\bar{C}+D\right)
\end{aligned}
$$

8.
$$
\begin{aligned}
AC+C\left(A+\bar{A}B\right) &= AC+C(A+B) && \left[\because \; A+\bar{A}B=A+B\right] \\
&= AC+AC+BC
\end{aligned}
$$

$$= AC + BC \qquad [\because \; A + A = A]$$
$$= C\,[A + B].$$

9. $$A\bar{B}C\bar{D} + A\bar{B}\bar{C}\bar{D} + \bar{A}\bar{B}\bar{D} = A\bar{B}\bar{D}\left[C + \bar{C}\right] + \bar{A}\bar{B}\bar{D}$$
$$= A\bar{B}\bar{D} + \bar{A}\bar{B}\bar{D} \qquad \left[\because \; A + \bar{A} = 1\right]$$
$$= \bar{B}\bar{D}\left[A + \bar{A}\right]$$
$$= \bar{B}\bar{D} \qquad \left[\because \; A + \bar{A} = 1\right].$$

10. $$\overline{\overline{AB} + \bar{A} + AB} = \overline{\bar{A} + \bar{B} + \bar{A} + AB} \qquad \left[\because \; \overline{A\cdot B} = \bar{A} + \bar{B}\right]$$
$$= \overline{\bar{A} + \bar{B} + \bar{A} + B} \qquad \left[\because \; A + \bar{A}B = A + B\right]$$
$$= \overline{\bar{A} + \bar{B} + B} \qquad [\because \; A + A = A]$$
$$= \overline{\bar{A} + 1} \qquad \left[\because \; A + \bar{A} = 1\right]$$
$$= \bar{1} \qquad [\because \; 1 + A = 1]$$
$$= 0.$$

11. $$x(\bar{x} + y) = x\bar{x} + xy \qquad \left[\because \; A\cdot\bar{A} = 0\right]$$
$$= 0 + xy$$
$$= xy$$

12. $$(A + B)(A + \bar{B}) = AA + AB + A\bar{B} + B\bar{B} \qquad \left[\because \; A\cdot A = A,\, A\cdot\bar{A} = 0\right]$$
$$= A + AB + A\bar{B} + 0$$
$$= A + A\bar{B} \qquad [A + AB = A]$$
$$= A\,[1 + B]$$
$$= A \qquad [\because \; A + 1 = 1]$$

13. $$ABC + \bar{A}B + AB\bar{C} = AB\left[C + \bar{C}\right] + \bar{A}B$$
$$= AB + \bar{A}B \qquad \left[\because \; A + \bar{A} = 1\right]$$
$$= B\left[A + \bar{A}\right]$$
$$= B \qquad \left[\because \; A + \bar{A} = 1\right]$$

14. $$\overline{x\bar{y} + \bar{x}y} = \overline{x\bar{y}}\cdot\overline{\bar{x}y} \qquad \left[\because \; \overline{A + B} = \bar{A}\cdot\bar{B}\right]$$
$$= \bar{x} + \bar{\bar{y}}\cdot\bar{\bar{x}} + \bar{y} \qquad \left[\because \; \overline{A\cdot B} = \bar{A} + \bar{B}\right]$$
$$= \bar{x} + y\cdot x + \bar{y} \qquad \left[\because \; \bar{\bar{A}} = A\right]$$
$$= x\bar{x} + xy + \bar{x}\bar{y} + y\bar{y}$$

$$= 0 + xy + \bar{x}\bar{y} + 0 \qquad [\because\ A\cdot\bar{A} = 0]$$

$$= xy + \bar{x}\bar{y}$$

15. $$\overline{(A\bar{B}+C)\bar{D}+E} = \overline{(A\bar{B}+C)\bar{D}}\cdot\bar{E} \qquad [\because\ \overline{A+B} = \bar{A}\cdot\bar{B}]$$

$$= \overline{(A\bar{B}+C)} + \bar{\bar{D}}\cdot\bar{E} \qquad [\because\ \overline{A\cdot B} = \bar{A}+\bar{B}]$$

$$= \overline{(A\bar{B}+C)} + D\cdot\bar{E} \qquad [\because\ \bar{\bar{A}} = A]$$

$$= \overline{A\bar{B}}\cdot\bar{C} + D\cdot\bar{E} \qquad [\because\ \overline{A+B} = \bar{A}\cdot\bar{B}]$$

$$= \bar{A} + \bar{\bar{B}}\cdot\bar{C} + D\cdot\bar{E} \qquad [\because\ \overline{A\cdot B} = \bar{A}+\bar{B}]$$

$$= \bar{A} + B\cdot\bar{C} + D\cdot\bar{E} \qquad [\because\ \bar{\bar{A}} = A]$$

$$= \left(\left(\bar{A}+B\right)\bar{C}+D\right)\bar{E}$$

16. $$\bar{A}\bar{C} + ABC + A\bar{C} = \bar{C}\left[A+\bar{A}\right] + ABC$$

$$= \bar{C}\cdot 1 + ABC \qquad [\because\ A+\bar{A} = 1]$$

$$= \bar{C} + ABC \qquad [\because\ A\cdot 1 = A]$$

$$= \bar{C} + AB \qquad [\because\ A+\bar{A}B = A+B]$$

17. $\bar{A}B\left(\bar{D}+\bar{C}D\right) + B\left(A+\bar{A}CD\right)$

$$= \bar{A}B\left(\bar{D}+\bar{C}\right) + B\left(A+CD\right) \qquad [\because\ A+\bar{A}B = A+B]$$

$$= \bar{A}B\bar{D} + \bar{A}B\bar{C} + AB + BCD$$

$$= \bar{A}B\bar{D} + B\left(A+\bar{A}\bar{C}\right) + BCD$$

$$= \bar{A}B\bar{D} + B\left(A+\bar{C}\right) + BCD \qquad (\because\ A+\bar{A}B = A+B)$$

$$= \bar{A}B\bar{D} + AB + B\bar{C} + BCD$$

$$= \bar{A}B\bar{D} + AB + B\left(\bar{C}+CD\right)$$

$$= \bar{A}B\bar{D} + AB + B\left(\bar{C}+D\right) \qquad (\because\ A+\bar{A}B = A+B)$$

$$= \bar{A}B\bar{D} + AB + B\bar{C} + BD$$

$$= B\left(D+\bar{A}\bar{D}\right) + AB + B\bar{C}$$

$$= B\left(D+\bar{A}\right) + AB + B\bar{C} \qquad (\because\ A+\bar{A}B = A+B)$$

$$= BD + \bar{A}B + AB + B\bar{C}$$

$$= BD + B(A + \bar{A}) + B\bar{C}$$

$$= BD + B \cdot 1 + B\bar{C} \qquad (\because\ A + \bar{A} = 1)$$

$$= BD + B + B\bar{C}$$

$$= B[1 + D + \bar{C}]$$

$$= B \cdot 1 \qquad [\because\ A + 1 = 1]$$

$$= B \qquad [\because\ A \cdot 1 = A]$$

18. $(\bar{A} + C)(\bar{A} + \bar{C})(A + B + \bar{C}D) = (\bar{A}\bar{A} + \bar{A}C + \bar{A}\bar{C} + C\bar{C})(A + B + \bar{C}D)$

$$= (\bar{A} + \bar{A}C + \bar{A}\bar{C} + 0)(A + B + \bar{C}D) \quad [\because\ A \cdot A = A, A \cdot \bar{A} = 0]$$

$$= \bar{A}[1 + C + \bar{C}]A + B + \bar{C}D$$

$$= (\bar{A} \cdot 1) \cdot (A + B + \bar{C}D) \qquad [\because\ A + 1 = 1]$$

$$= \bar{A}(A + B + \bar{C}D) \qquad [\because\ A \cdot 1 = A]$$

$$= A\bar{A} + \bar{A}B + \bar{A}\bar{C}D$$

$$= 0 + \bar{A}B + \bar{A}\bar{C}D \qquad [\because\ A \cdot \bar{A} = 0]$$

$$= \bar{A}[B + \bar{C}D]$$

19. $\overline{(\bar{x}\bar{y} + z)} + z + xy + wz = \overline{\bar{x}\bar{y}} \cdot \bar{z} + z + xy + wz \qquad [\because\ \overline{A + B} = \bar{A} \cdot \bar{B}]$

$$= (\bar{\bar{x}} + \bar{\bar{y}})\bar{z} + z + xy + wz \qquad [\because\ \overline{A \cdot B} = \bar{A} + \bar{B}]$$

$$= (x + y)\bar{z} + z + xy + wz \qquad [\because\ \bar{\bar{A}} = A]$$

$$= x\bar{z} + y\bar{z} + z + xy + wz$$

$$= x\bar{z} + z + y + xy + wz \qquad [\because\ A + \bar{A}B = A + B]$$

$$= z + x + y + xy + wz \qquad [\because\ A + \bar{A}B = A + B]$$

$$= z + y + x + wz \qquad [\because\ A + AB = A]$$

$$= z + y + x \qquad [\because\ A + AB = A]$$

20. $xy + x(wz + w\bar{z}) = xy + xw(z + \bar{z})$

$$= xy + xw \cdot 1 \qquad (\because\ A + \bar{A} = 1)$$

$$= xy + xw \qquad (\because\ A \cdot 1 = A)$$

$$= x(y + w)$$

21. $\overline{(x+y)}(\bar{x}+\bar{y}) = (\bar{x}\cdot\bar{y})(\bar{x}+\bar{y})$ $\quad [\because \overline{A+B}=\bar{A}\cdot\bar{B}]$

$= \bar{x}\bar{x}\bar{y} + \bar{x}\bar{y}\bar{y}$

$= \bar{x}\bar{y} + \bar{x}\bar{y}$ $\quad [\because \bar{\bar{A}} = A]$

$= \bar{x}\bar{y}$ $\quad [\because \bar{\bar{A}} = A]$

22. $\bar{x}yz + xz = z[x + \bar{x}y]$

$= z[x+y]$ $\quad [\because A + \bar{A}B = A + B]$

23. $A + \bar{A}B + \bar{A}\bar{B}C + \bar{A}\bar{B}\bar{C}D = A + B + \bar{A}\bar{B}C + \bar{A}\bar{B}\bar{C}D$ $\quad [\because A + \bar{A}B = A + B]$

$= A + \bar{B}C + B + \bar{A}\bar{B}\bar{C}D$ $\quad [\because A + \bar{A}B = A + B]$

$= A + B + C + \bar{A}\bar{B}\bar{C}D$ $\quad [\because A + \bar{A}B = A + B]$

$= A + C + B + \bar{A}\bar{C}D$ $\quad [\because A + \bar{A}B = A + B]$

$= A + \bar{C}D + C + B$ $\quad [\because A + \bar{A}B = A + B]$

$= A + B + C + D$ $\quad [\because A + \bar{A}B = A + B]$

24. $ABC + A\bar{B}C + AB\bar{C} = AB(C + \bar{C}) + A\bar{B}C$

$= AB + A\bar{B}C$ $\quad [\because A + \bar{A} = 1]$

$= A[B + \bar{B}C]$

$= A[B + C]$ $\quad [\because A + \bar{A}B = A + B]$

25. $\bar{A}\bar{B}C + BC + AC = C[B + \bar{B}\bar{A}] + AC$

$= C[B + \bar{A}] + AC$ $\quad [\because A + \bar{A}B = A + B]$

$= BC + \bar{A}C + AC$

$= BC + C[A + \bar{A}]$

$= BC + C\cdot 1$ $\quad [\because A + \bar{A} = 1]$

$= BC + C$ $\quad [\because A\cdot 1 = A]$

$= C[1 + B] = C\cdot 1 = C$ $\quad [\because A + 1 = 1, A\cdot 1 = A]$

3.6 COMBINATIONAL CIRCUITS

Introduction

A digital system consists of two types of circuits namely

(*i*) Combinational circuits

(*ii*) Sequential circuits

In a combinational circuit, the output at any time depends only on the input values at that time. It consists of input variables, logic gates and output variables. The logic gates accept signals from the input variables and generate output signals. Figure 3.36 shows the block diagram of a combinational circuit. It accepts *n* input binary variables and generates output variables depending on the logical combination of gates. In a sequential circuit, the output at any time depends on the present input values as well as the past output values. Let us see in detail about the basic building blocks of an arithmetic unit in a digital computer such as half adder and full adder, are combinational circuits.

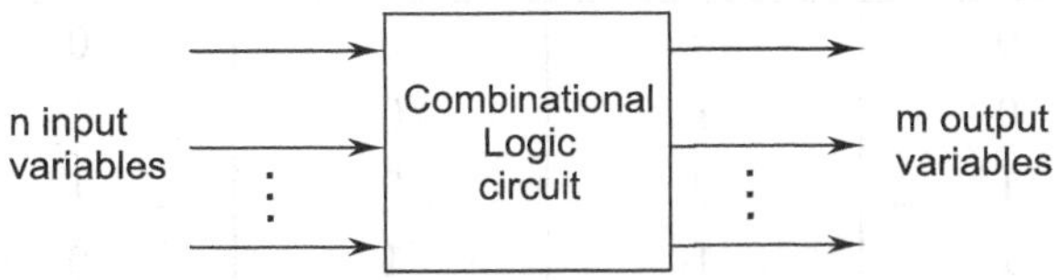

Fig. 3.36 Block diagram of a combinational circuit

Design of Combinational Circuits

The design of the combinational circuits involves following steps :

1. Identify the inputs and outputs and draw a block diagram.
2. Draw a truth table such that it completely describes the operation of the circuit for different combinations of inputs.
3. Write down the switching expression (s) for the output (s).
4. Simplify the switching expression using either algebraic or K-map method.
5. Implement the simplified expression using logic gates.

3.7 ADDERS

Digital computers perform various arithmetic operations. The most basic operation is the addition of two binary digits. This simple addition consists of four possible operations namely,

$$0 + 0 = 0$$

$$0 + 1 = 1$$

$$1 + 0 = 1$$

$$1 + 1 = 10_2$$

For the last operation, the sum is of two digits. The higher significant bit of this result is called a carry, and lower significant bit is called sum. Let us see the half adder and full adder in detail.

Half adder

The simplest combinational circuit which performs the arithmetic addition of two binary digits is called a half adder. Its operation needs two binary inputs and two outputs. Figure 3.37 shows the block diagram of half adder. The two inputs are the two 1-bit numbers A and B, and the two outputs are the sum (S) of A and B and the carry bit denoted by C.

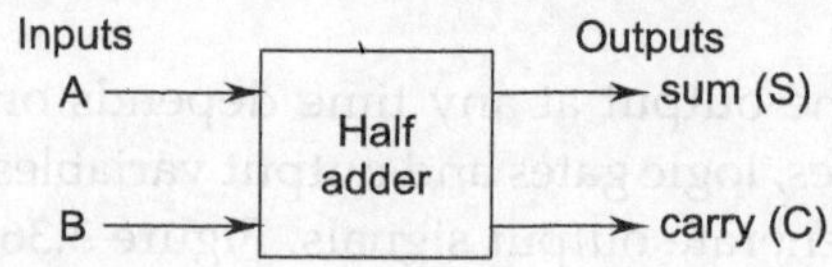

Fig. 3.37 Block diagram of half adder

Table 3.15. Truth table of half adder

Inputs		Outputs	
A	B	Carry (C)	Sum (S)
0	0	0	0
0	1	0	1
1	0	0	1
1	1	1	0

The truth table shown in Table 3.15 gives the relation between input and output variables for half adder. Here, the sum output is 1 when either of the inputs (A or B) is 1, and the carry output is 1 when both the inputs (A and B) are 1.

From the Table 3.15, the logic expression for the sum output can be written. In this, the sum output is 1 when AB = 01, and AB = 10. Therefore, the expression for sum is

$$S = \overline{A}B + A\overline{B} = A \oplus B$$

Similarly, the logic expression for carry is

$$C = AB$$

To simplify the logic expression, we can also use the K-map method. Figure 3.38 shows the K-map for carry and Fig. 3.39 shows the K-map for sum. In this, the expression for carry cannot be simplified.

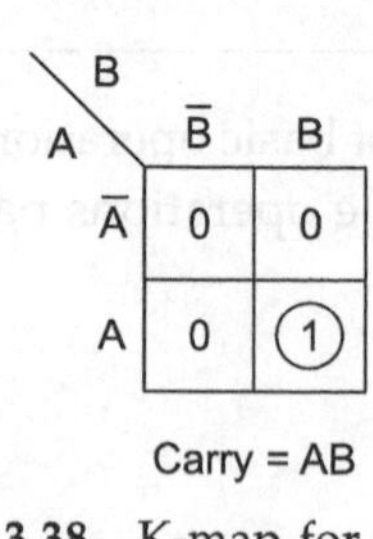

Fig. 3.38 K-map for carry

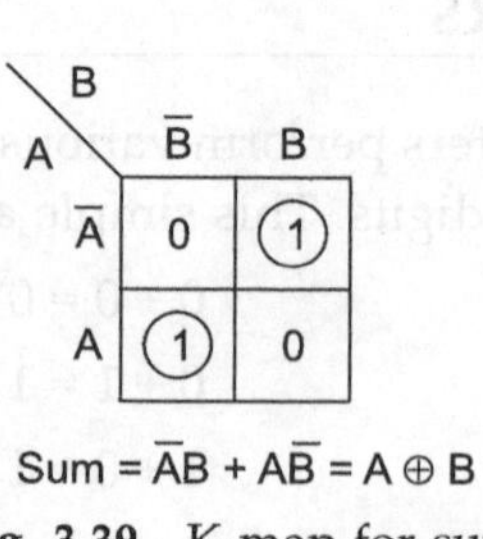

Fig. 3.39 K-map for sum

The sum output corresponds to an Ex-OR function while the carry output corresponds to an AND function. So the half adder circuit can be implemented using Ex-OR and AND gates as shown in Fig. 3.40. The implementation of the half adder circuit using basic gates AND, OR and NOT is shown in Fig. 3.41.

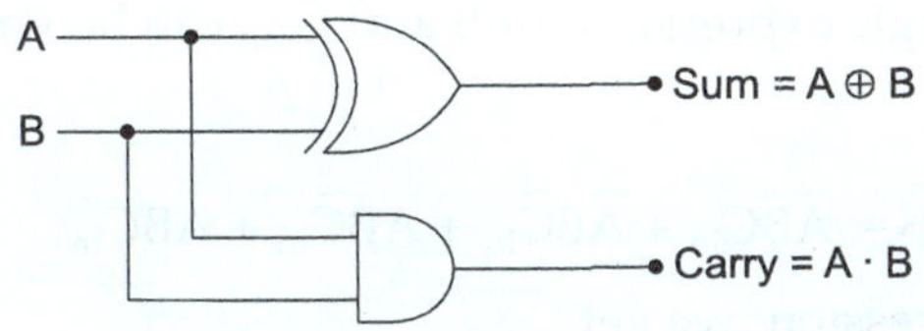

Fig. 3.40 Logic diagram for half adder

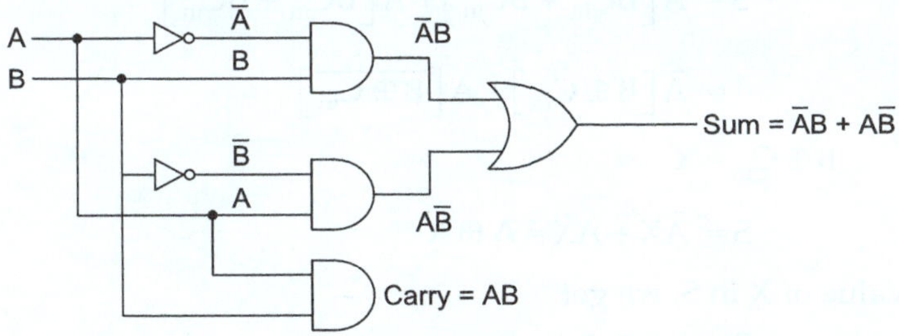

Fig. 3.41 Half adder using basic AND, OR and NOT gates

Full Adder

A full adder is a combinational circuit that performs the arithmetic sum of three input bits and produces a sum output and a carry. In half adder, there is no provision to add a carry coming from the lower order bits when multibit addition is performed. For this purpose only, a full adder is designed. The block diagram of the full adder is shown in Fig. 3.42. It consists of 3 inputs A, B and C_{in}. The third input, C_{in} represents the carry from the previous lower significant position. The outputs are designated by symbols S (for sum) and C_{out} (for carry). The truth table for the full adder circuit is shown in Table 3.16.

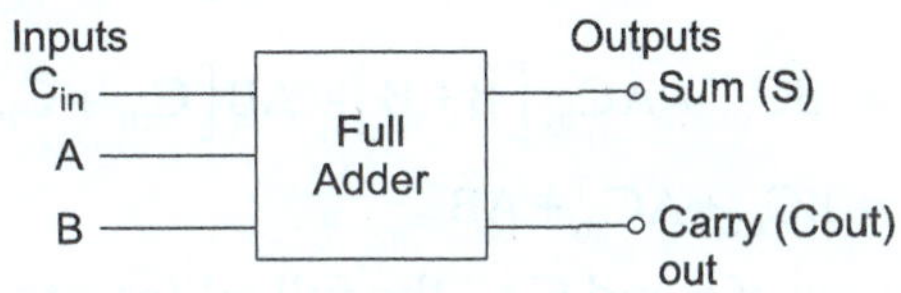

Fig. 3.42 Block diagram of full adder

As shown in Table 3.16, there are eight possible input combinations for the three inputs and for each case the S and C_{out} values are listed.

Table 3.16. Truth table of full adder

Inputs			Outputs	
A	B	C_{in}	Carry out (C_{out})	Sum (S)
0	0	0	0	0
0	0	1	0	1
0	1	0	0	1
0	1	1	1	0
1	0	0	0	1
1	0	1	1	0
1	1	0	1	0
1	1	1	1	1

From the truth table, the logic expression for S and C_{out} can be written as follows :

For Sum S :

$$S = \bar{A}\bar{B}C_{in} + \bar{A}B\bar{C}_{in} + A\bar{B}\bar{C}_{in} + ABC_{in}$$

Simplifying the above expression, we get

$$S = \bar{A}\left[\bar{B}C_{in} + B\bar{C}_{in}\right] + A\left[\bar{B}\bar{C}_{in} + BC_{in}\right]$$

$$= \bar{A}\left[B \oplus C_{in}\right] + A\left[\overline{B \oplus C_{in}}\right]$$

Let $\quad B \oplus C_{in} = X$

Now, $\quad S = \bar{A}X + A\bar{X} = A \oplus X$

Substituting the value of X in S, we get

$$S = A \oplus B \oplus C_{in}$$

For Carry out C_{out} :

$$C_{out} = \bar{A}BC_{in} + A\bar{B}C_{in} + AB\bar{C}_{in} + ABC_{in}$$

$$= BC_{in}\left[A + \bar{A}\right] + A\bar{B}C_{in} + AB\bar{C}_{in}$$

$$= BC_{in} + A\bar{B}C_{in} + AB\bar{C}_{in}$$

Now, the ABC_{in} term is added for simplification

$$C_{out} = BC_{in} + A\bar{B}C_{in} + AB\bar{C}_{in} + ABC_{in} + ABC_{in}$$

$$= BC_{in} + AC_{in}\left[\bar{B} + B\right] + AB\left[C_{in} + \bar{C}_{in}\right] \qquad \left[\because \; A + \bar{A} = 1\right]$$

$$= BC_{in} + AC_{in} + AB$$

From the simplified expressions of S and C_{out}, the full adder circuit can be implemented using one 3 input Ex-OR gate and three 2 input AND gates and one 3 input OR gate as shown in Fig. 3.43.

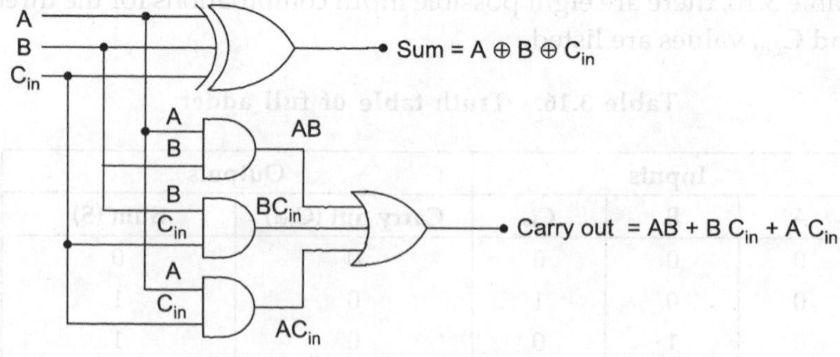

Fig. 3.43 Logic diagram for full adder

K-map method can also be used for simplifying the logic expressions for S and C_{out}. The K-maps for the outputs S and C_{out} are given in Fig. 3.44 and Fig. 3.45 respectively.

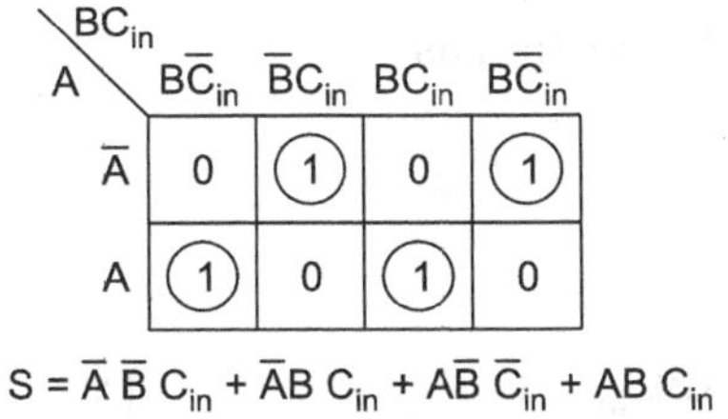

Fig. 3.44 K-map for sum

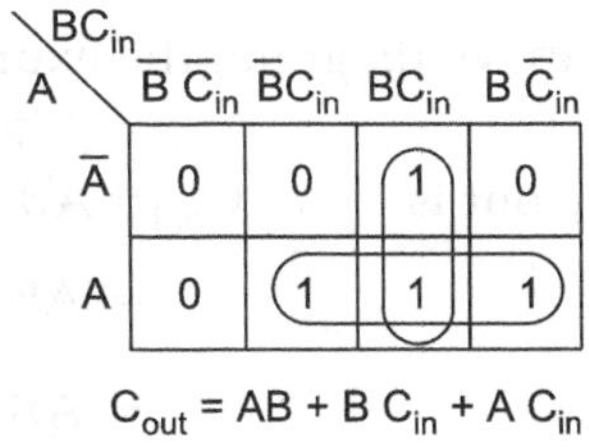

Fig. 3.45 K-map for carry

From the K-map, the simplified expressions for S and C_{out} can be written as

$$S = \bar{A}\bar{B}C_{in} + \bar{A}BC_{in} + A\bar{B}\bar{C}_{in} + ABC_{in}$$

$$C_{out} = AB + BC_{in} + AC_{in}$$

Using the above expressions, the full adder can be implemented using basic gates as shown in Fig. 3.46.

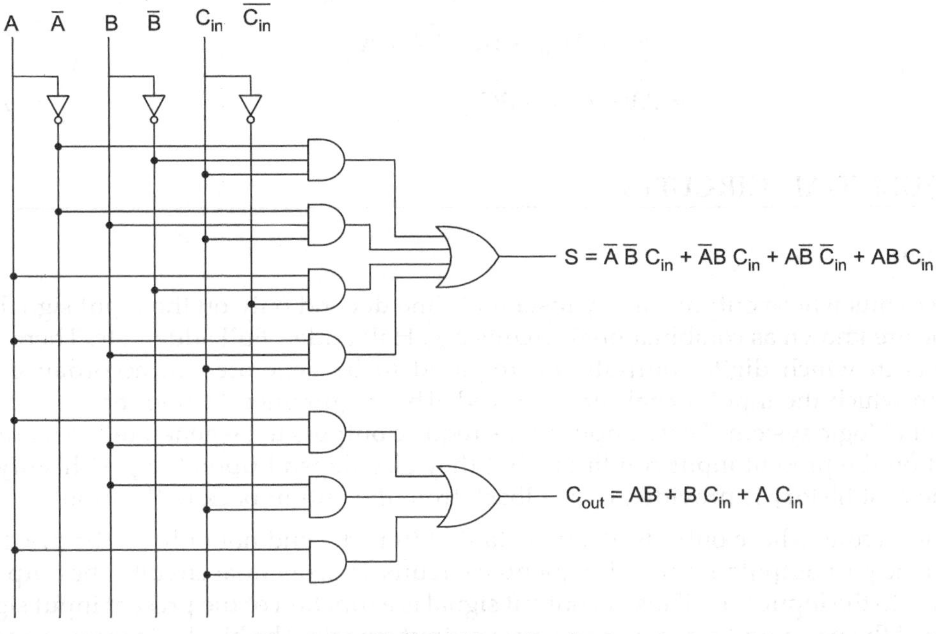

Fig. 3.46 Full adder using basic gates

We can also implement a full adder using two half adders and an OR gate as shown in Fig. 3.47.

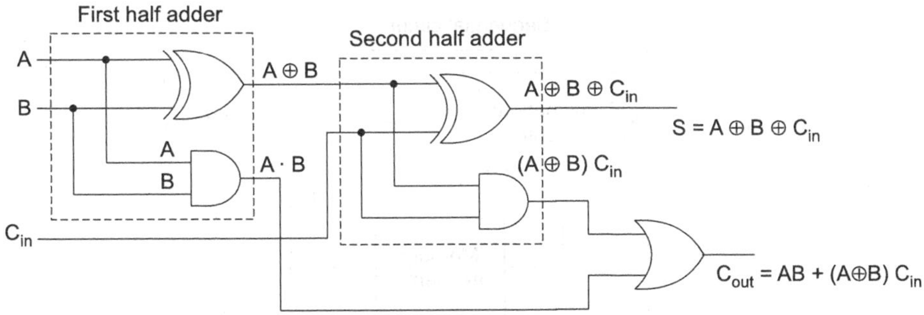

Fig. 3.47 Implementation of a full adder with two half adders and an OR gate

From the above diagram, the expression for the S can be written as

$$S = A \oplus B \oplus C_{in}$$

and the carry out is

$$\begin{aligned} C_{out} &= AB + (A \oplus B)\, C_{in} \\ &= AB + \left(A\bar{B} + \bar{A}B\right) C_{in} \\ &= AB\left(C_{in} + 1\right) + A\bar{B}C_{in} + \bar{A}BC_{in} && [\because\ A + 1 = 1] \\ &= ABC_{in} + AB + A\bar{B}C_{in} + \bar{A}BC_{in} \\ &= AB + AC_{in}\left(B + \bar{B}\right) + \bar{A}BC_{in} \\ &= AB + AC_{in} + \bar{A}BC_{in} && \left[\because\ A + \bar{A} = 1\right] \\ &= AB\left(C_{in} + 1\right) + AC_{in} + \bar{A}BC_{in} && [\because\ A + 1 = 1] \\ &= ABC_{in} + AB + AC_{in} + \bar{A}BC_{in} \\ &= AB + AC_{in} + BC_{in}\left[A + \bar{A}\right] \\ &= AB + AC_{in} + BC_{in} && \left[\because\ A + \bar{A} = 1\right] \end{aligned}$$

3.8 SEQUENTIAL CIRCUITS

Introduction

The logic circuits whose outputs at any instant of time depend only on the input signals present at that time are known as combinational circuits *e.g.*, Half adder, Full adder, etc. There are many applications in which digital outputs are required to be generated in accordance with the sequence in which the input signals are received. This requirement cannot be satisfied using a combinational logic system. These applications require outputs to be generated that are not only dependent on the present input conditions but they also depend upon the past history of these inputs. The past history is provided by feedback from the output back to the input.

The logic circuits whose outputs at any instant of time depend not only on the present inputs but also on the past outputs are called sequential circuits. In sequential circuits, the output signals are feedback to the input side. Thus, an output signal is a function of the present input signals and a sequence of the past input signals *i.e.*, the past output signals. The block diagram of a sequential circuit is shown in Fig. 3.48. It consists of a combinational circuit to which memory elements are connected to form a feedback path.

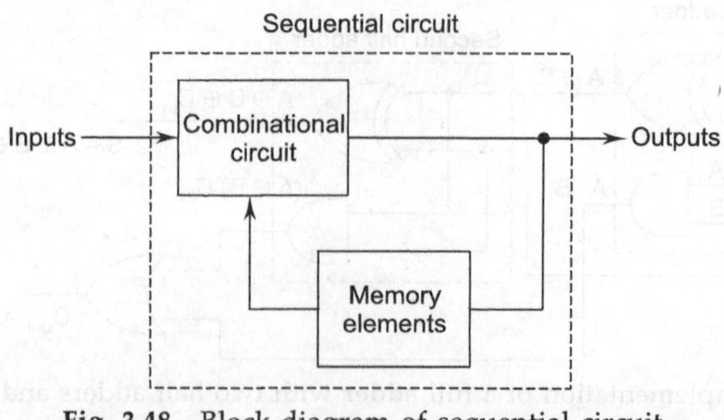

Fig. 3.48 Block diagram of sequential circuit

The information stored in the memory elements at any time defines the present state of the sequential circuit. The present state and the external inputs determine the outputs and the next state of the sequential circuit.

Combinational circuits are often faster than sequential circuits since the combinational circuits do not require memory elements whereas the sequential circuits need memory devices to perform their operations in sequence. However, modern digital computers, must have memories to function properly. Thus, sequential circuits are of prime importance in modern digital devices.

Sequential circuits are of two types, *viz.*, (*i*) synchronous or clocked and (*ii*) asynchronous or unclocked. In synchronous sequential circuits, signals can affect the memory elements only at discrete instants of time. In asynchronous sequential circuits change in input signals can affect memory element at any instant of time. In general, asynchronous circuits are considerably faster than synchronous sequential circuits. However, in an asynchronous circuit, events are allowed to occur without any synchronization. In such a case, the system becomes unstable which results in difficulties.

To have a sequential circuit, a storage device is required to know what has happened in the past. The basic unit of storage is the flip-flop.

3.9 FLIP-FLOPS

The simplest kind of sequential circuit is a memory cell that has only two states. It can be either 1 (+5 V) or 0 (0 V). Such two state sequential circuits are called flip-flops because they flip from one state to another and then flop back. It is a 1-bit memory unit. A flip-flop is also known as bistable multivibrator, latch or toggle.

There are many types of flip-flops. The major differences among various types of flip-flops are in the number of inputs they process and manner in which the inputs affect their output state. The most common types of flip-flops are discussed below.

Types of flip-flops

There are four basic types of flip-flops. They are

(*i*) S-R flip-flop
(*ii*) D flip-flop
(*iii*) J-K flip-flop
(*iv*) T flip-flop

3.10 SET-RESET (S-R) FLIP-FLOP

The simplest type of flip-flop is the Set-Reset (S-R) flip-flop. It has two inputs, namely SET (S) and RESET (R), and two outputs Q and $\bar{Q}$. The two outputs are complement to each other. It can be constructed from either two NAND gates or two NOR gates. The block diagram of S-R flip-flop is shown in Fig. 3.49.

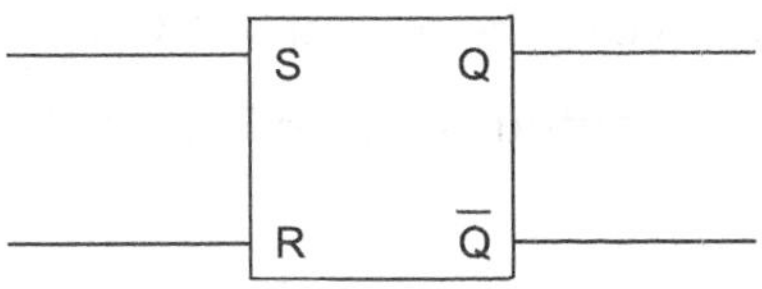

Fig. 3.49 Block diagram of S-R flip-flop

S-R flip-flop using NOR gates

The S-R flip-flop can be easily constructed using two NOR gates connected back to back as shown in Fig. 3.50. The cross coupled connections from the output of one gate to the input of the other gate constitute a feedback path. For this reason, the circuits are classified as asynchronous sequential circuits. The truth table for the NOR based S-R flip-flop is shown in Table 3.17.

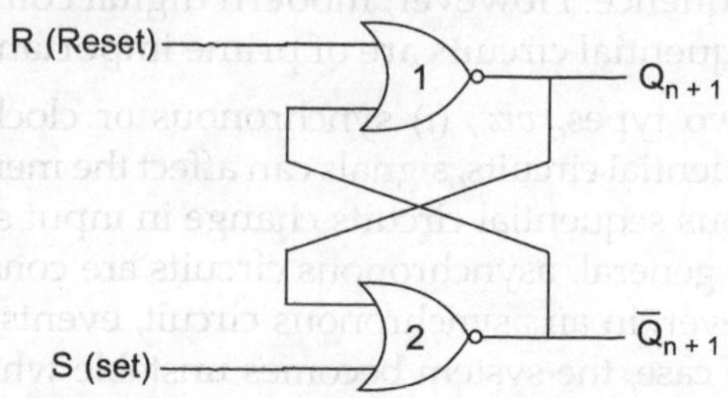

Fig. 3.50 S-R flip-flop using NOR gates

Table 3.17. Truth table of S-R flip-flop using NOR gates

Inputs		Outputs		Action
S	R	Q_{n+1}	$\bar{Q}_{n+1}$	
0	0	Q_n	$\bar{Q}_n$	No change
0	1	0	1	Reset
1	0	1	0	Set
1	1	?	?	Forbidden

To analyze the circuit, one must know that the output of a NOR gate is 0 if any input is 1 and the output is 1 only when all inputs are 0. The outputs for the four possible input combinations are described below.

Case 1. S = 0 and R = 0

Let us assume that $Q_n = 0$ and $\bar{Q}_n = 1$. The inputs of NOR gate 1 are 1 and 0, and therefore its output $Q_{n+1} = 0$. This $Q_{n+1} = 0$ is fed back to NOR gate 2 input, thereby producing a 1 at its output, So, $\bar{Q}_{n+1} = 1$, as originally assumed.

Next, let us assume that $Q_n = 1$ and $\bar{Q}_n = 0$. This 1 is applied to the input of NOR gate 2 and therefore the output $\bar{Q}_{n+1}$ becomes 0. This $Q_{n+1} = 0$ is fed to the input of NOR gate 1, thereby producing a 1 at its output. So, $Q_{n+1} = 1$, as originally assumed. Thus, the condition S = 0 and R = 0 will not affect the outputs of flip-flop *i.e.*, the flip-flop simply remains in its present state (Q_n).

Case 2. S = 0 and R = 1

In this case, 1 at the RESET input makes the output of NOR gate 1 low (*i.e.*, $Q_{n+1} = 0$). Now both the inputs of NOR gate 2 are 0 and its output $\bar{Q}_{n+1} = 1$. Thus, the input condition S = 0 and R = 1 will always reset the flip-flop to 0.

Case 3. S = 1 and R = 0

In this case, 1 at the SET input makes the output of NOR gate 2 low (*i.e.*, $\bar{Q}_{n+1} = 0$). Now both the inputs of NOR gate 1 are 0 and its output $Q_{n+1} = 1$. Thus, the input condition S = 1 and R = 0 will always set the flip-flop to 1.

Case 4. S = 1 and R = 1

In this case, both the NOR gates will produce 0 at the output. Hence $Q_{n+1} = 0$ and $\bar{Q}_{n+1} = 0$. This condition violates the fact that the outputs Q_{n+1} and $\bar{Q}_{n+1}$ are the complements of each other. So, we call this an indeterminate or forbidden state. In normal operation, this condition must be avoided by making sure that 1s are not applied to both inputs simultaneously.

$\bar{S}$-$\bar{R}$ flip-flop using NAND gate

The S-R flip-flop can also be constructed using two NAND gates connected back to back as shown in Fig. 3.51. The truth table for the NAND based $\bar{S}$-$\bar{R}$ flip-flop is shown in Table 3.18. To analyze the circuit, one must know that the output of NAND gate is 1 if any input is 0 and the output is 0 only when all the inputs are 1. The outputs for the four possible input combinations are described below.

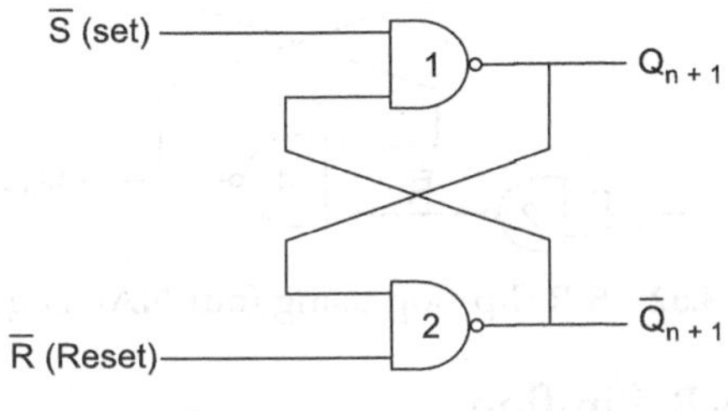

Fig. 3.51 $\bar{S}$-$\bar{R}$ flip-flop using NAND gate

Table 3.18. Truth table of $\bar{S}$-$\bar{R}$ flip-flop using NAND gates

Inputs		Outputs		Action
$\bar{S}$	$\bar{R}$	Q_{n+1}	$\bar{Q}_{n+1}$	
0	0	?	?	Forbidden
0	1	1	0	Set
1	0	0	1	Reset
1	1	Q_n	$\bar{Q}_n$	No change

Case 1. $\bar{S} = 0$ and $\bar{R} = 0$.

In this case, both the NAND gates will produce 1 at the output *i.e.*, $Q_{n+1} = 1$ and $\bar{Q}_{n+1} = 1$. This condition is indeterminate or forbidden state. This should not be used.

Case 2. $\bar{S} = 0$ and $\bar{R} = 1$

In this case, 0 at the set input makes the output of NAND gate 1 HIGH (*i.e.*, $Q_{n+1} = 1$). Now both the inputs of NAND gate 2 are 1 and its output $\bar{Q}_{n+1} = 0$. Thus, the input condition $\bar{S} = 0$ and $\bar{R} = 1$ will always set the flip-flop to 1.

Case 3. $\bar{S} = 1$ and $\bar{R} = 0$

In this case, 0 at the reset input, makes the output of NAND gate 2 HIGH (*i.e.,* $\bar{Q}_{n+1} = 1$). Now both the inputs of NAND gate 1 are 1 and its output $Q_{n+1} = 0$. Thus, the input condition $\bar{S} = 1$ and $\bar{R} = 0$ will always reset the flip-flop to 0.

Case 4. $\bar{S} = 1$ and $\bar{R} = 1$

In this case, $\bar{S} = 1$ and $\bar{R} = 1$ does not affect the state of the flip-flop. It remains in its prior state.

We can convert a NAND based $\bar{S}$-$\bar{R}$ flip-flop into an S-R flip-flop by placing an inverter at each input of the flip-flop. The resulting four NAND gate circuits, shown in Fig. 3.52, will behave in the same manner as an S-R flip-flop.

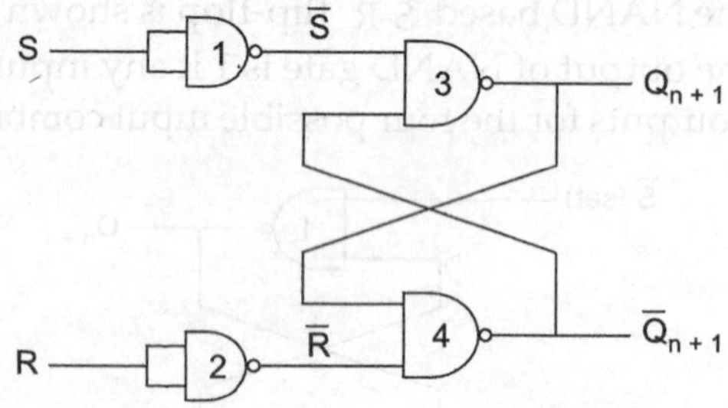

Fig. 3.52 S-R flip-flop using four NAND gates

Characteristic Equation of S-R flip-flop

The logic expression which describes the operation of the S-R flip-flop is called as characteristic equation of the S-R flip-flop. We can derive the equation from the truth table as described below.

From the truth table of S-R flip-flop (Table 3.17), the Present state-Next state table and application or excitation table for S-R flip-flop can be drawn, as shown in Table 3.19 and Table 3.20 respectively.

Table 3.19. Present state-next state table for S-R flip-flop

Present State	Inputs		Next State
Q_n	S	R	Q_{n+1}
0	0	0	0
0	0	1	0
0	1	0	1
0	1	1	X
1	0	0	1
1	0	1	0
1	1	0	1
1	1	1	X

X : don't care

Table 3.20. Application or excitation table for S-R flip-flop

Q_n	Q_{n+1}	Excitation Inputs	
		S	R
0	0	0	X
0	1	1	0
1	0	0	1
1	1	X	0

Using the Table 3.19, a Karnaugh map for the next state transition (Q_{n+1}) can be drawn as shown in Fig. 3.53.

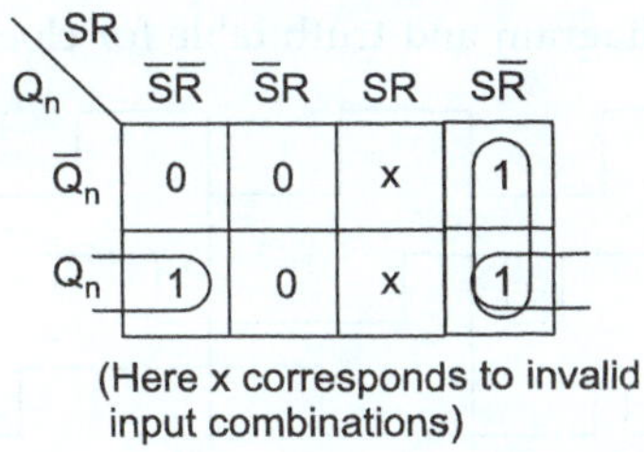

Fig. 3.53 Next state (Q_{n+1}) map for S-R flip-flop

From the above K-map,

$$Q_{n+1} = S\overline{R} + Q_n\overline{R}$$

$$Q_{n+1} = \overline{R}\left[S + Q_n\right]$$

This logic expression is called the characteristic equation of the S-R flip-flop.

Clocked S-R flip-flop

Synchronous circuits change their states only when clock pulses are present. The clocked S-R flip-flop is shown in Fig. 3.54. It consists of a basic S-R flip-flop circuit and two additional NAND gates. The pulse input acts as an enable signal for the other two inputs. The outputs of NAND gates 1 and 2 stay at the logic 1 as long as the clock pulse (CLK) input remains 0. When the pulse input goes to 1, information from the S or R input is allowed to reach the output. Fig. 3.55 shows the Block diagram and truth table of clocked S-R flip-flop and Fig. 3.56 shows input and output waveforms.

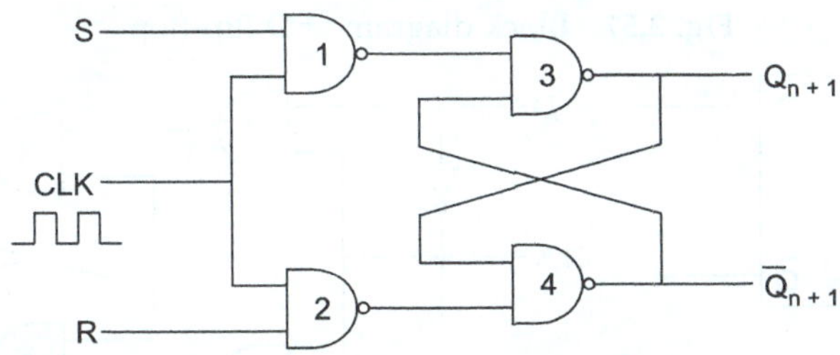

Fig. 3.54 Clocked S-R flip-flop

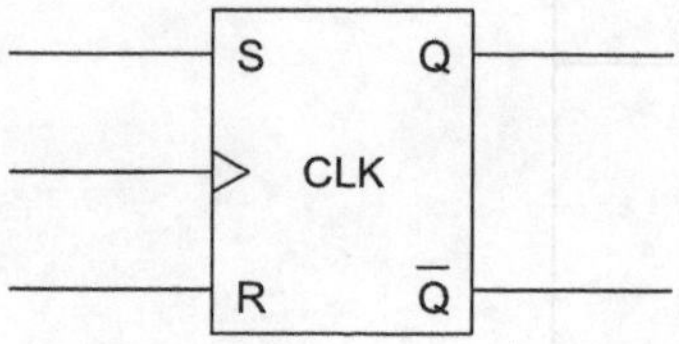

(a) Block diagram

CP	S	R	Q_{n+1}	$\bar{Q}_{n+1}$	State
0	0	0	No change	No change	No change
0	0	1	No change	No change	No change
0	1	0	No change	No change	No change
0	1	1	No change	No change	No change
1	0	0	Q	$\bar{Q}$	No change
1	0	1	0	1	Reset
1	1	0	1	0	Set
1	1	1	?	?	Forbidden

(b) Truth table

Fig. 3.55 Block diagram and truth table for clocked S-R flip-flop

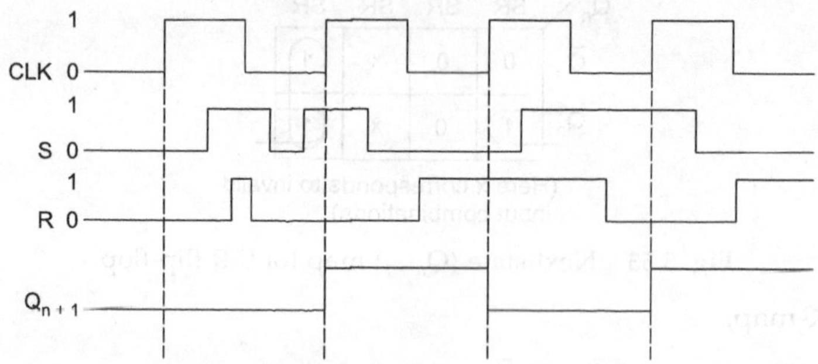

Fig. 3.56 Input and output waveform for clocked S-R flip-flop

3.11 D FLIP-FLOP

The D (delay) flip-flop has only one input called the Delay (D) input and two outputs Q and $\bar{Q}$. By inserting an inverter between S and R and assigning the symbol D to the S input, we can construct the D flip-flop from an S-R flip-flop. The block diagram of D flip-flop is shown in Fig. 3.57. The D flip-flop using S-R flip-flop is shown in Fig. 3.58.

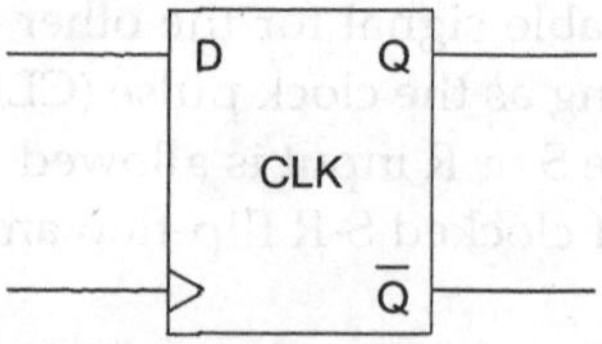

Fig. 3.57 Block diagram of D flip-flop

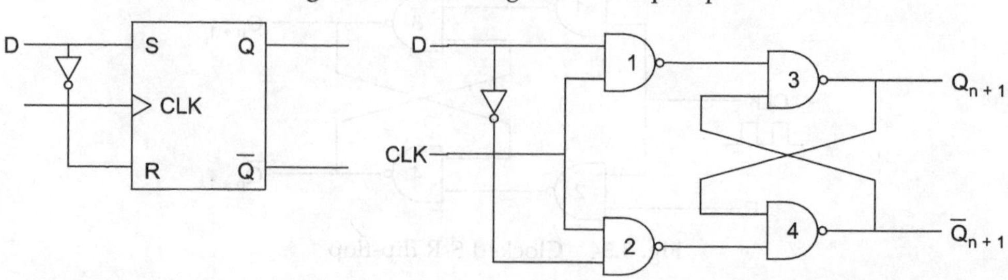

Fig. 3.58 D flip-flop using S-R flip-flop

The operation of the D flip-flop is as follows :

1. When the CLK input is LOW, the D input has no effect, since the set and reset inputs of the NAND flip-flops are kept HIGH.
2. When the CLK goes HIGH, the Q output will take on the value of the D input. Therefore, during the occurrence of Clock pulse if D = 1, the Q output is set and if D = 0, the Q output is reset.

The truth table of D flip-flop is given in Table 3.21 from which it is clear that the next state of the flip-flop at time (Q_{n+1}) follows the value of the input D when the clock pulse is applied. As transfer of data from the input to the output is delayed, it is known as delay (D) flip-flop. It is either used as a delay device or as a latch to store 1 bit of binary information. Figure 3.59 shows the input and output waveforms of D flip-flop.

Table 3.21. Truth table of D flip-flop

CLK	Input D	Output Q_{n+1}	Action
0	X	No change	No change
1	0	0	Reset
1	1	1	Set

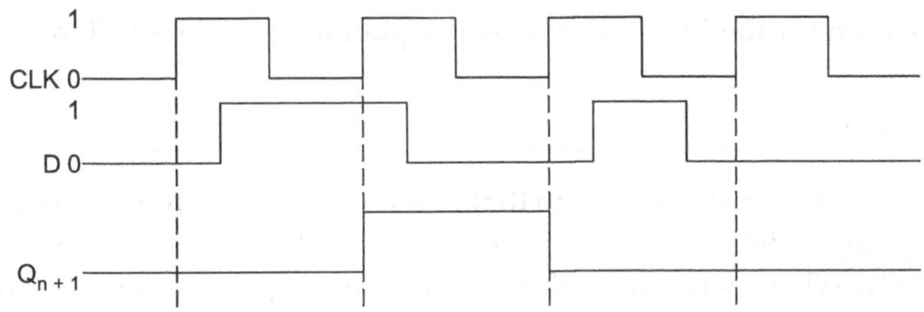

Fig. 3.59 Input and output waveforms of D flip-flop

Characteristic equation of D flip-flop :

From the truth table of D flip-flop (Table 3.21), the present state-next state table and application or excitation table for D flip-flop can be drawn as shown in Table 3.22 and Table 3.23 respectively.

Table 3.22. Present State-Next state table for D flip-flop

Present State Q_n	Delay Input D	Next State Q_{n+1}
0	0	0
0	1	1
1	0	0
1	1	1

Table 3.23. Application-excitation table for D flip-flop

Q_n	Q_{n+1}	Excitation Input D
0	0	0
0	1	1
1	0	0
1	1	1

Using the Table 3.22, a Karnaugh map for the next state transition (Q_{n+1}) can be drawn as shown in Fig. 3.60.

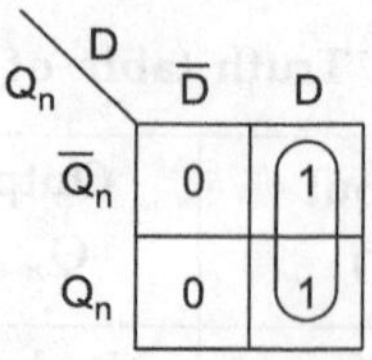

Fig. 3.60 Next state (Q_{n+1}) map for D flip-flop

From the above K-map,

$$Q_{n+1} = D$$

This logic expression is called the characteristic equation of the D flip-flop.

3.12 J-K FLIP-FLOP

A J-K flip-flop has a characteristic similar to that of an S-R flip-flop. In J-K flip-flop the forbidden state in the S-R flip-flop is defined. Inputs J and K behave like inputs S and R to set and reset the flip-flop, respectively. When J = K = 1, the flip-flop output toggles, *i.e.*, switches to its complement state ; if Q = 0, it switches to Q = 1 and *vice-versa*. The block diagram of J-K flip-flop is shown in Fig. 3.61.

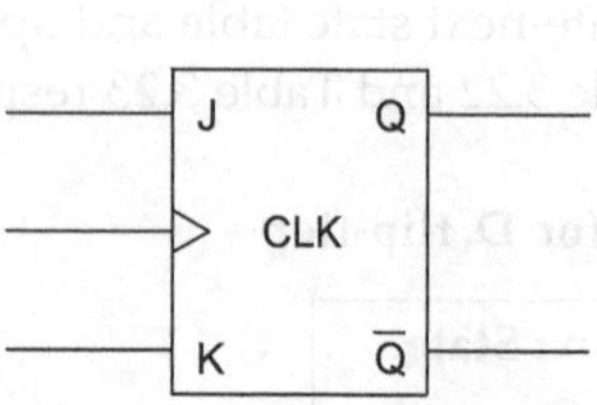

Fig. 3.61 Block diagram of J-K flip-flop

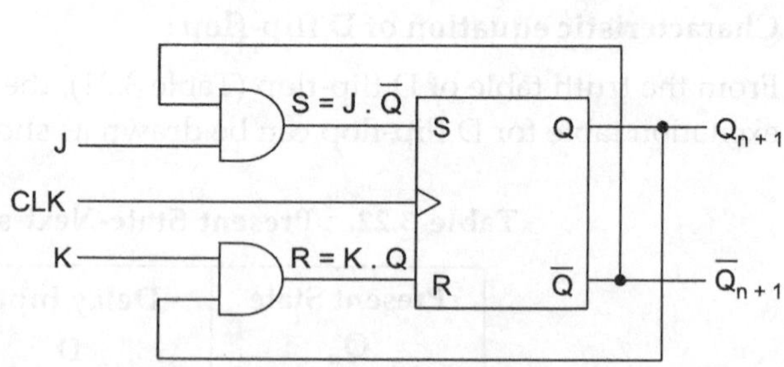

Fig. 3.62 J-K flip-flop using S-R flip-flop

A J-K flip-flop can be obtained from the clocked S-R flip-flop by augmenting two AND gates as shown in Fig. 3.62. The operation of the J-K flip-flop is as follows :

Case 1. J = 0 and K = 0

When J and K are both 0 (low), both AND gates are disabled. Therefore, the inputs to the basic flip-flop are S = 0 and R = 0. This condition forces the flip-flop into the same state.

Case 2. J = 0 and K = 1

When J = 0, K = 1 and the previous state of flip-flop is SET (*i.e.*, $Q_n = 1$, $\bar{Q}_n = 0$), then $S = J\bar{Q}_n = 0$ and $R = KQ_n = 1$. Since S = 0 and R = 1, the flip-flop RESETS on the application of a clock pulse. But if the flip-flop is already RESET (*i.e.*, $Q_n = 0$ and $\bar{Q}_n = 1$), then the application of J = 0 and K = 1 will not alter the state of the flip-flop ($S = J\bar{Q}_n = 0$ and R = 0) and it remains in the RESET state.

Case 3. J = 1 and K = 0

When J = 1, K = 0 and the previous state of flip-flop is RESET ($Q_n = 0$ and $\bar{Q}_n = 1$), then $S = J\bar{Q}_n = 1$ and $R = KQ_n = 0$. Since S = 1 and R = 0, the flip-flop will SET on the application of a clock pulse. But, if the flip-flop is already SET (*i.e.*, $Q_n = 1$, $\bar{Q}_n = 0$), then the inputs J = 1, K = 0 will make S = 0 and R = 0, and the flip-flop remains in the SET state.

Case 4. J = 1 and K = 1

This condition deserves special attention. When J = 1, K = 1 and the previous state is a SET state (*i.e.*, $Q_n = 1$ and $\bar{Q}_n = 0$), then $S = J\bar{Q}_n = 0$ and $R = KQ_n = 1$. Since S = 0 and R = 1, the flip-flop RESETS on the application of a clock pulse, *i.e.*, the flip-flop toggles from SET to RESET state.

Again for J = 1, K = 1, if the previous state is a RESET state (*i.e.*, $Q_n = 0$, $\bar{Q}_n = 1$), then $S = J\bar{Q}_n = 1$ and $R = KQ_n = 0$. Since S = 1 and R = 0, the flip-flop on the application of a clock pulse will toggle to the SET state. Hence, when J = 1 and K = 1 the flip-flop toggles on the application of the next clock pulse. Toggle means to switch to the opposite state.

Table 3.24 shows the truth table for clocked J-K flip-flop. Figure 3.63 shows the input and output waveforms of clocked J-K flip-flop.

Table 3.24. Truth table of clocked J-K flip-flop

CLK	Inputs		Output		Action
	J	K	Q_{n+1}	$\bar{Q}_{n+1}$	
X	0	0	Q_n	$\bar{Q}_n$	No change
1	0	1	0	1	Reset
1	1	0	1	0	Set
1	1	1	$\bar{Q}_n$	Q_n	Toggle

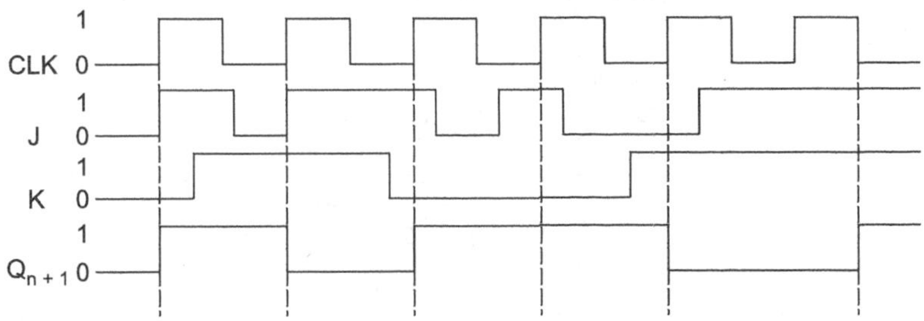

Fig. 3.63 Input and Output waveforms of clocked J-K flip-flop

Race Around Condition

In J-K flip-flop output is feed back to the input, and therefore change in the output results change in the input. Due to this in the positive half of the clock pulse if J and K are both high then output toggles continuously. This condition is known as 'race around condition' and must be avoided. To avoid this problem, the positive pulse width of the clock must be kept shorter than the propagation delay through the flip-flop. The restriction on the clock pulse width and thus the race around condition can be eliminated with a master-slave flip-flops, as discussed in the next discussion.

Characteristic Equation of J-K Flip-flop

From the truth table of J-K flip-flop, the Present state-Next state table and application or excitation table for J-K flip-flop are shown in Table 3.25 and Table 3.26 respectively.

Table 3.25. Present State-Next State table for J-K flip-flop

Present State Q_n	Inputs J	Inputs K	Next State Q_{n+1}
0	0	0	0
0	0	1	0
0	1	0	1
0	1	1	1
1	0	0	1
1	0	1	0
1	1	0	1
1	1	1	0

Table 3.26. Application or excitation table for J-K flip-flop

Q_n	Q_{n+1}	Excitation Inputs J	Excitation Inputs K
0	0	0	X
0	1	1	X
1	0	X	1
1	1	X	0

Using the Table 3.25, a Karnaugh map for the next state transition (Q_{n+1}) can be drawn as shown in Fig. 3.64.

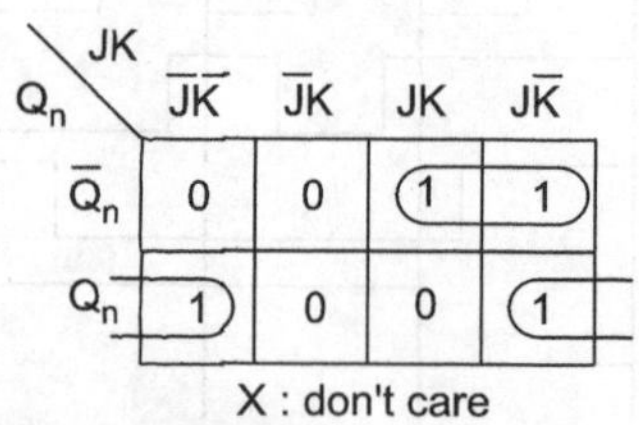

Fig. 3.64 Next state (Q_{n+1}) map for J-K flip-flop

From the above K-map,

$$Q_{n+1} = J\bar{Q}_n + \bar{K}Q_n$$

This logic expression is called the characteristic equation of the J-K flip-flop.

3.13 T FLIP-FLOP

Another basic flip-flop, called the T or Trigger or toggle flip-flop, has only a single data (T) input, a clock input and two outputs Q and $\bar{Q}$. The T flip-flop is obtained from a J-K flip-flop by connecting its J and K inputs together. The block diagram of a T flip-flop and its circuit implementation using a J-K flip-flop are shown in Fig. 3.65 and Fig. 3.66 respectively.

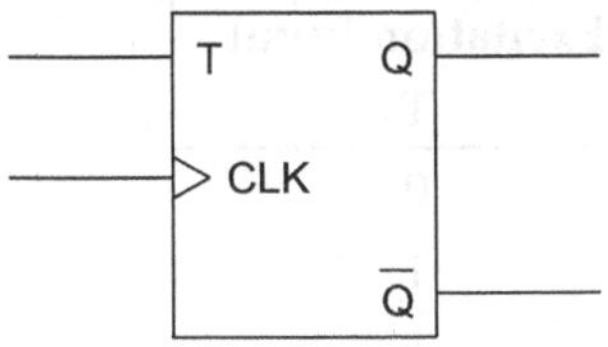

Fig. 3.65 Block diagram of T flip-flop

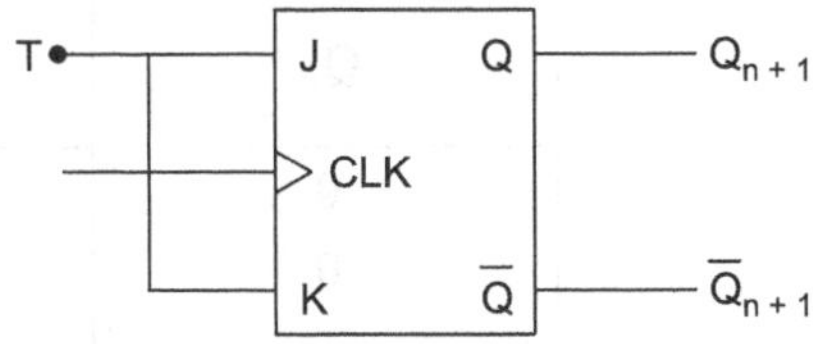

Fig. 3.66 T flip-flop using J-K flip-flop

The operation of the T flip-flop is as follows:

1. When the T input is in the 0 state (*i.e.*, J = K = 0), both the AND gates are disabled, and hence there is no change in the output.
2. When the T input is at a 1 (*i.e.*, J = K = 1) level, output toggles.

The truth table of T flip-flop is shown in Table 3.27. Figure 3.67 shows the present state input and output waveforms of T flip-flop.

Table 3.27. Truth table of T flip-flop

Present State Q_n	Input T	Output Q_{n+1}
0	0	0
0	1	1
1	0	1
1	1	0

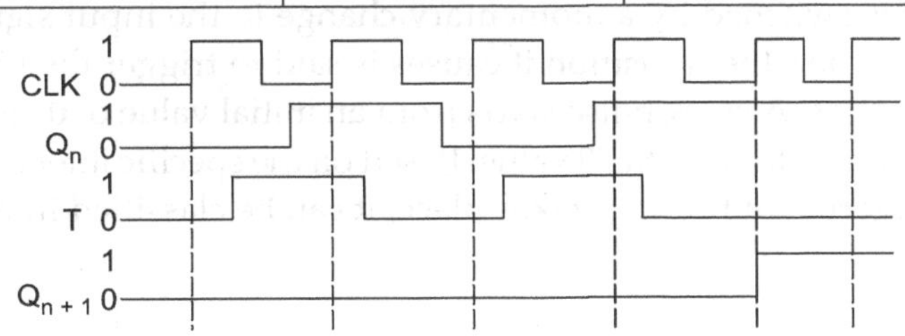

Fig. 3.67 Present state, input and output waveforms of T flip-flops

Characteristic Equation of T flip-flop

From the truth table of T flip-flop, the present state-next state table and application or excitation table for T flip-flop are shown in Table 3.28 and Table 3.29 respectively.

Table 3.28. Present State - Next State table for T flip-flop

Present State Q_n	Input T	Next State Q_{n+1}
0	0	0
0	1	1
1	0	1
1	1	0

Table 3.29. Application or excitation table for T flip-flop

Q_n	Q_{n+1}	Excitation Input T
0	0	0
0	1	1
1	0	1
1	1	0

Using the Table 3.28, a Karnaugh map for the next state transition (Q_{n+1}) can be drawn as shown in Fig. 3.68.

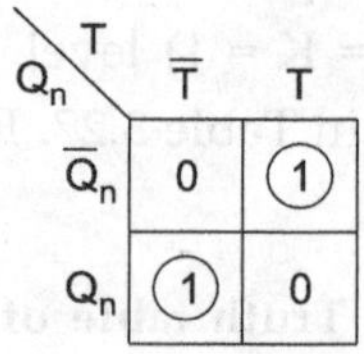

Fig. 3.68 Next state (Q_{n+1}) map for T flip-flop

From the above K-map,

$$Q_{n+1} = \bar{Q}_n T + Q_n \bar{T}$$

This logic expression is called the characteristic equation of the T flip-flop.

Triggering of Flip-Flops

The state of a flip-flop is switched by a momentary change in the input signal. This momentary change is called a trigger and the transition it causes is said to trigger the flip-flop. Clocked flip-flops are triggered by pulses. A clock pulse starts from an initial value of 0, goes momentarily to 1, and after a short time, returns to its initial 0 value. Based on the specific interval or point in the clock during or at which triggering of flip-flop takes place, it can be classified into two different types:

1. Level triggering
2. Edge triggering

1. **Level Triggering :** Level triggering means, the flip-flops are enabled in HIGH (+ve level) or LOW (–ve level). The flip-flop action is dependent on the entire period of the pulse.

   Positive level triggering : If the flip-flop changes its state when the clock is positive, it is called positive level triggering.

Negative level triggering : If the flip-flop changes its state when the clock is negative, it is called negative level triggering.

2. **Edge Triggering :** A clock pulse may be either negative or positive. A positive clock source remains at 0 during the interval between pulses and goes to 1 during the occurrence of a pulse. The pulse goes through two signal transitions, from 0 to 1 and returns from 1 to 0. The positive transition is defined as the positive edge and the negative transition is defined as the negative edge. Figure 3.69 shows the positive pulse and Fig. 3.70 shows the negative pulse with +ve edge and –ve edge.

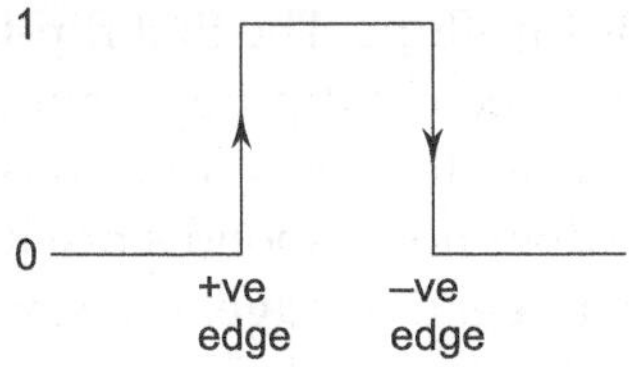

Fig. 3.69 Positive pulse

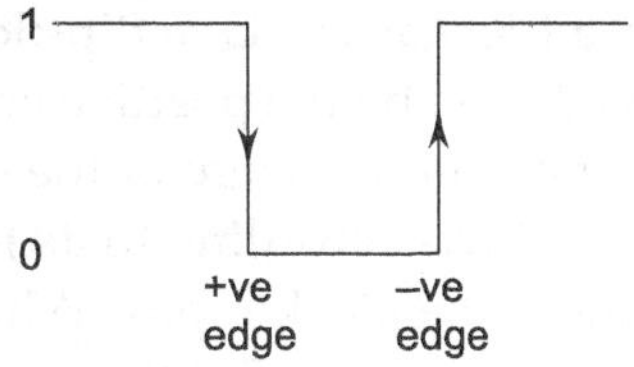

Fig. 3.70 Negative pulse

The flip-flop changes its state either at the positive edge or at the negative edge of the clock pulse and is sensitive to its inputs only at this transition of the clock.

To make the flip-flop respond only to a pulse transition is to use capacitive coupling. An RC (resistor-capacitor) circuit is inserted in the clock input of the flip-flop. The RC circuit is shown in Fig. 3.71. This circuit generates a spike in response to a momentary change of input signal.

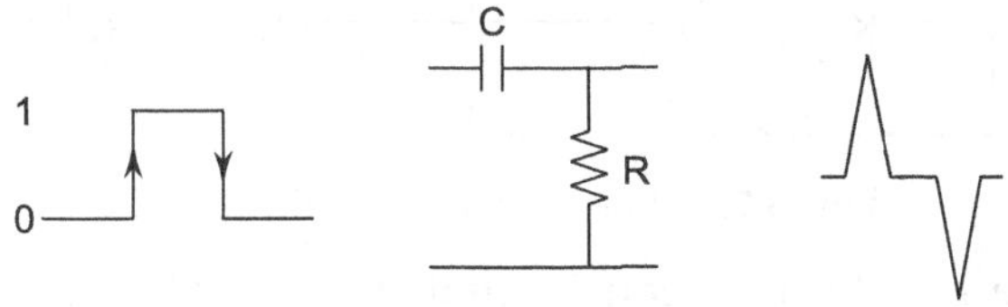

Fig. 3.71 RC circuit

A positive edge emerges from the RC circuit with a positive spike and a negative edge emerges with a negative spike. Edge triggering is achieved by designing the flip-flop to neglect one spike and trigger on the occurrence of the other spike.

The graphic symbols of level triggered and edge triggered flip-flops are given in Fig. 3.72.

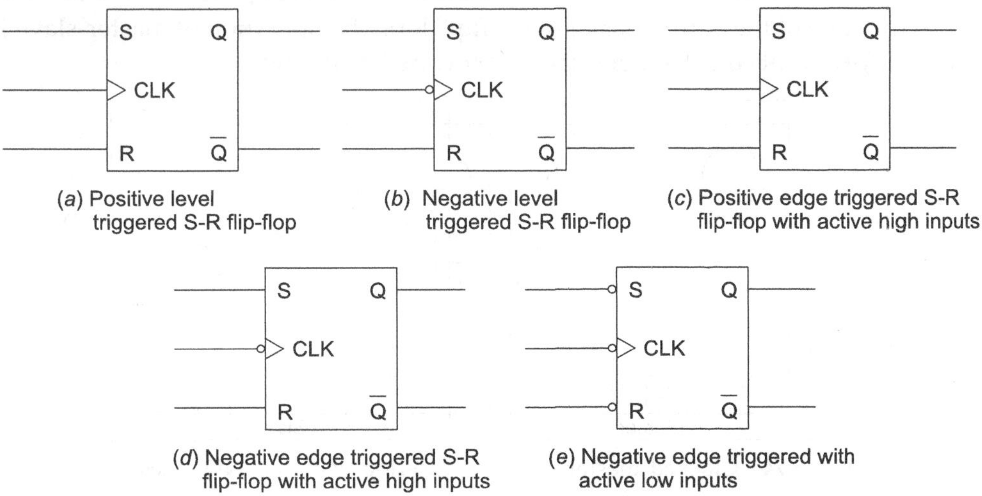

Fig. 3.72 Graphic symbols

3.14 MASTER-SLAVE FLIP-FLOPS

The race around condition can be avoided by using Master-Slave flip-flops. A master-slave flip-flop is constructed using two separate flip-flops connected serially. The first flip-flop serves as the master and the second as a slave, and the overall circuit is referred to as a Master-Slave flip-flop.

Master-Slave J-K flip-flop

The master-slave combination can be constructed for any type of flip-flop. Figure 3.73 shows one way to build a J-K master-slave flip-flop. It consists of two J-K flip-flops. The first flip-flop, called the master, is driven by the positive edge of the clock pulse ; the second flip-flop, called the slave, is driven by the negative edge of the clock pulse. Therefore, when the clock input has a positive edge, the master acts according to its J-K inputs, but the slave does not respond since it requires a negative edge at the clock input. When the clock input has a negative edge, the slave flip-flop copies the master outputs. But the master does not respond to the feedback from Q and $\overline{Q}$, since it requires a positive edge at its clock input. Thus, the master-slave flip-flop does not have race-around problem.

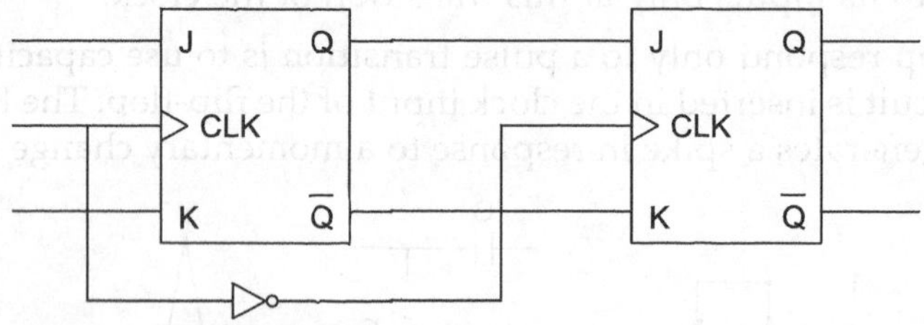

Fig. 3.73 Master-slave J-K flip-flop

A clocked master-slave J-K flip-flop constructed using NAND gates is shown in Fig. 3.74. It consists of two flip-flops connected in series. NAND gates-1 through 4 form the master flip-flop and NAND gates-5 through 8 form the slave flip-flop. When the clock is positive, change in J and K inputs cause a change of state in the master flip-flop. During this period, the slave retains its previous state and serves as a buffer between the master and the output. When the clock goes negative, the master flip-flop does not respond, *i.e.*, it maintains its previous state, while the slave flip-flop is enabled and changes its state to that of the master flip-flop. The new state of the slave then becomes the state of the entire master-slave flip-flop. The operation of master-slave J-K flip-flop for different J-K input combinations can be explained as follows :

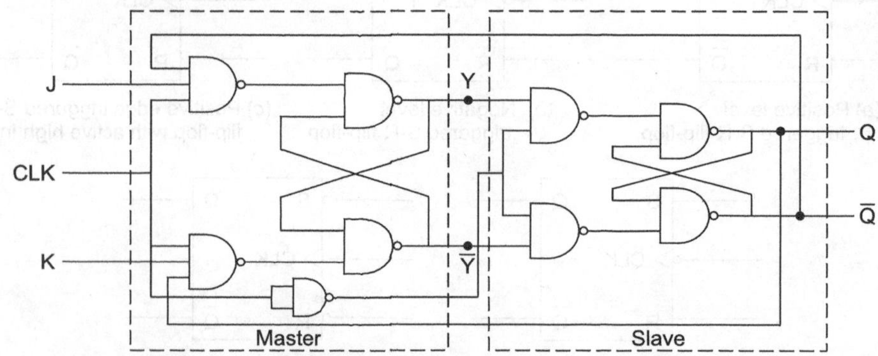

Fig. 3.74 Clocked master-slave J-K flip-flop using NAND gates

Case 1. J = 0 and K = 0

This condition does not produce any change at the output.

Case 2. J = 0 and K = 1

When J = 0 and K = 1, the master resets on the positive clock. The high $\overline{Y}$ output of the master goes to the R input of the slave. Therefore, at the negative clock slave resets, again copying the action of the master.

Case 3. J = 1 and K = 0.

When J = 1 and K = 0, the master sets on the positive clock. The high Y output of the master drives the S input of the slave, so at negative clock, slave sets, copying the action of the master.

Case 4. J = 1 and K = 1.

When J = 1 and K = 1, master toggles on the positive clock and slave then copies the output of master on the negative clock. At this instant, feedback inputs to the master flip-flop are complemented but as it is negative half of the clock pulse master flip-flop is inactive. This prevents race around condition. Figure 3.75 shows input and output waveforms of master-slave J-K flip-flop.

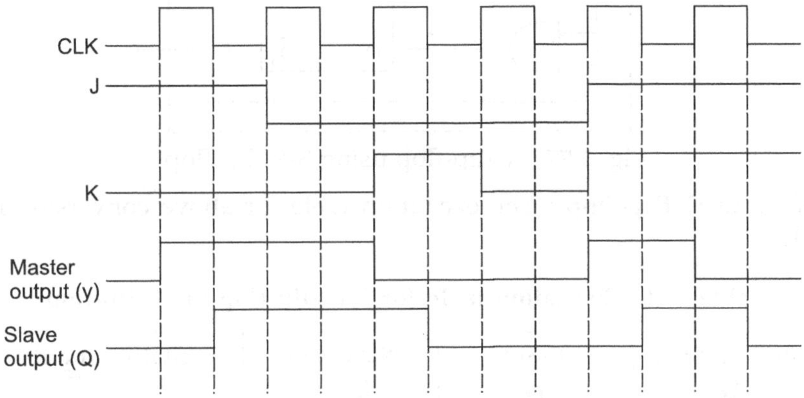

Fig. 3.75 Input and output waveforms of Master-Slave J-K flip-flop

Flip-flop Conversions

It is possible to convert one flip-flop into another the flip-flop with some additional gates or simply doing some extra connection. Let us see few conversions among flip-flops.

1. **S-R flip-flop to T flip-flop :** The excitation table for above conversion is as shown in Table 3.30.

Table 3.30. Excitation table for S-R flip-flop to T flip-flop conversion

Present State Q_n	Input T	Next State Q_{n+1}	Excitation Inputs	
			S	R
0	0	0	0	X
0	1	1	1	0
1	0	1	X	0
1	1	0	0	1

K-map Simplification:

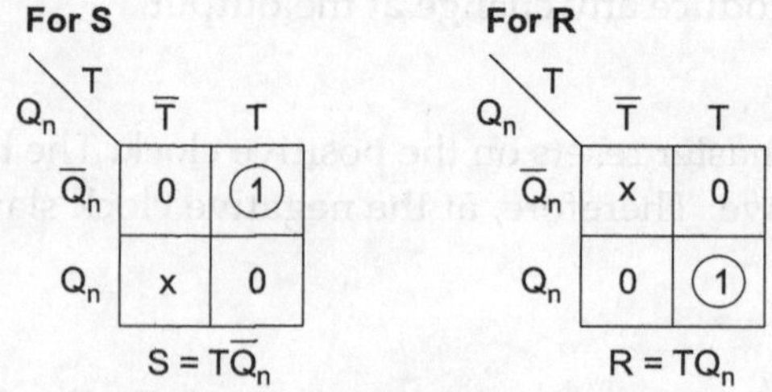

Fig. 3.76 Excitation map for S and R

Logic Diagram:

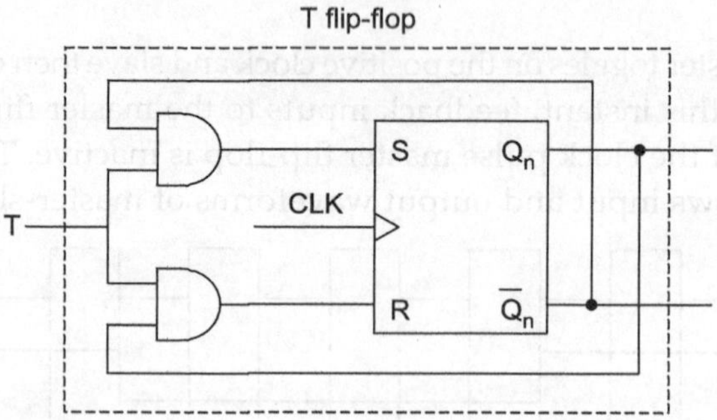

Fig. 3.77 T flip-flop using S-R flip-flop

2. **S-R flip-flop to D flip-flop** : The excitation table for above conversion is as shown in Table 3.31.

Table 3.31. Excitation table for S-R flip-flop to D flip-flop

Present State Q_n	Input D	Next State Q_{n+1}	Excitation Inputs S	Excitation Inputs R
0	0	0	0	X
0	1	1	1	0
1	0	0	0	1
1	1	1	X	0

K-map Simplification :

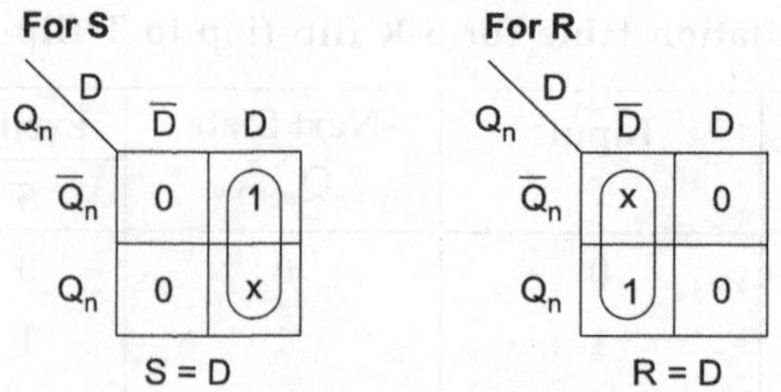

Fig. 3.78 Excitation map for S and R

Logic Diagram :

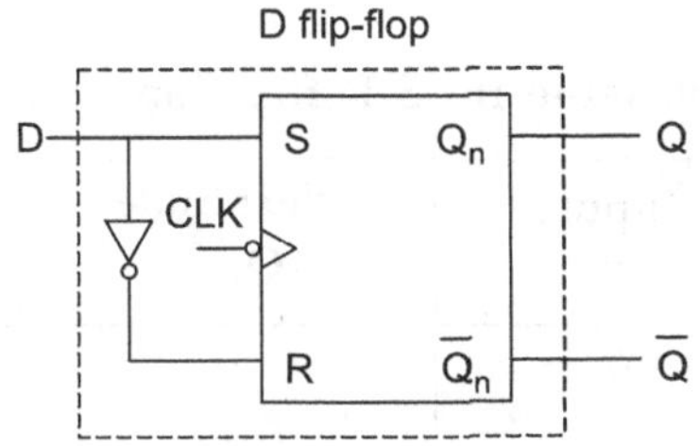

Fig. 3.79 D flip-flop using S-R flip-flop

3. **S-R flip-flop to J-K flip-flop:** The excitation table for above conversion is as shown in Table 3.32.

Table 3.32. Excitation table for S-R flip-flop to J-K flip-flop

Present State Q_n	Inputs		Next State Q_{n+1}	Excitation Inputs	
	J	K		S	R
0	0	0	0	0	X
0	0	1	0	0	X
0	1	0	1	1	0
0	1	1	1	1	0
1	0	0	1	X	0
1	0	1	0	0	1
1	1	0	1	X	0
1	1	1	0	0	1

K-map Simplification :

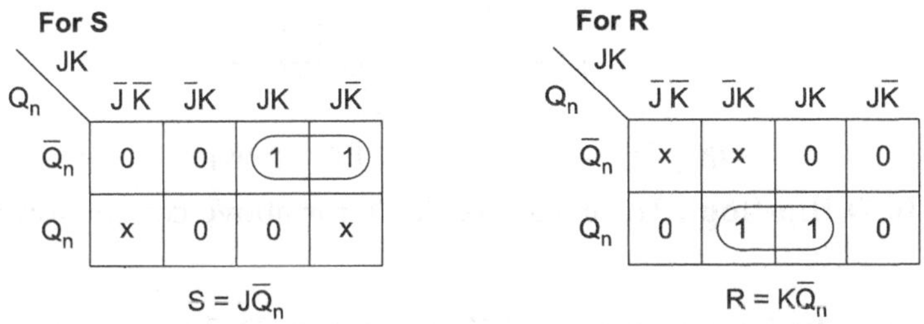

Fig. 3.80 Excitation map for S and R

Logic Diagram :

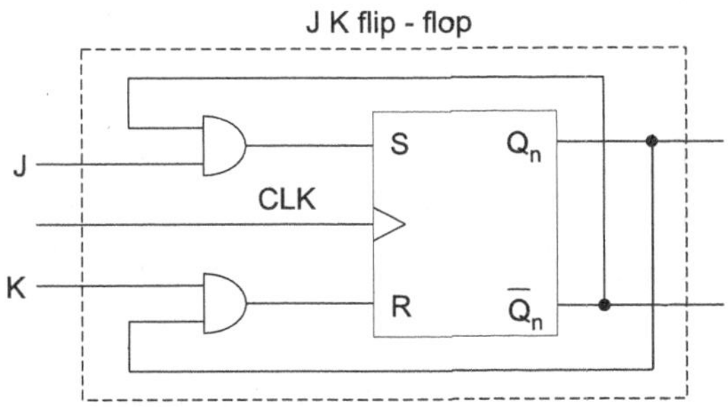

Fig. 3.81 J-K flip-flop using S-R flip-flop

4. **J-K flip-flop to T flip-flop :** The excitation table for above conversion is as shown in the Table 3.33.

Table 3.33. Excitation table for S-R flip-flop to T flip-flop conversion

Present State Q_n	Input T	Next State Q_{n+1}	Excitation Inputs	
			J	K
0	0	0	0	X
0	1	1	1	X
1	0	1	X	0
1	1	0	X	1

K-map Simplification :

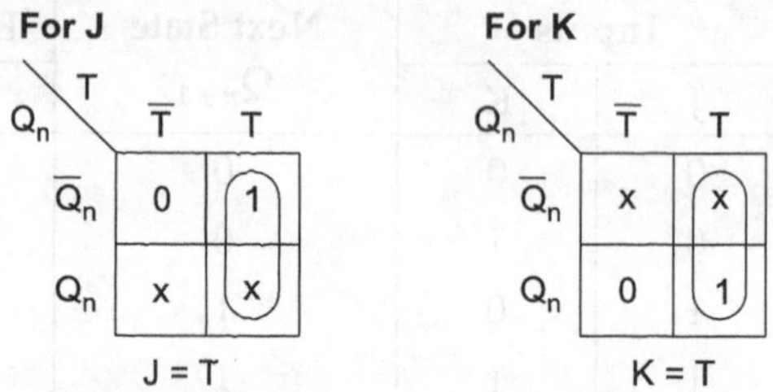

Fig. 3.82 Excitation map for S and R

Logic Diagram :

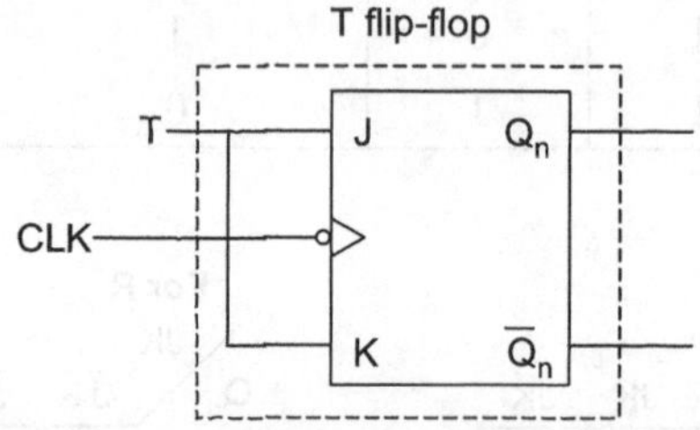

Fig. 3.83 T flip-flop using J-K flip-flop

5. **J-K flip-flop to D flip-flop :** The excitation table for above conversion is as shown in the Table 3.34.

Table 3.34. Excitation table for J-K flip-flop to D flip-flop conversion

Present State Q_n	Input D	Next State Q_{n+1}	Excitation Inputs	
			J	K
0	0	0	0	X
0	1	1	1	X
1	0	0	X	1
1	1	1	X	0

K-map Simplification :

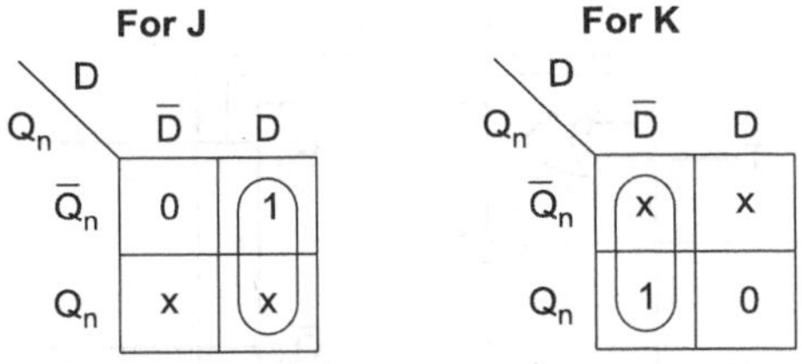

$J = D$ $\qquad$ $K = \overline{D}$

Fig. 3.84 Excitation map for S and R

Logic Diagram :

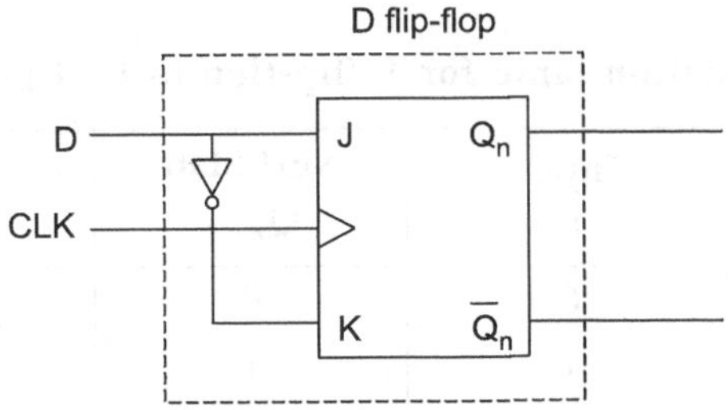

Fig. 3.85 D flip-flop using J-K flip-flop

6. **D flip-flop to T flip-flop :** The excitation table for above conversion is as shown in the Table 3.35.

Table 3.35. Excitation table for D flip-flop to T flip-flop conversion

Present State Q_n	Input T	Next State Q_{n+1}	Excitation Inputs D
0	0	0	0
0	1	1	1
1	0	1	1
1	1	0	0

K-map Simplification :

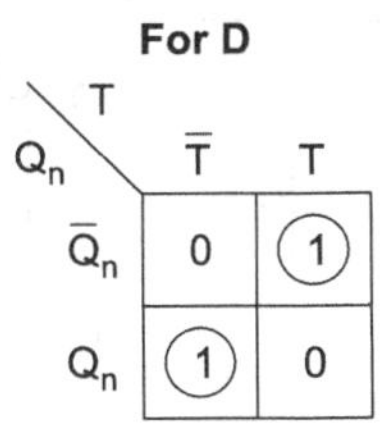

$D = \overline{T}Q_n + T\overline{Q}_n = T \oplus Q_n$

Fig. 3.86 Excitation map for D

Logic Diagram :

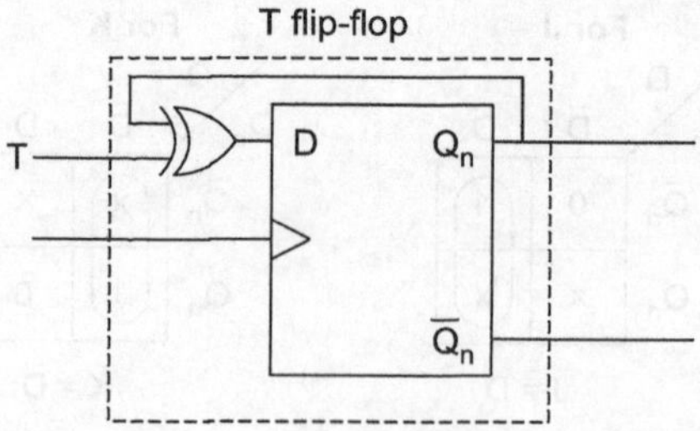

Fig. 3.87 T flip-flop using D flip-flop

7. **T flip-flop to D flip-flop :** The excitation table for above conversion is as shown in the Table 3.36.

Table 3.36. Excitation table for T flip-flop to D flip-flop conversion

Present State Q_n	Input D	Next State Q_{n+1}	Excitation Inputs T
0	0	0	0
0	1	1	1
1	0	0	1
1	1	1	0

K-map Simplification :

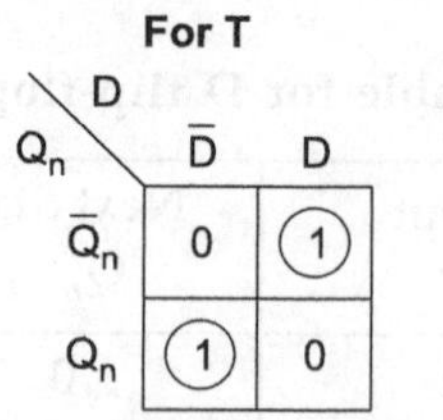

$T = D\bar{Q}_n + \bar{D}Q_n = D \oplus Q_n$

Fig. 3.88 Excitation map for T

Logic Diagram :

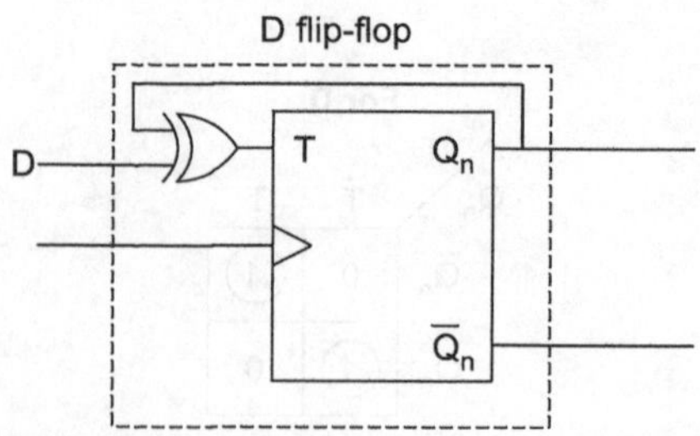

Fig. 3.89 D flip-flop using T flip-flop

8. **D flip-flop to J-K flip-flop :** The excitation table for above conversion is as shown in the Table 3.37.

Table 3.37. Excitation table for D flip-flop to J-K flip-flop conversion

Present State Q_n	Input		Next State Q_{n+1}	Excitation Inputs D
	J	K		
0	0	0	0	0
0	0	1	0	0
0	1	0	1	1
0	1	1	1	1
1	0	0	1	1
1	0	1	0	0
1	1	0	1	1
1	1	1	0	0

K-map Simplification :

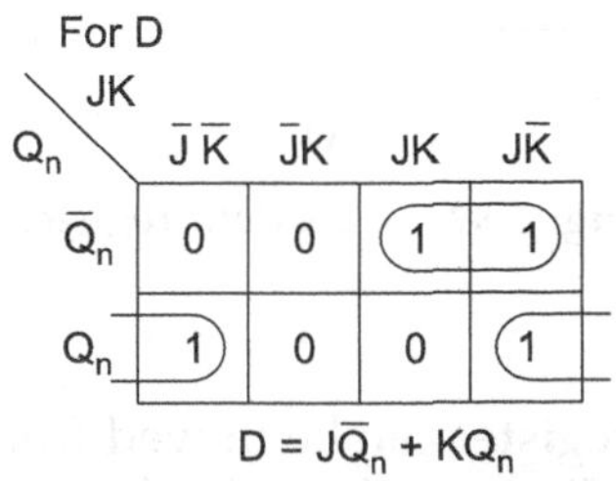

Fig. 3.90 Excitation map for D

Logic Diagram :

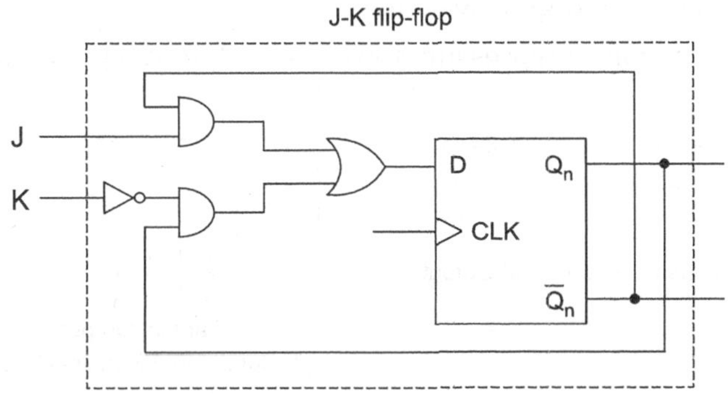

Fig. 3.91 J-K flip-flop using D flip-flop

3.15 REGISTERS

Introduction

A register is a group of flip-flops suitable for storing binary information. Each flip-flop is a binary cell capable of storing one bit of information. So an *n*-bit register has a group of *n* flip-flops and is capable of storing any binary information containing *n*-bits. The register is mainly used for storing and shifting binary data entered into it from an external source.

The register is a type of sequential circuit and an important building block used in digital systems like multipliers, dividers, memories, microprocessors, etc. A register has no specific sequence of states except in certain specialised applications. This is the main difference between a register and a counter.

Data Register

The simplest type of register is a data register which is used for the temporary storage of data. In its simplest form, it consists of a set of N D flip-flops, all sharing a common clock. All of the digits in the N bit data word are connected to the data register by an N line "data bus". Figure 3.92 shows a four bit data register, implemented with four D flip-flops. The data register is said to be a synchronous device, because all the flip-flops change state at the same time.

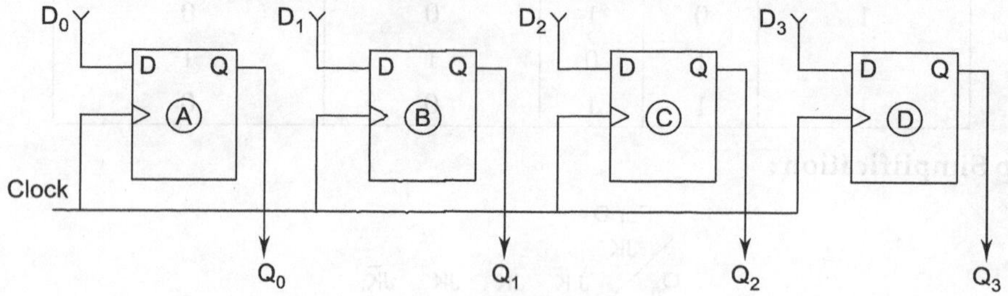

Fig. 3.92 4 bit data register

Shift Registers

The binary information (data) in a register can be moved from stage to stage within the register or into or out of the register upon application of clock pulses. This type of bit movement or shifting is essential for certain arithmetic and logic operations used in microprocessors. This gives rise to a group of registers called "shift registers". They are very important in applications involving the storage and transfer of data in a digital system.

Figure 3.93 gives the symbolical representation of the different types of data movement in shift register operations.

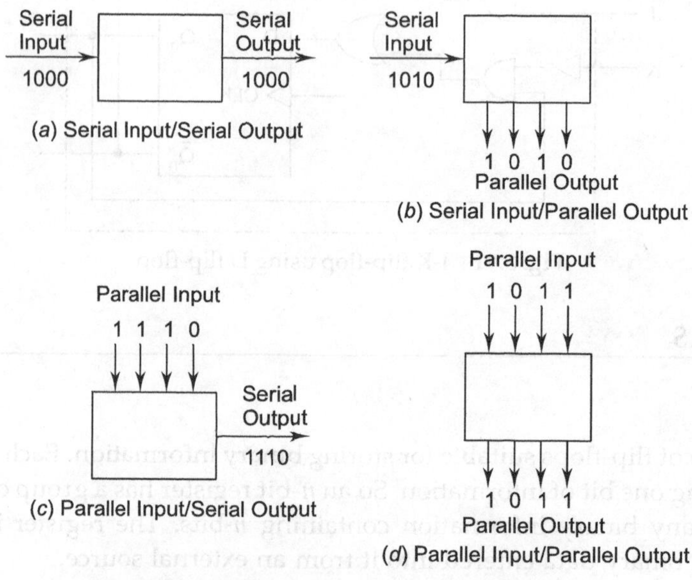

Fig. 3.93 Basic data movement in registers

Shift registers have found considerable application in arithmetic operations. Since moving a binary number one bit to the left is equivalent to multiplying the number of 2 and moving the number one bit position to the right amounts to dividing the number by 2. Thus, multiplication and division can be accomplished by shifting data bits. Shift register find considerable application in generating a sequence of control pulses.

According to the data movement in a register, let us see some of the types of shift registers.

Types of Shift Registers

Shift registers are classified into the following four types based on how binary information is entered or shifted out.

1. Serial-in-Serial-out Shift Registers
2. Serial-in-Parallel-out Shift Registers
3. Parallel-in-Serial-out Shift Registers
4. Parallel-in-Parallel-out Shift Registers

Let us discuss in detail about the types of shift registers.

1. Serial-in-Parallel-out Shift Registers

Shift Left Register

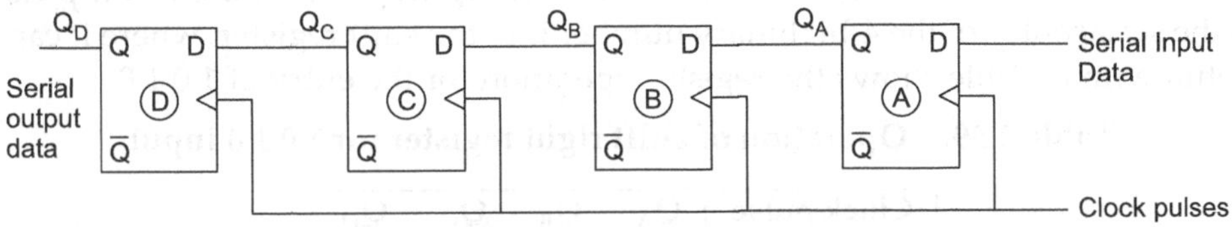

Fig. 3.94 Serial-in-serial-out shift left register

In the above shift register delay (D) input of the rightmost flip-flop is used as a serial input line. To input data 1, one should apply a 1 at D input and to input data 0, a 0 at the D input should be applied.

As shown in Fig. 3.94, the clock pulse is applied to all the flip-flops simultaneously. When the clock pulse occurs, each flip-flop is set or reset according to the data at the respective flip-flop input. Thus, the input data bit at serial input line is entered into stage A by the first clock pulse. At the same time, the data of stage A is shifted into stage B and so on for the following stages. For each clock pulse, data stored in the register stage shifts to the left by one stage. New data is entered into stage A, whereas the data present in stage D are shifted out (to the left) for use by some other shift register or other building block of digital system.

For example, consider that all stages are reset and a steady logical 1 is applied at the serial input line connected to stage A. The data in each stage after each of the four clock pulses is shown in Table 3.38. Note that the logical 1 input enters into stage A and then shifts left to stage D after four shift pulses.

Table 3.38. Operation of shift left register for 1 1 1 1 input

Clock Pulse	Q_D	Q_C	Q_B	Q_A
0	0	0	0	0
1	0	0	0	1
2	0	0	1	1
3	0	1	1	1
4	1	1	1	1

Shift-Right Register

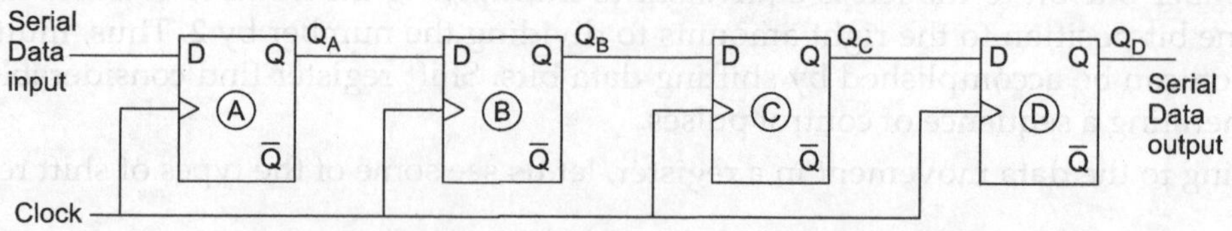

Fig. 3.95 Serial-In-Serial-Out Shift Right Register

A shift-right register is shown in Fig. 3.95. Let us illustrate the entry of the 4-bit binary number 1 0 1 0 into the register beginning with the rightmost bit A is applied at the serial input line *i.e.*, at the delay (D) input of the first flip-flop. When the first clock pulse is applied, flip-flop A is RESET, thus storing the 0. Next a 1 is applied to the serial input, making D = 1 for flip-flop A and D = 0 for flip-flop B because D input of flip-flop B is connected to the Q_A output.

When the second clock pulse occurs, the 1 on the data input is 'shifted' to flip-flop A and the 0 in flip-flop A is shifted to flip-flop B. The next 0 in the binary number is now applied at the serial input line, and a clock pulse is applied. This 0 is entered into flip-flop A, and the 1 stored in flip-flop A is shifted to flip-flop B and the 0 stored in flip-flop B is shifted to flip-flop C. The last bit in the binary number, *i.e.*, 1, is now applied at the serial input and a clock pulse is applied. Now, this 1 is entered into flip-flop A, and the 0 stored in flip-flop A is shifted to flip-flop B, the 1 stored in flip-flop B is shifted to flip-flop C, and the 0 stored in flip-flop C is shifted to flip-flop D. This completes the serial entry of the 4-bit binary number into the shift register, where it can be stored for an indefinite time. Table shows the register operation for the entry of 1 0 1 0.

Table 3.39. Operation of shift right register for 1 0 1 0 input

Clock pulse	Q_A	Q_B	Q_C	Q_D
0	0	0	0	0
1	0	0	0	0
2	1	0	0	0
3	0	1	0	0
4	1	0	1	0

2. Serial-in-Parallel-out Shift Registers

A 4-bit serial-in-parallel-out shift register is shown in Fig. 3.96. It consists of one serial input, and outputs are taken from all the flip-flops parallel. In this register, data is shifted in serially but shifted out in parallel. In order to shift the data out in parallel, it is necessary to have all the data available at the outputs at the same time. Once the data is stored, each bit appears on its respective output line and all the bits are available simultaneously, rather than on a bit-by-bit basis as with the serial output.

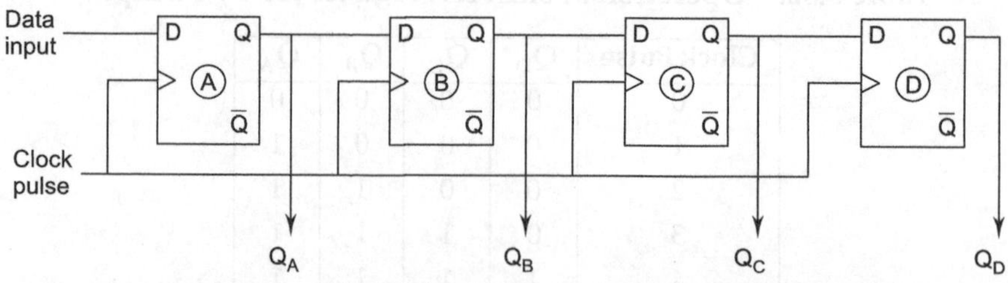

Fig. 3.96 4-bit serial-in-parallel-out shift register

Table 3.40 shows the register operation for the entry of 1 1 1 0.

Table 3.40. Operation of serial-in-parallel-out shift register for 1 1 1 0 input

Clock pulse	Q_A	Q_B	Q_C	Q_D
0	0	0	0	0
1	0	0	0	0
2	1	0	0	0
3	1	1	0	0
4	1	1	1	0

3. Parallel-in-Serial-out Shift Registers

In this type, the bits are entered in parallel *i.e.*, simultaneously into their respective stages on parallel lines.

A four bit parallel-in-serial-out shift register is illustrated shown in Fig. 3.97. Let A, B, C and D be the four parallel data-input lines and SHIFT/$\overline{\text{LOAD}}$ is a control input that allows the four bits of data at A, B, C and D inputs to enter into the register in parallel or shift the data in serial. When SHIFT/$\overline{\text{LOAD}}$ is LOW, AND gates G_1 through G_3 are enabled, allowing the data at parallel inputs *i.e.*, B, C and D to the D input of its respective flip-flop. The A input is directly connected to the D input of the first flip-flop. When a clock pulse is applied, the flip-flops with D = 1 will be SET and the flip-flops with D = 0 will be RESET, thereby storing all four bits simultaneously.

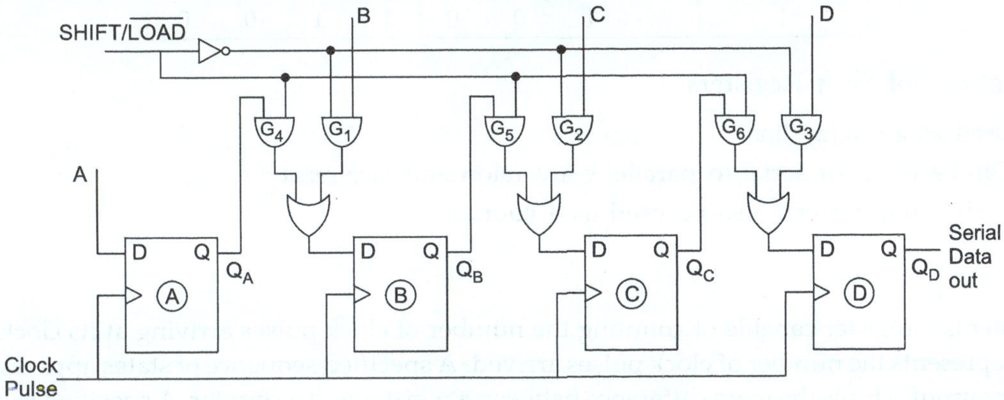

Fig. 3.97 A 4-bit parallel-in-serial-out shift register

When SHIFT/$\overline{\text{LOAD}}$ is HIGH, AND gates G_1 through G_3 are disabled and the remaining AND gates G_4 through G_6 are enabled, allowing the data bits to shift right from one stage to the next. The OR gates allow either the normal shifting operation or the parallel data-entry operation, depending on which the AND gates are enabled by the level on the SHIFT/$\overline{\text{LOAD}}$ input.

4. Parallel-in-Parallel-out Shift Registers

In this type of register, data inputs can be shifted either in or out of the register in parallel. The parallel entry of the data is carried out as discussed in the previous section of parallel-in-serial-out shift register. Also, in this register, there is no interconnection between successive flip-flops since no serial shifting is required. Therefore, the moment the parallel entry of the input data is accomplished, the respective bits will appear at the parallel outputs.

A simple 4-bit parallel-in-parallel-out shift register using D flip-flops is shown in Fig. 3.98. Here the parallel inputs to be entered should be applied at A, B, C and D inputs which are directly connected to delay (D) inputs of respective flip-flops. Now on applying a clock pulse, these inputs are entered into the register and are immediately available at the outputs Q_A, Q_B, Q_C and Q_D. Table 3.41 shows the register operation for the entry of 1 1 0 0.

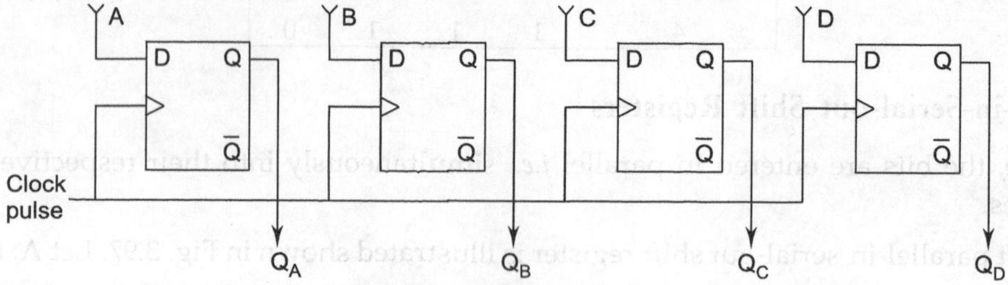

Fig. 3.98 A 4-bit parallel-in-parallel-out shift register

Table 3.41. Operation of parallel-in-parallel-out shift register for 1 1 0 0 input

Clock pulse	Inputs				Outputs			
	A	B	C	D	Q_A	Q_B	Q_C	Q_D
0	X	X	X	X	0	0	0	0
1	1	1	0	0	1	1	0	0

Applications of Shift Registers

1. Used as an delay line.
2. Can be used for serial to parallel conversion and *vice-versa.*
3. A shift register can also be used as a counter.

Counters

A counter is a register capable of counting the number of clock pulses arriving at its clock input count represents the number of clock pulses arrived. A specified sequence of states appears as the counter output. This is the main difference between a register and a counter. A specified sequence of states is different for different types of counters.

There are two types of counters, namely:

1. Synchronous (or) Parallel counters
2. Asynchronous (or) Ripple (or) serial counters

Synchronous Counters

In synchronous counter, the common clock input is connected to all of the flip-flops and thus they are clocked simultaneously.

Asynchronous Counter

In asynchronous counter, the first flip-flop is clocked by the external clock pulse and then each successive flip-flop is clocked by the Q or $\bar{Q}$ output of the previous flip-flop. Therefore, in an asynchronous counter, the flip-flops are not clocked simultaneously. This is also called as ripple counters.

Let us start with asynchronous counters

2-bit Asynchronous/Ripple Counters

A 2-bit asynchronous counter using JK flip-flops is shown in Fig. 3.99. In this, the clock signal is connected to the clock input of only first stage flip-flop. The clock input of the second stage flip-flop is triggered by the Q_A output of the first stage and thus the counter has a cumulative setting time. Hence, its speed of operation is limited. The first flip-flop switches first on the application of a clock pulse and the second stage change their state in turn causing a ripple-through effect of the count pulses. As the triggers move through the flip-flops like a ripple it is called a ripple counter.

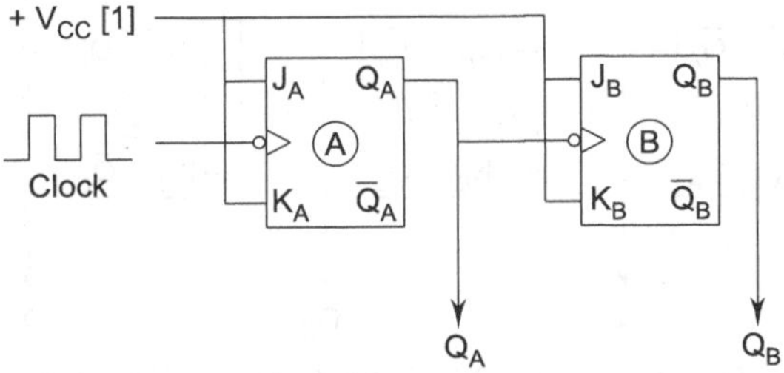

Fig. 3.99 2-bit Asynchronous/Ripple Counter

In the above diagram, (Fig. 3.99) all the J and K inputs are connected to V_{CC} (1), which means that each flip-flop toggles on the negative edge of its clock input. Consider initially all flip-flops to be in the logical 0 state (*i.e.*, $Q_A = Q_B = 0$). A negative transition (1 to 0) in clock input which drives flip-flop A causes Q_A to change from logical 0 to logical 1. Flip-flop B does not change its state since it also requires negative transition at its clock input, *i.e.*, it requires its clock input (Q_A) to change from logical 1 to logical 0. With the arrival of the second clock pulse to flip-flop A, Q_A goes from 1 to 0. This change of state creates the negative going edge needed to trigger flip-flop B, and thus Q_B goes to 1 from 0. During the third clock pulse to flip-flop A, Q_A goes from 0 to 1. Flip-flop B does not change its state since it also requires negative transition at its clock input, *i.e.*, it requires its clock input (Q_A) to change from logical 1 to logical 0. Clock pulse 4 causes Q_A and Q_B to go to logical 0 state in turn. Figure 3.100 shows the timing waveforms for 2-bit asynchronous counter. It illustrates the changes in the state of the flip-flop outputs in response to the clock. Table 3.42 shows the sequence of binary states that the flip-flops will follow as clock pulses are applied continuously.

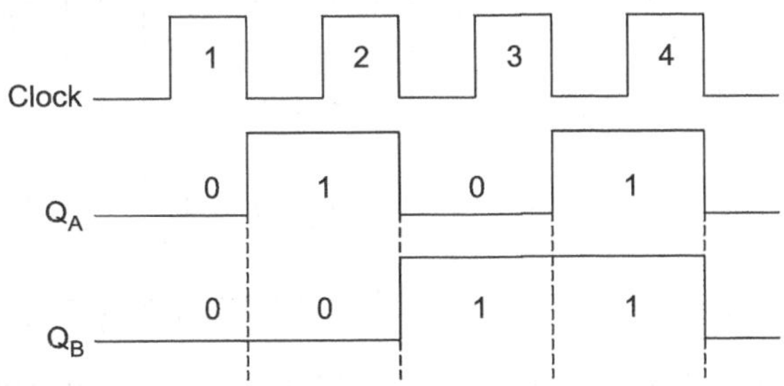

Fig. 3.100 Timing waveforms for the 2-bit asynchronous counter

Table 3.42. Truth table of 2-bit asynchronous counter

CLOCK	Q_A	Q_B
0	0	0
1	0	1
2	1	0
3	1	1
0	0	0

Four-bit Asynchronous/Ripple Counter

Figure 3.101 shows a 4-bit Asynchronous/Ripple counter. As explained above, a ripple counter is constructed using clocked JK flip-flops. The system clock, a square wave, drives flip-flop A. The output of A drives flip-flop B, output of B drives flip-flop C, and the output of C drives flip-flop D.

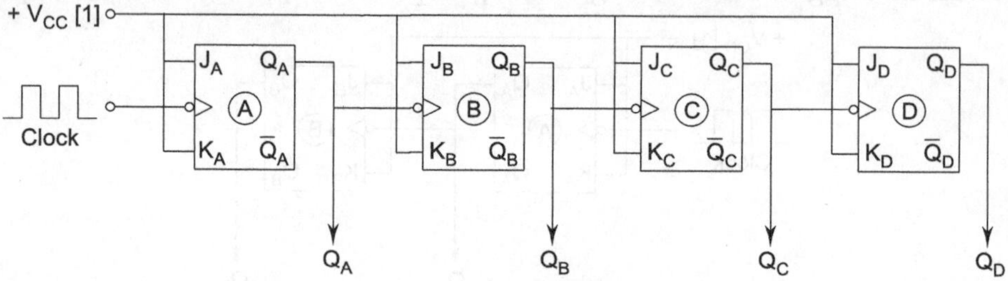

Fig. 3.101 A 4-bit Asynchronous Counter

Consider initially all flip-flops to be in the logical 0 state (*i.e.*, $Q_A = Q_B = Q_C = Q_D = 0$). A negative transition (1 to 0) in clock input which drives flip-flop A causes Q_A to change from logical 0 to logical 1. Flip-flop B does not change its state. Since it also requires negative transition at its clock input, *i.e.*, it requires its clock input (Q_A) to change from logical 1 to logical 0. With the arrival of the second clock pulse to flip-flop A, Q_A goes from 1 to 0. This change of state creates the negative going edge needed to trigger flip-flop B, and thus Q_B goes from 0 to 1. Before the arrival of the sixteenth clock pulse, all the flip-flops are in the logical 1 state, clock pulse 16 causes Q_A, Q_B, Q_C and Q_D to go to logical 0 state in turn.

Figure 3.102 shows the timing waveform for the 4-bit Asynchronous counter. Table 3.43 shows the sequence of binary states that the flip-flops will follow as clock pulses are applied continuously.

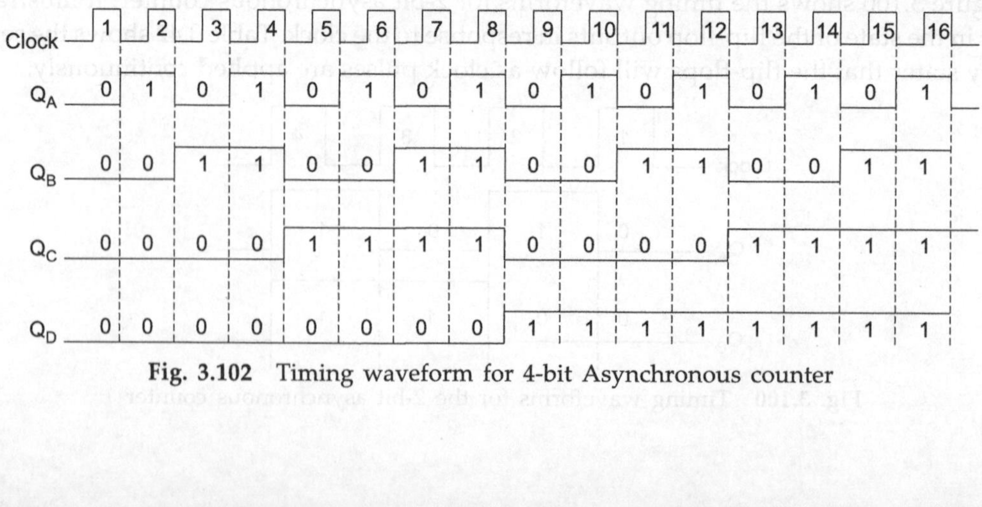

Fig. 3.102 Timing waveform for 4-bit Asynchronous counter

Table 3.43. Truth table of 4-bit Asynchronous counter

CLOCK	Q_D	Q_C	Q_B	Q_A
0	0	0	0	0
1	0	0	0	1
2	0	0	1	0
3	0	0	1	1
4	0	1	0	0
5	0	1	0	1
6	0	1	1	0
7	0	1	1	1
8	1	0	0	0
9	1	0	0	1
10	1	0	1	0
11	1	0	1	1
12	1	1	0	0
13	1	1	0	1
14	1	1	1	0
15	1	1	1	1

Asynchronous/Ripple Down Counter

The down counter using *n* flip-flops counts downward from a maximum count of $(2^n - 1)$ to zero. The Fig. 3.103 shows the 3-bit asynchronous down counter using JK flip-flops. Let Q_A, Q_B and Q_C represent the flip-flop output states as the counter goes through its sequence. Here also, the output Q_A (LSB) change its state (toggle) at each negative transition of clock as it does in the up counter. The Q_B output changes state every time Q_A goes from LOW to HIGH *i.e.*, when $\bar{Q}_A$ goes from HIGH to LOW ; Q_C changes state each time Q_B goes from LOW to HIGH, *i.e.*, when $\bar{Q}_B$ goes from HIGH to LOW. Thus, in a down counter, each flip-flop except the LSB flip-flop, must toggle when the inverted output $(\bar{Q})$ of the preceding flip-flop goes from HIGH to LOW. The count down sequence for a 3-bit down counter is shown in Table 3.44.

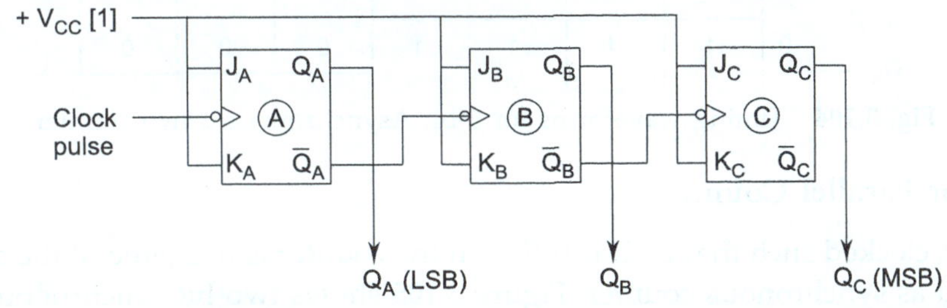

Fig. 3.103 3-bit Asynchronous down counter

Table 3.44. Truth table for 3-bit down counter

CLOCK	Q_C	Q_B	Q_A
0	0	0	0
1	1	1	1
2	1	1	0
3	1	0	1
4	1	0	0
5	0	1	1
6	0	1	0
7	0	0	1

The input pulses are applied to the clock input of A flip-flop, the $\bar{Q}_A$ output serves as the clock input for the B flip-flop ; the $\bar{Q}_B$ output serves as the clock input for the C flip-flop. The waveforms at Q_A, Q_B and Q_C of 3-bit Asynchronous down counter is shown in Fig. 3.104. It shows that Q_A toggles at the negative transition of the clock, Q_B toggles whenever Q_A goes from LOW to HIGH, and Q_C toggles whenever Q_B goes from LOW to HIGH. Whenever Q_A and Q_B go HIGH, $\bar{Q}_A$ and $\bar{Q}_B$ go LOW, and it is this negative transition at $\bar{Q}_A$ and $\bar{Q}_B$ that triggers outputs Q_B and Q_C respectively.

As shown in truth Table 3.44, if the initial counter content is 0 0 0, at the first negative transition of the clock, the counter content becomes 1 1 1, at the second negative transition, the content becomes 1 1 0 ; at the third negative transition of clock, the content changes to 1 0 1, and so on. Thus, in the down counter, the counter content is decremented by one for every negative transition in clock.

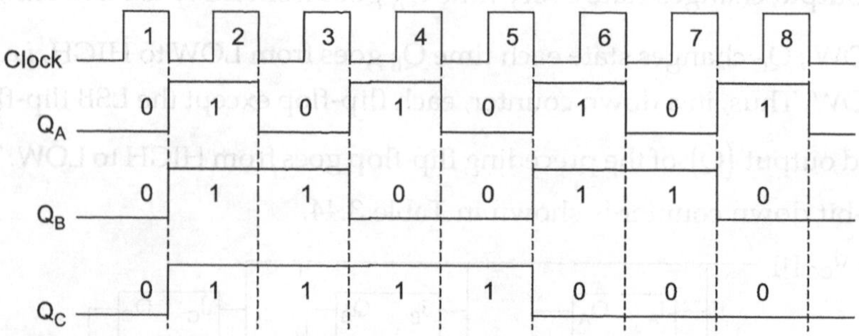

Fig. 3.104 Timing waveforms for 3-bit Asynchronous down counter

Synchronous or Parallel Counter

When counter is clocked such that each flip-flop in the counter is triggered at the same time, the counter is called as synchronous counter. Figure 3.105 shows two-bit synchronous counter.

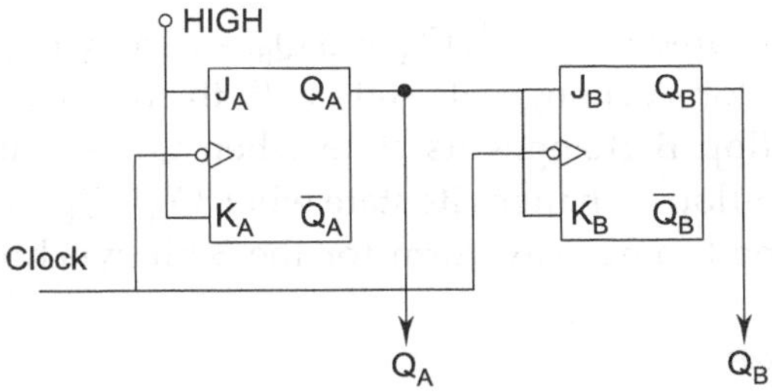

Fig. 3.105 A two-bit synchronous counter

Here, clock signal is connected in parallel to clock inputs of both the flip-flops. But the Q_A output of first stage is used to drive the J and K inputs of the second stage. Let us see the operation of the circuit.

Initially, we assume that Q the $Q_A = Q_B = 0$. When positive edge of the first clock pulse is applied, flip-flop A will toggle because $J_A = K_A = 1$, whereas flip-flop B output will remain zero because $J_B = K_B = 0$. After first clock pulse $Q_A = 1$ and $Q_B = 0$. At negative going edge of the second clock pulse both flip-flops will toggle because they both have a toggle condition on their J and K inputs ($J_A = K_A = J_B = K_B = 1$). Thus after second clock pulse, $Q_A = 0$ and $Q_B = 1$. At negative going edge of the third clock pulse flip-flop A toggles making $Q_A = 1$, but flip-flop B remains set *i.e.*, $Q_B = 1$. Finally, at the leading edge of the fourth clock pulse both flip-flops toggle as their J-K inputs are at logic 1. This results $Q_A = Q_B = 0$ and counter recycled back to its original state. The timing waveform and the truth table for the 2-bit synchronous counter is shown in Fig. 3.106 and Table 3.45 respectively.

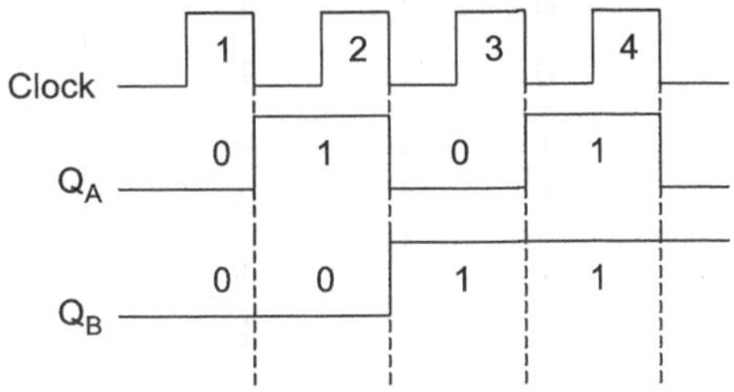

Fig. 3.106. Timing waveform for 2-bit synchronous counter

Table 3.45. Truth table for 2-bit synchronous counter

CLOCK	Q_B	Q_A
0	0	0
1	0	1
2	1	0
3	1	1

3-Bit Synchronous Counter

Figure 3.107 shows the 3-bit synchronous counter. In this counter, the clock inputs of all the flip-flops are connected together so that the input clock signal is applied simultaneously to each flip-flop. Also, only the LSB flip-flop A has its J and K inputs connected permanently to V_{CC} while the J and K inputs of the other flip-flops are driven by some combination of flip-flop outputs. The J and K inputs of flip-flop B are connected with Q_A output of flip-flop A ; the J and K inputs of flip-flop C are connected with AND operated output of Q_A and Q_B. As shown in Fig. 3.107 and according to truth table (Table 3.46), flip-flop A changes its state with the occurrence of negative transition at each clock pulse. The flip-flop B changes its state when $Q_A = 1$ and when there is negative transition at clock input. Flip-flop C changes its state when $Q_A = Q_B = 1$ and when there is negative transition at clock input. The timing waveform for the 3-bit synchronous counter is shown in Fig. 3.108.

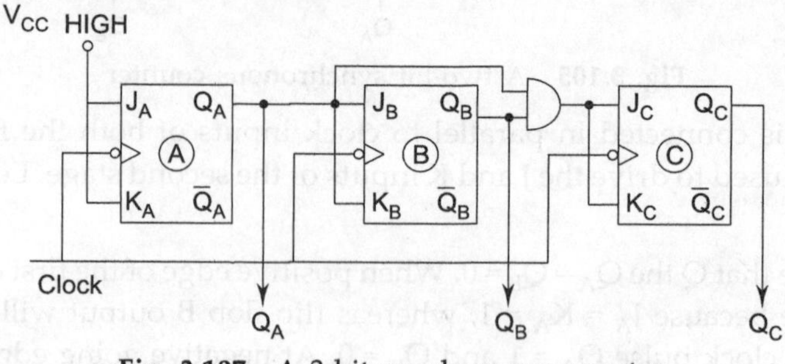

Fig. 3.107 A 3-bit synchronous counter

Table 3.46. Truth table for 3-bit synchronous counter

CLOCK	Q_C	Q_B	Q_A
0	0	0	0
1	0	0	1
2	0	1	0
3	0	1	1
4	1	0	0
5	1	0	1
6	1	1	0
7	1	1	1

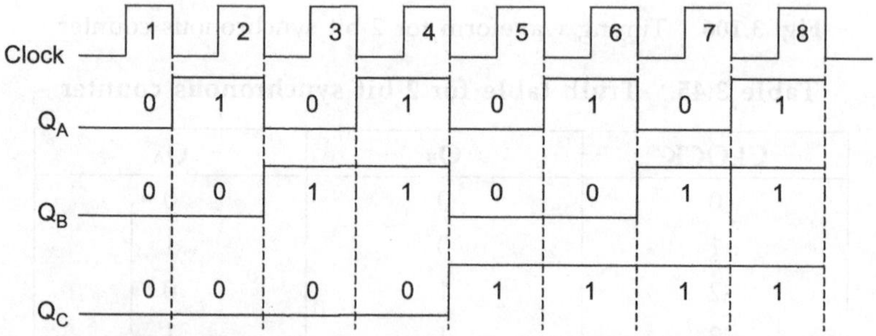

Fig. 3.108. Timing waveforms for 3-bit synchronous counter

Synchronous Down Counter

A synchronous down counter can be made to count down by using the inverted outputs of flip-flops to feed the various logic gates. The 4-bit synchronous up counter can be converted to a down counter by connecting the $\bar{Q}_A$, $\bar{Q}_B$ and $\bar{Q}_C$ outputs to the AND gates as shown in Fig. 3.109. The truth table for 4-bit synchronous down counter is shown in Table 3.47.

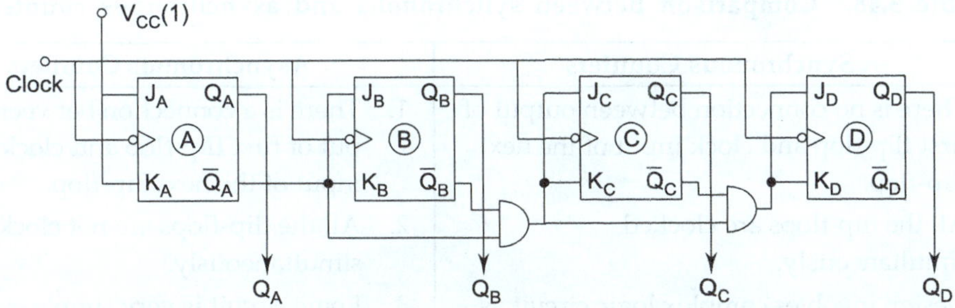

Fig. 3.109. 4-bit synchronous down counter

Table 3.47. Truth table for 4-bit synchronous down counter

CLOCK	Q_D	Q_C	Q_B	Q_A
0	0	0	0	0
1	1	1	1	1
2	1	1	1	0
3	1	1	0	1
4	1	1	0	0
5	1	0	1	1
6	1	0	1	0
7	1	0	0	1
8	1	0	0	0
9	0	1	1	1
10	0	1	1	0
11	0	1	0	1
12	0	1	0	0
13	0	0	1	1
14	0	0	1	0
15	0	0	0	1

Fig. 3.110 shows the timing waveform for the 4-bit synchronous down counter.

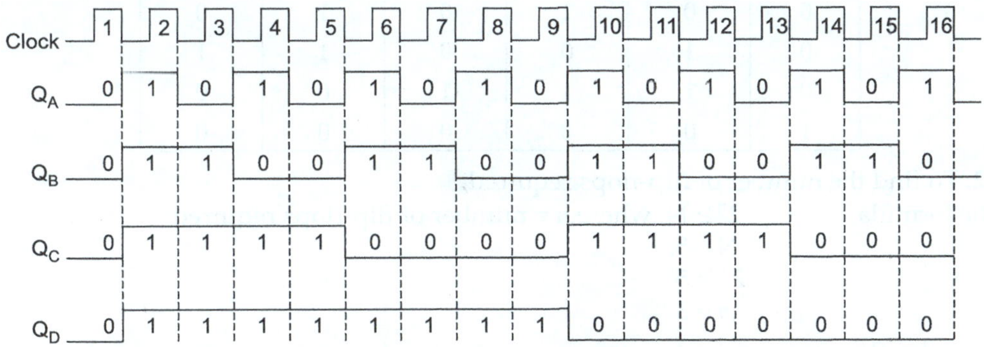

Fig. 3.110. Timing waveform for a 4-bit synchronous down counter

Synchronous Vs Asynchronous Counters

The Table 3.48 shows the comparison between synchronous and asynchronous counters.

Table 3.48. Comparison between synchronous and asynchronous counters

S.No.	Synchronous Counters	Asynchronous Counters
1.	There is no connection between output of first flip-flop and clock input of the next flip-flop.	1. There is a connection between output of first flip-flop and clock input of the next flip-flop.
2.	All the flip-flops are clocked simultaneously.	2. All the flip-flops are not clocked simultaneously.
3.	Design involves complex logic circuit as number of states increases.	3. Logic circuit is very simple even for more number of states.
4.	Speed is high (*i.e.,* there is no problem of propagation delay).	4. Speed is low (*i.e.,* there is a propagation delay problem).

Design of Synchronous Counters

The steps involved in the design of synchronous counter are listed below.

1. Obtain the PS-NS table from the given circuit information.
2. Determine the number of flip-flops needed.
3. Decide the type of memory element to be used.
4. From PS-NS table obtain the excitation table using the application table of the flip-flop.
5. Draw the excitation maps for various excitation inputs of flip-flops and simplify the excitation functions.
6. Draw the schematic diagram.

Design of MOD-5 Synchronous Counter

Step 1. PS-NS table

Table 3.49. PS-NS table for MOD-5 counter

Present State			Next State		
Q_C	Q_B	Q_A	Q_{C+1}	Q_{B+1}	Q_{A+1}
0	0	0	0	0	1
0	0	1	0	1	0
0	1	0	0	1	1
0	1	1	1	0	0
1	0	0	0	0	0

Step 2. To find the number of flip-flops required,

Use the formula, $2^n \geq N$, where n = number of flip-flops required.

Here $N = 5$,

$\therefore$ $2^n \geq 5$

$\therefore$ $n = 3$

i.e., three flip-flops are required.

Step 3. Let us use JK flip-flop.

Step 4. Excitation table

Table 3.50. Excitation table for MOD-5 counter

Present State			Next State			Excitation Inputs					
Q_C	Q_B	Q_A	Q_{C+1}	Q_{B+1}	Q_{A+1}	J_C	K_C	J_B	K_B	J_A	K_A
0	0	0	0	0	1	0	X	0	X	1	X
0	0	1	0	1	0	0	X	1	X	X	1
0	1	0	0	1	1	0	X	X	0	1	X
0	1	1	1	0	0	1	X	X	1	X	1
1	0	0	0	0	0	X	1	0	X	0	X
1	0	1	X	X	X	X	X	X	X	X	X
1	1	0	X	X	X	X	X	X	X	X	X
1	1	1	X	X	X	X	X	X	X	X	X

Step 5. Excitation maps.

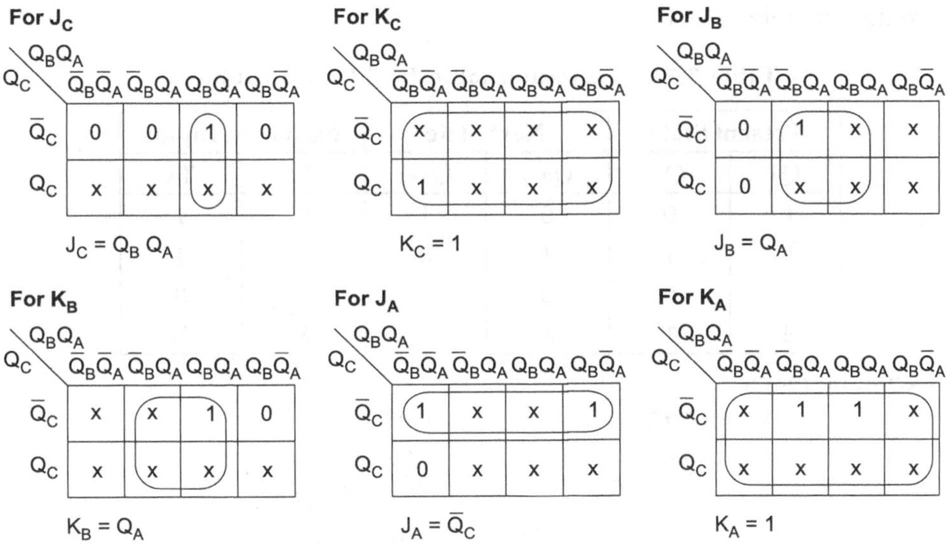

Fig. 3.111. Excitation maps

Step 6. Schematic diagram.

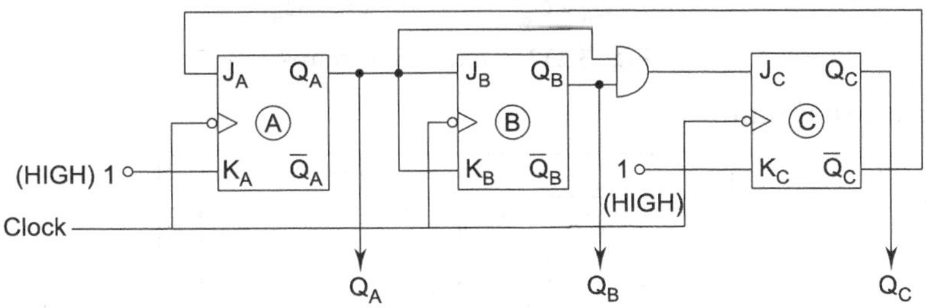

Fig. 3.112. Schematic diagram for MOD-5 synchronous counter

Design of Mod-3 Synchronous Counter

Step 1. PS-NS table.

Table 3.51. PS-NS table for MOD = 3

Present State		Next State	
Q_B	Q_A	Q_{B+1}	Q_{A+1}
0	0	0	1
0	1	1	0
1	0	0	0

Step 2. To find the number of flip-flops required use the formula,

$2^n \geq N$, where n = number of flip-flops required.

Here $N = 3$,

$2^n \geq 3$

$\therefore$ $n = 2$

i.e., two flip-flops are required

Step 3. Let us use T flip-flops.

Step 4. Excitation table

Table 3.52. Excitation table for MOD-3 counter

Present State		Next State		Excitation Inputs	
Q_B	Q_A	Q_{B+1}	Q_{A+1}	T_B	T_A
0	0	0	1	0	1
0	1	1	0	1	1
1	0	0	0	1	0
1	1	X	X	X	X

Step 5. Excitation maps

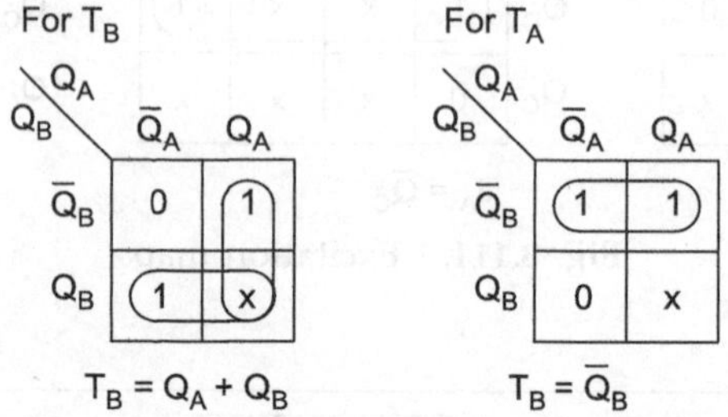

Fig. 3.113. Excitation maps

Step 6. Schematic diagram

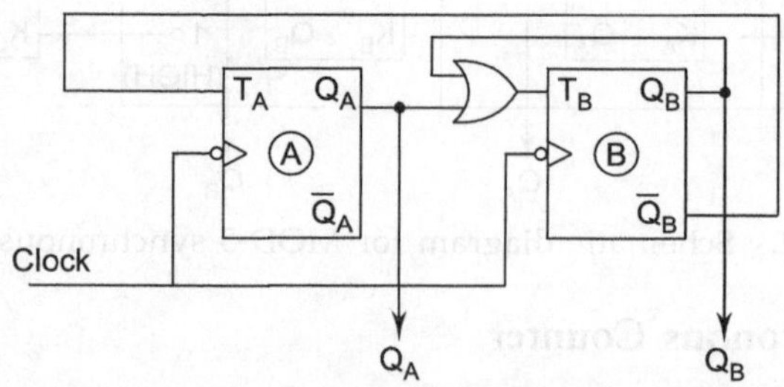

Fig. 3.114. Schematic diagram for MOD-3 synchronous counter

MOD-10 (Decade) Synchronous Counter

Step 1. PS-NS table

Table 3.53. PS-NS table for MOD-10 synchronous counter

Present State				Next State			
Q_D	Q_C	Q_B	Q_A	Q_{D+1}	Q_{C+1}	Q_{B+1}	Q_{A+1}
0	0	0	0	0	0	0	1
0	0	0	1	0	0	1	0
0	0	1	0	0	0	1	1
0	0	1	1	0	1	0	0
0	1	0	0	0	1	0	1
0	1	0	1	0	1	1	0
0	1	1	0	0	1	1	1
0	1	1	1	1	0	0	0
1	0	0	0	1	0	0	1
1	0	0	1	0	0	0	0

Step 2. To find the number of flip-flops required, use the formula,

$$2^n \geq N$$

Here $N = 10$

$\therefore$ $2^n \geq 10$

$\therefore$ $n = 4$

i.e., four flip-flops are required.

Step 3. Let us use T flip-flops.

Step 4. Excitation table

Table 3.54. Excitation table

Present State				Next State				Excitation Inputs			
Q_D	Q_C	Q_B	Q_A	Q_{D+1}	Q_{C+1}	Q_{B+1}	Q_{A+1}	T_D	T_C	T_B	T_A
0	0	0	0	0	0	0	1	0	0	0	1
0	0	0	1	0	0	1	0	0	0	1	1
0	0	1	0	0	0	1	1	0	0	0	1
0	0	1	1	0	1	0	0	0	1	1	1
0	1	0	0	0	1	0	1	0	0	0	1
0	1	0	1	0	1	1	0	0	0	1	1
0	1	1	0	0	1	1	1	0	0	0	1
0	1	1	1	1	0	0	0	1	1	1	1
1	0	0	0	1	0	0	1	0	0	0	1
1	0	0	1	0	0	0	0	1	0	0	1
1	0	1	0	X	X	X	X	X	X	X	X
1	0	1	1	X	X	X	X	X	X	X	X
1	1	0	0	X	X	X	X	X	X	X	X
1	1	0	1	X	X	X	X	X	X	X	X
1	1	1	0	X	X	X	X	X	X	X	X
1	1	1	1	X	X	X	X	X	X	X	X

Step 5. Excitation maps

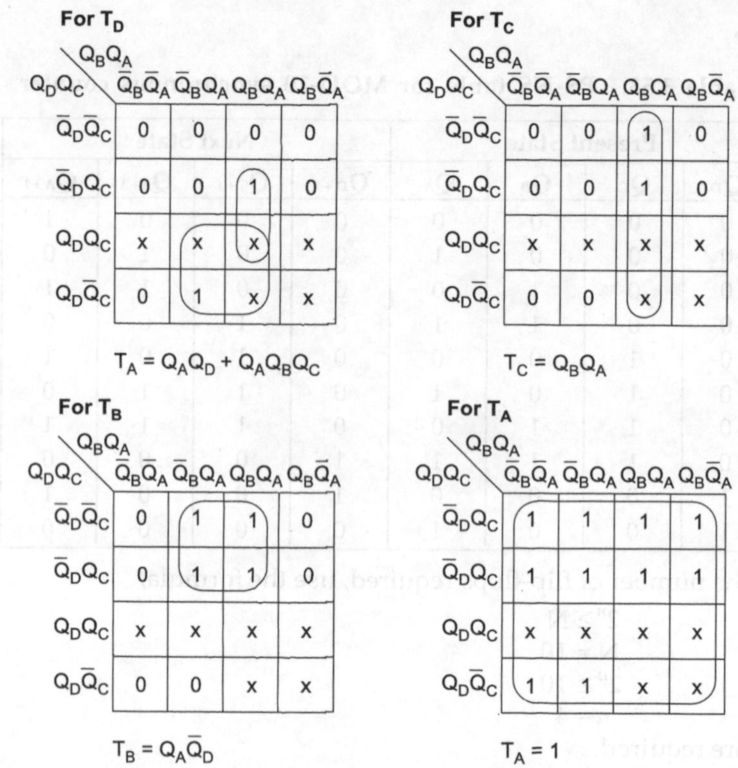

Fig. 3.115. Excitation maps

Step 6. Schematic diagram

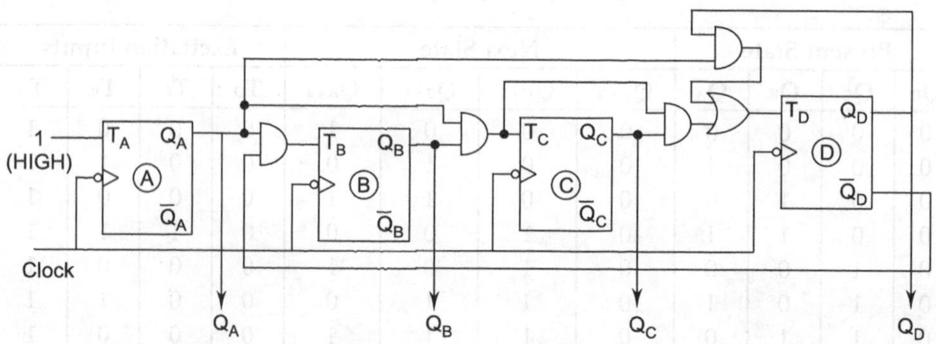

Fig. 3.116. Schematic Diagram for MOD-10 (decade) counter

3-bit Synchronous UP/$\overline{\text{DOWN}}$ counter

An up/down counter is one that is capable of progressing in increasing order or decreasing order through a certain sequence. Usually up/down operation of the counter is controlled by UP/$\overline{\text{DOWN}}$ signal. When this signal is HIGH, counter goes through UP sequence, *i.e.*, 0, 1, 2, ..., *n*. When this signal is LOW, counter follows reverse sequence *i.e.*, *n*, *n* – 1, ..., 0. Let us design the 3-bit synchronous UP/$\overline{\text{DOWN}}$ counter.

Step 1. PS-NS table

Table 3.55. PS-NS table for 3-bit synchronous up-down counter

Control input UP/$\overline{\text{DOWN}}$	Present State			Next State		
	Q_C	Q_B	Q_A	Q_{C+1}	Q_{B+1}	Q_{A+1}
0	0	0	0	1	1	1
0	0	0	1	0	0	0
0	0	1	0	0	0	1
0	0	1	1	0	1	0
0	1	0	0	0	1	1
0	1	0	1	1	0	0
0	1	1	0	1	0	1
0	1	1	1	1	1	0
1	0	0	0	0	0	1
1	0	0	1	0	1	0
1	0	1	0	0	1	1
1	0	1	1	1	0	0
1	1	0	0	1	0	1
1	1	0	1	1	1	0
1	1	1	0	1	1	1
1	1	1	1	0	0	0

Step 2. To find the number of flip-flops required, use the formula

$2^n \geq N$, where n = number of flip-flops required

Here, $N = 8$

$2^n \geq 8$

$\therefore$ $n = 3$.

Step 3. Using T flip-flop

Step 4. Excitation table

Table 3.56. Excitation table for 3-bit synchronous up-down counter

Control Input UP/$\overline{\text{DOWN}}$ (UD)	Present State			Next State			Excitation Inputs		
	Q_C	Q_B	Q_A	Q_{C+1}	Q_{B+1}	Q_{A+1}	T_C	T_B	T_A
0	0	0	0	1	1	1	1	1	1
0	0	0	1	0	0	0	0	0	1
0	0	1	0	0	0	1	0	1	1
0	0	1	1	0	1	0	0	0	1
0	1	0	0	0	1	1	1	1	1
0	1	0	1	1	0	0	0	0	1
0	1	1	0	1	0	1	0	1	1
0	1	1	1	1	1	0	0	0	1
1	0	0	0	0	0	1	0	0	1
1	0	0	1	0	1	0	0	1	1
1	0	1	0	0	1	1	0	0	1

1	0	1	1	1	0	0	1	1	1
1	1	0	0	1	0	1	0	0	1
1	1	0	1	1	1	0	0	1	1
1	1	1	0	1	1	1	0	0	1
1	1	1	1	0	0	0	1	1	1

Step 5. Excitation maps.

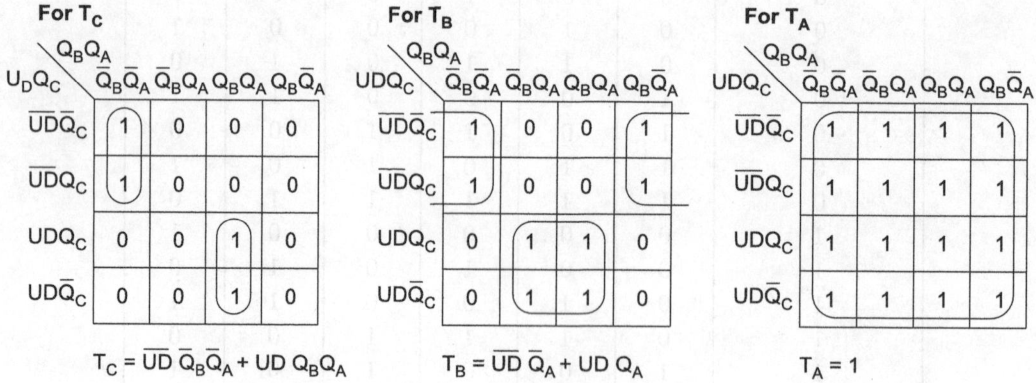

$T_C = \overline{UD}\,\overline{Q}_B\overline{Q}_A + UD\,Q_BQ_A$ $\quad T_B = \overline{UD}\,\overline{Q}_A + UD\,Q_A$ $\quad T_A = 1$

Fig. 3.117. Excitation maps

Step 6. Schematic diagram

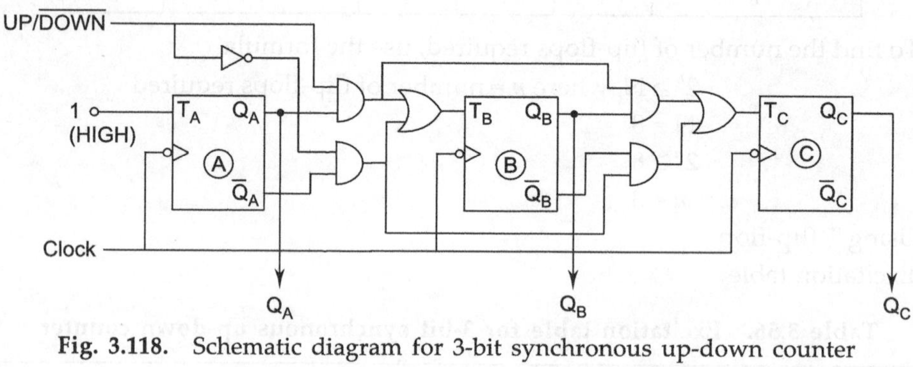

Fig. 3.118. Schematic diagram for 3-bit synchronous up-down counter

Or

Step 4. Using J-K flip-flop.

Table 3.57. Excitation table using J-K flip-flops

Control Input	**Present State**			**Next State**			**Excitation Inputs**					
$\overline{\text{UP/DOWN}}$ (UD)	**Q_C**	**Q_B**	**Q_A**	**Q_{C+1}**	**Q_{B+1}**	**Q_{A+1}**	**J_C**	**K_C**	**J_B**	**K_B**	**J_A**	**K_A**
0	0	0	0	1	1	1	1	X	1	X	1	X
0	0	0	1	0	0	0	0	X	0	X	X	1
0	0	1	0	0	0	1	0	X	X	1	1	X
0	0	1	1	0	1	0	0	X	X	0	X	1
0	1	0	0	0	1	1	X	1	1	X	1	X
0	1	0	1	1	0	0	X	0	0	X	X	1

0	1	1	0	1	0	1	X	0	X	1	1	X
0	1	1	1	1	1	0	X	0	X	0	X	1
1	0	0	0	0	0	1	0	X	0	X	1	X
1	0	0	1	0	1	0	0	X	1	X	X	1
1	0	1	0	0	1	1	0	X	X	0	1	X
1	0	1	1	1	0	0	1	X	X	1	X	1
1	1	0	0	1	0	1	X	0	0	X	1	X
1	1	0	1	1	1	0	X	0	1	X	X	1
1	1	1	0	1	1	1	X	0	X	0	1	X
1	1	1	1	0	0	0	X	1	X	1	X	1

Step 5. Excitation maps

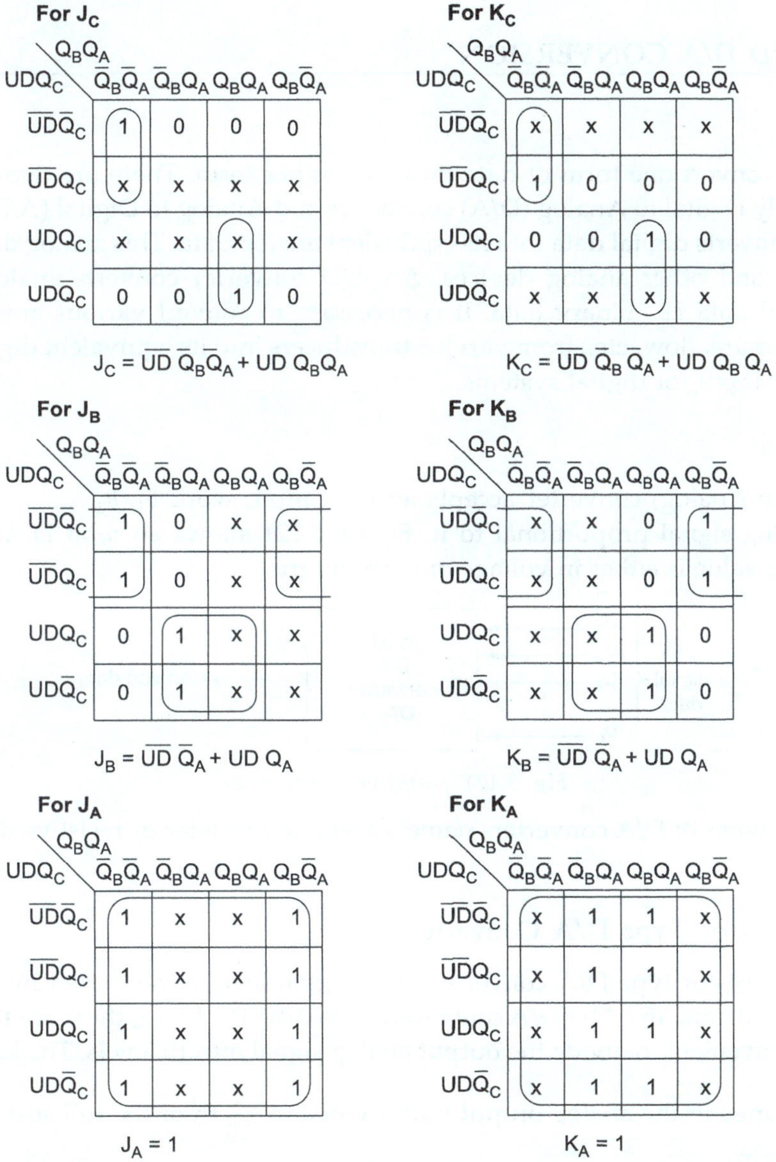

Fig. 3.119. Excitation maps

Step 6. Schematic diagram

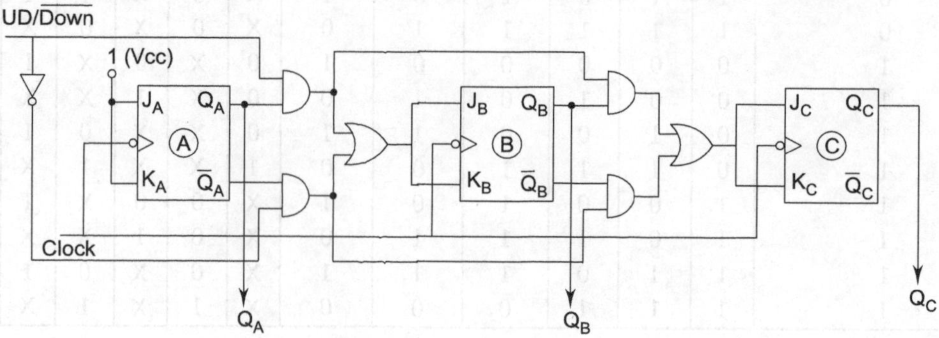

Fig. 3.120. Schematic diagram using J-K flip-flop

3.16 A/D AND D/A CONVERSION

Introduction

Data converters convert one form of data into an another form. There are two important data converters namely Digital to Analog (D/A) converters and Analog to Digital (A/D) converters. A D/A converter converts digital data into its equivalent analog data. This analog data are required to drive motors and other analog devices. An A/D converter converts analog data into its equivalent digital data *i.e.,* binary data. It is necessary to convert various analog signals like temperature, pressure, flow, etc., from various transducers into its equivalent digital data, which in turn act as the input for digital systems.

D/A Converters

A D/A (Digital to Analog) converter accepts an n-bit input word b_1, b_2, ..., b_n in binary and produce an analog signal proportional to it. Figure 3.121 shows an n-bit D/A converter. The converted analog value is either in voltage or current form.

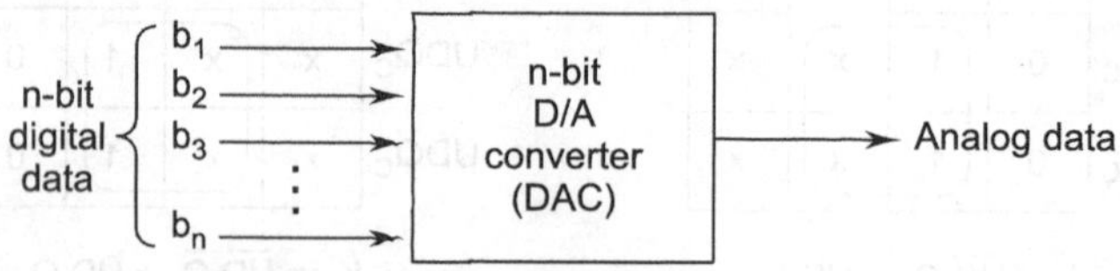

Fig. 3.121 n-bit D/A converter

There are two types of D/A converters namely weighted resistor or resistive divider type and R-2R ladder type.

1. Weighted Resistor Type D/A Converter

In the weighted resistor type D/A converter each digital level is converted into an equivalent analog voltage or current. In a 4 bit D/A converter, from 0 0 0 0 to 1 1 1 1, there are 15 discrete levels and hence it is convenient to divide the output analog signal into 15 levels. The LSB of the digital data causes a change in the analog output that is equal to $\frac{1}{15}$th of the full-scale analog output voltage. The weighted resistor network should then be designed in such a way that a '1' in LSB (2) position results in $15 \times \frac{1}{15} = 1$ V at the output. A '1' in the 2^1 bit position must cause a change in the analog output voltage that is equal to $\frac{2}{15}$ th of the full-scale analog output voltage (*i.e.*, twice the size of the LSB). The weighted resistor network should then be designed in such a way that a '1' in the 2^1 position results in $15 \times \frac{2}{15} = 2$ V at the output. Similarly, a '1' in 2^2 and 2^3 bit positions must cause a change of 4 V and 8 V respectively at the analog output. It is important to note that the sum of the weights assigned to various bit positions must be equal to '1'. In general, the weight assigned to the LSB is $\frac{1}{(2^n - 1)}$, where n is the number of bits in the digital input.

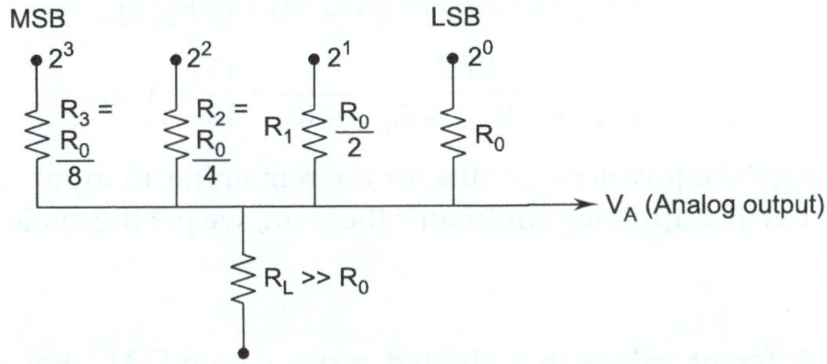

Fig. 3.122. 4 bit weighted resistor type D/A converter

Thus, the 4-bit weighted resistor network shown in Fig. 3.122 performs the following D/A conversion :

1. The 2^0 bit is changed to 1 V, 2^1 bit to 2 V, 2^2 bit to 4 V, and 2^3 bit to 8 V.
2. These four voltages are added together to form the analog output voltage.

The resistor R_0, R_1, R_2 and R_3 form the divider network and R_L acts as the load resistor. The value of R_L should be large enough so as not to load the divider network. The LSB should be connected with the highest input resistance (R_o) while 2^1 bit with a resistance of half that of LSB $\left(\frac{R_o}{2}\right)$ so that its output contribution will be twice that of LSB. The 2^3 bit is connected with a quarter of LSB resistance $\left(\frac{R_o}{4}\right)$ while the MSB with only one eighth of the LSB resistance $\left(\frac{R_o}{8}\right)$. The output is the sum of these four attenuated voltages.

Apply the digital data 0 0 0 1 to the network ('0' means 0 V , '1' means '15 V'). The equivalent circuit of Fig. 3.122 is shown in Fig. 3.123. The analog output voltage V_A can be calculated using Millman's theorem which states that the voltage at any node in a resistive network is equal to the sum of the currents entering the node divided by the sum of the conductance connected to the node, all determined by assuming that the voltage at the node is zero.

$$V_A = \frac{V_1/R_1 + V_2/R_2 + V_3/R_3 + \cdots}{1/R_1 + 1/R_2 + 1/R_3 + \cdots}$$

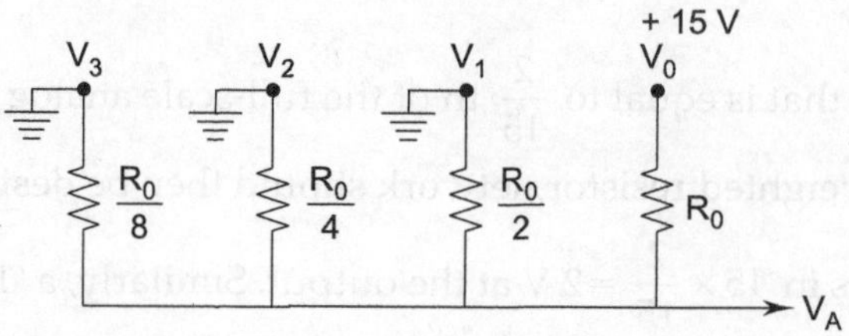

Fig. 3.123. Equivalent circuit of a 4-bit weighted resistor type D/A converter for 0 0 0 1 input

Applying the Millman's theorem to the circuit of Fig. 3.123, we get

$$V_A = \frac{\frac{V_0}{R_0} + \frac{V_1}{(R_0/2)} + \frac{V_2}{(R_0/4)} + \frac{V_3}{(R_0/8)}}{1/R_0 + 1/(R_0/2) + 1/(R_0/4) + 1/(R_0/8)}$$

$$= \frac{15/R_0}{1/R_0 + 2/R_0 + 4/R_0 + 8/R_0} = \frac{15}{15} = 1\text{ V}.$$

Similarly, by drawing the equivalent circuits for the remaining 15 input combinations (*i.e.*, 0 0 0 0 ; 0 0 1 0 ; .... 1 1 1 1) and applying Millman's theorem, we get the analog conversions.

Drawbacks

1. Resistors have different values in weighted resistor type DAC. For perfect D/A conversion, these resistors should be of precision type and this increases cost.
2. Resistors have to handle different currents.

Hence, the R-2R ladder type D/A converter that does not have such drawback is a better choice for practical applications.

2. R-2R Ladder Type D/A Converter

R-2R ladder D/A converter uses only two resistor values *i.e.*, R and 2R are used. The ladder type network for 4 bit D/A conversion is shown in Fig. 3.124 with 4 digital inputs (A, B, C, D) one analog output (V_A) and one terminating resistor (2R).

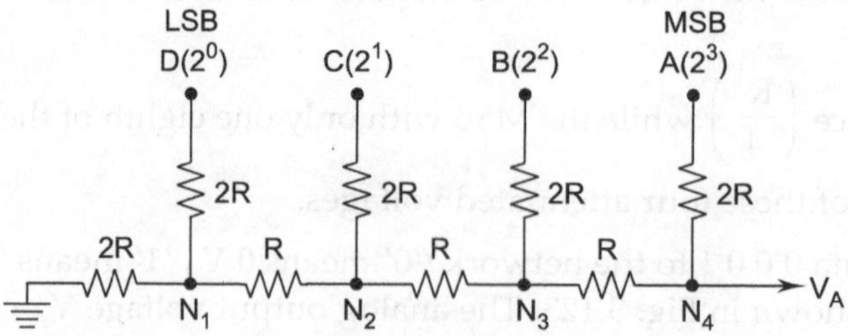

Fig. 3.124. 4 bit R-2R ladder type D/A converter

In this ladder circuit, the output voltage is a weighted sum of digital inputs. For example, if 4-bit digital input ABCD is 1000 *i.e.*, if MSB is 1 (+ V) while the other three inputs are 0 (ground), the circuit shown in Fig. 3.124 can be modified as shown in Fig. 3.125 (*a*).

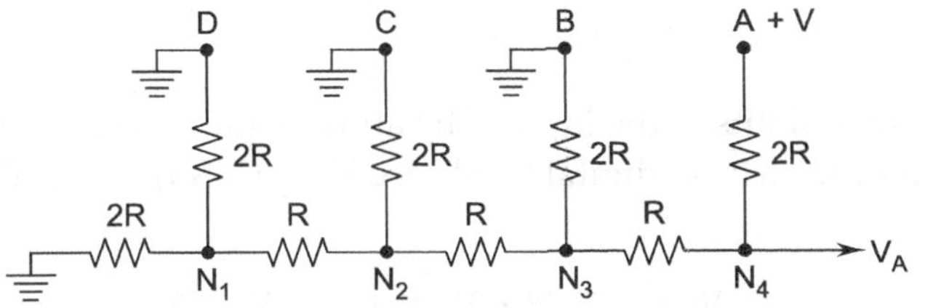

Fig. 3.125 (*a*)

Here, the terminating resistor (2R) and the resistor connected to D input (2R) are combined at Node (N_1), as they are parallel, to form an equivalent resistor (R) as shown in Fig. 3.125 (*b*).

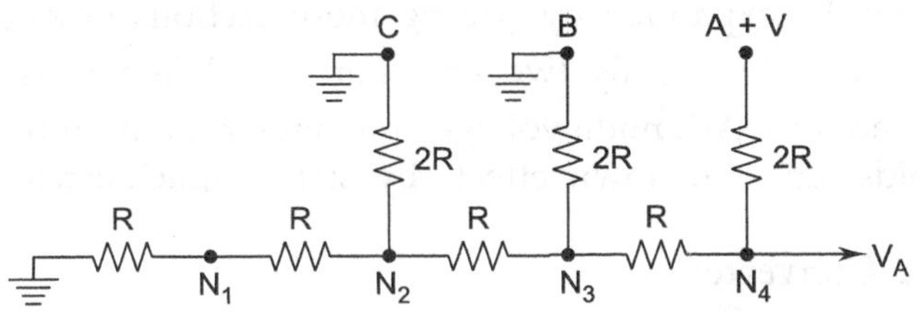

Fig. 3.125 (*b*)

Then, at Node 2 (N_2), the resistor connected with C input (2R) can be combined with R + R = 2R resistor to form an equivalent resistor R as shown in Fig. 3.125 (*c*).

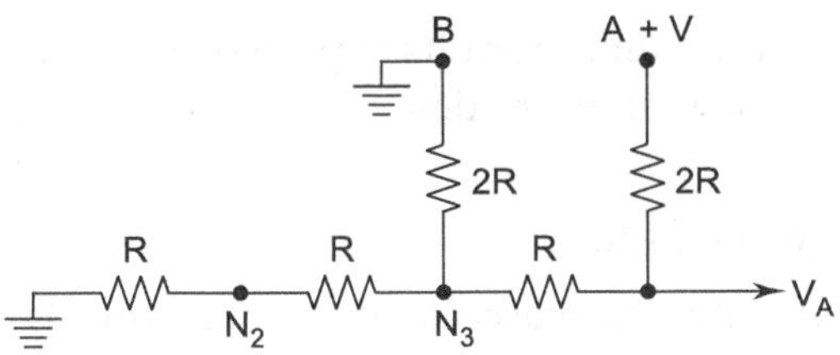

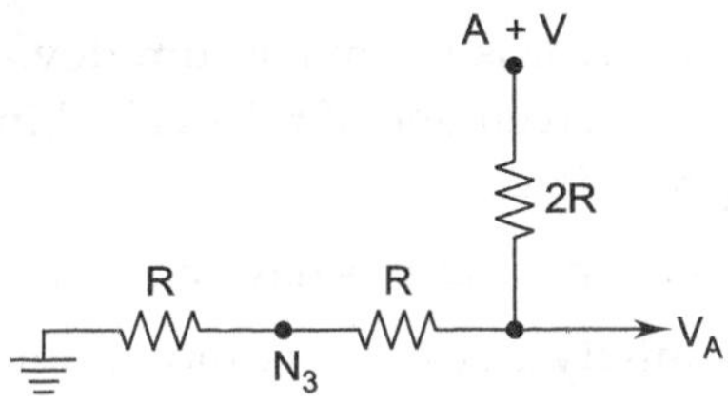

Fig. 3.125 (*c*) **Fig. 3.125** (*d*)

Similarly, at Node 3 (N_3), the equivalent resistor is R as shown in Fig. 3.125 (*d*). From the equivalent circuit shown in Fig. 3.125 (*d*), the analog output voltage V_A is given by

$$V_A = \frac{V \times 2R}{R + R + 2R} = \frac{V}{2}$$

Thus, for digital input ABCD = 1 0 0 0 *i.e.*, when MSB = 1, the output is $\frac{V}{2}$. Similarly, it can be found that for digital input ABCD = 0 1 0 0 *i.e.*, when second MSB = 1, the output is $\frac{V}{4}$, for ABCD = 0 0 1 0, the output is $\frac{V}{8}$; for ABCD = 0 0 0 1, *i.e.*, when LSB = 1, the output is $\frac{V}{16}$.

Since the ladder is a linear network, the principle of superposition can be used to find the total analog output voltage for a particular digital input by adding the output voltages caused by the individual digital inputs. This can be represented for an *n*-bit D/A converter as follows :

$$V_A = \frac{V}{2} + \frac{V}{4} + \frac{V}{8} + \cdots\cdots + \frac{V}{2^n}$$

where n is the total number of bits at the input. The above equation can be written in generalized form for an n-bit digital input whose digital levels are V_{n-1} (MSB), V_{n-2}, V_{n-3}, ..., V_3, V_2, V_1, V_0 (LSB) as follows :

$$V_A = \frac{V_0 2^0 + V_1 2^1 + V_2 2^2 + \cdots\cdots V_{n-1} 2^{n-1}}{2^n}$$

Advantages

1. Number of bits can be expanded by adding more sections of same R-2R values.
2. Easier to build accurately as only two precision metal film resistors are required.
3. In inverted R-2R ladder DAC, node voltages remain constant with changing input binary words. This avoids any slowdown effects by stray capacitances.

Specifications of D/A Converter

The various performance parameters of D/A converter are :

Accuracy : Accuracy indicates how close the measured value is to the true value. An accuracy of ± 1% means that 100 volts could be in the range 99 – 101 volts. DAC manufactures have many ways of specifying accuracy. The most common are full scale error and linearity error.

Full scale error is the maximum deviation of the output value from its expected (ideal) value expressed in percentage of full-scale. Linearity error is the maximum deviation in step size from the ideal step size.

General purpose DACs have accuracies in the range of 0.01 to 0.1%.

Monotonicity : A DAC is monotonic if its output value increases as the binary inputs are incremented from one value to the next *i.e.*, the staircase output having no downward steps as the binary input is incremented from zero to full scale.

Offset Voltage : Ideally, the output of a DAC will be zero volts when the binary inputs are all 0's. In practice, however, there will be a very small output voltage called offset voltage or offset error.

Resolution : Resolution of DAC is defined as the smallest change that can occur in the analog output as a result of a change in the digital input.

Resolution can also expressed as the amount of voltage or current per step, it is more useful to express it as a percentage of the full scale output. If the maximum full-scale output of the 4-bit DAC is 15 V, and the step size is 1 V, then the percentage resolution is given by

$$\text{Percentage resolution} = \frac{\text{Step size}}{\text{Full size}} \times 100$$

$$= \frac{1\text{V}}{15\text{V}} \times 100 = 6.67\%.$$

Percentage resolution can also be calculated as

$$\% \text{ resolution} = \frac{1}{\text{total number of steps}} \times 100$$

For an n-bit digital input, the total number of steps is $2^n - 1$. Then

$$\% \text{ resolution} = \frac{1}{2^n - 1} \times 100$$

Settling Time : The time required for the output of the DAC to settle to within $\pm \left(\frac{1}{2}\right)$ LSB of the final value for a given digital input is known as settling time.

A/D Converters

The A/D conversion is a quantizing process whereby an analog signal is converted into equivalent binary word. Thus the A/D converter is exactly opposite function that of the D/A converter. Figure 3.126 shows the n-bit A/D converter.

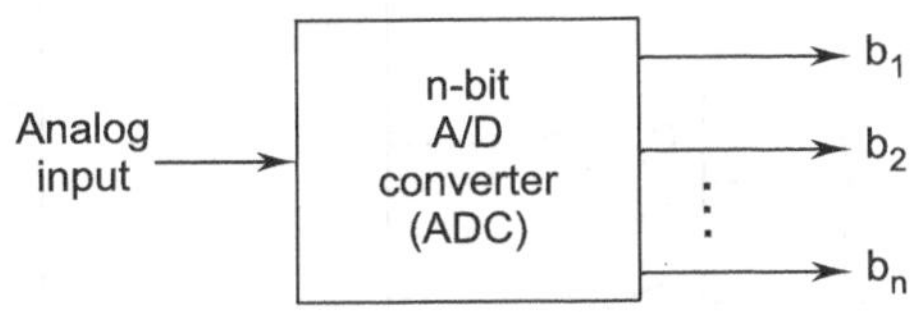

Fig. 3.126. n-bit A/D converter

There are different types of A/D converters as listed below :

1. Simultaneous or flash type
2. Counter type
3. Continuous type
4. Single slope type
5. Dual slope type
6. Successive approximation type.

Let us see in detail about the flash type, dual slope type and successive approximation type of A/D converters.

Simultaneous or Parallel Comparator or Flash Type

When system designs call for the highest speed available, flash type A/D converters (ADCs) are likely to be the right choice. They get their names from their ability to do the conversion very rapidly. Flash A/D converters, also known as a simultaneous or parallel comparator ADC, because the fast conversion speed is accomplished by providing $2^n - 1$ comparators and simultaneously comparing the input signal with unique reference levels spaced 1 LSB apart.

Figure 3.127 shows 3 bit flash A/D converter. For this ADC, seven $(2^3 - 1)$ comparators are required. As shown in Fig. 3.127, one input of each comparator is connected to the input signal and other input to the reference voltage level generated by the reference voltage divider. The reference voltage (V_{REF}) is equal to the full scale input signal voltage. The manner in which the flash A/D converter performs a quantization is relatively simple.

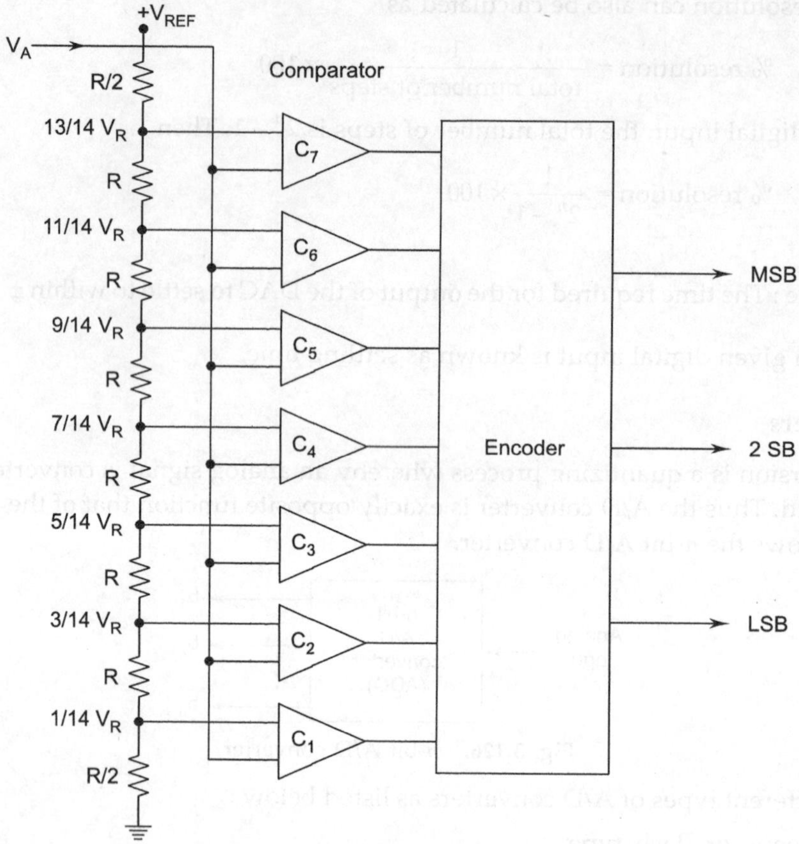

Fig. 3.127. 3-bit flash converter

The comparators give output '1' or '0' state depending on whether the input signal is above or below the reference level at that instant. Those comparators referred above the input signal, remain turned-off, representing a "0" state. The comparators at or below the input signal conversely become a '1' state. The code resulting from this comparator is converted to a binary code by the encoder.

The number of comparators required for n-bit resolution is

Number of comparators = $2^n - 1$

Here, the quantization error is $\pm \frac{1}{2}$ LSB. Thus, for an ADC, the maximum frequency for a sine wave V_{in} to be digitised within an accuracy of $\pm \frac{1}{2}$ LSB is,

$$f_{max} \cong \frac{1}{2\pi (T_c) 2^n}$$

where f_{max} = maximum input frequency

T_c = conversion time

n = number of bits

Dual Slope Type

Dual slope conversion is an indirect method for A/D conversion where an analog voltage and a reference voltage are converted into time periods by an integrator, and then measured by a counter. The speed of this conversion is slow but the accuracy is high.

Figure 3.128 shows a typical dual slope converter circuit. It consists of integrator (ramp generator), comparator, binary counter, output latch and reference voltage. The ramp generator input is switched between the analog input voltage V_i and a negative reference voltage, $-V_{REF}$. The analog switch is controlled by the MSB of the counter. When the MSB is a logic 0, the voltage being measured is connected to the ramp generator input. When MSB is logic 1, the negative reference voltage is connected to the ramp generator.

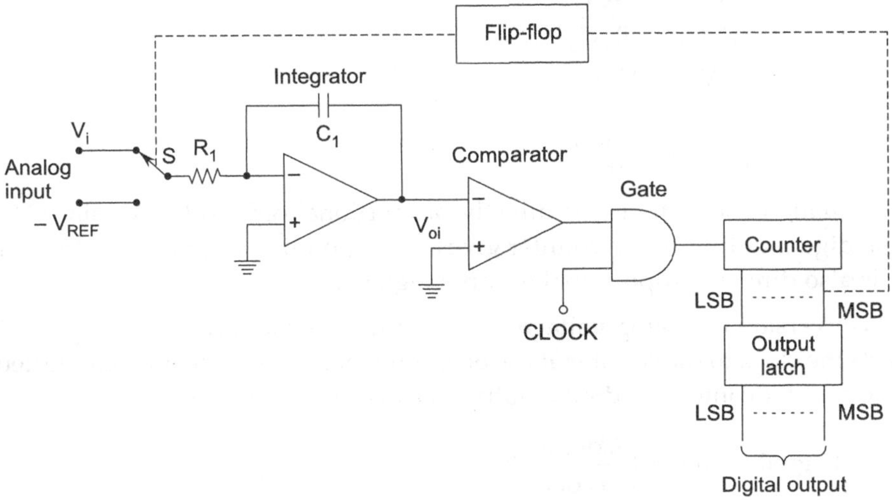

Fig. 3.128. Dual slope A/D converter

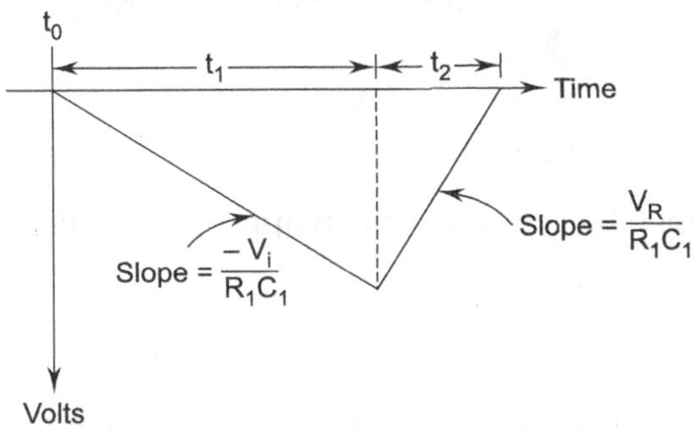

Fig. 3.129. Integrator output voltage

Figure 3.129 shows the output voltage of the integrator. In this, at time $t = 0$, analog switch S is connected to the analog input voltage V_i, so that the analog input voltage integration begins. The output voltage of the integrator can be given as

$$V_{oi} = \frac{-1}{R_1 C_1} \int_0^t V_i \, dt = \frac{-V_i t}{R_1 C_1}$$

where R_1C_1 is the integrator time constant and V_i is assumed constant over the integration time period. At the end of 2^N clock periods MSB of the counter goes high. As a result the output of the flip-flop goes high, which causes analog switch S to be switched from V_i to $-V_R$. At this very same time the counter which has gone through its entire count sequence is reset.

The negative input voltage ($-V_R$) connected to the input of integrator causes the integrator output to ramp positive. When integrator output reaches zero, the comparator output voltage goes low, which disables the clock AND gate. This stops the clock pulses reaching the counter, so that the counter will be stopped at a count corresponding to the number of clock pulses in time t_2.

The integrator output ramp down to a voltage V and get back upto 0. Therefore, the charge voltage is equal to discharge voltage and we can write,

$$\frac{V_i t_1}{R_1 C_1} = \frac{V_R t_2}{R_1 C_1}$$

$$V_i t_1 = V_R t_2$$

$$t_2 = \frac{V_i}{V_R} \cdot t_1$$

The above equation shows that t_2 is directly proportional only to the V_i, since V_R and t_1 are constants. The digital output of the counter gives corresponding digital value for time period t_2 and hence it is also directly proportional to input signal V_i.

The actual conversion of analog voltage V_{in} into a digital count occurs during t_2. The control circuit connects the clock to the counter at the beginning of t_2. The clock is disconnected at the end of t_2. Thus, the counter contents is digital output. Hence, we can write

$$\text{Digital output} = \left(\frac{\text{Counts}}{\text{Second}}\right) t_2$$

Substituting $t_2 = \frac{V_i t_1}{V_R}$ in the above equation, we get

$$\text{Digital output} = \left(\frac{\text{Counts}}{\text{Second}}\right) t_1 \left(\frac{V_i}{V_R}\right)$$

The counter output can then be connected to an appropriate digital display.

Advantages

1. It is highly accurate
2. It is immune to temperature caused variations in R_1 and C_1.
3. Cost is low.

Successive Approximation Type

The successive approximation technique uses a very efficient code search strategy to complete n-bit conversion in just n-clock periods. An eight bit converter would require eight clock pulses to obtain a digital output. Figure 3.130 shows an eight bit converter. It consists of a DAC, a comparator, and a successive approximation register (SAR). The circuit uses a successive approximation register (SAR) to find the required value of each bit by trial and error.

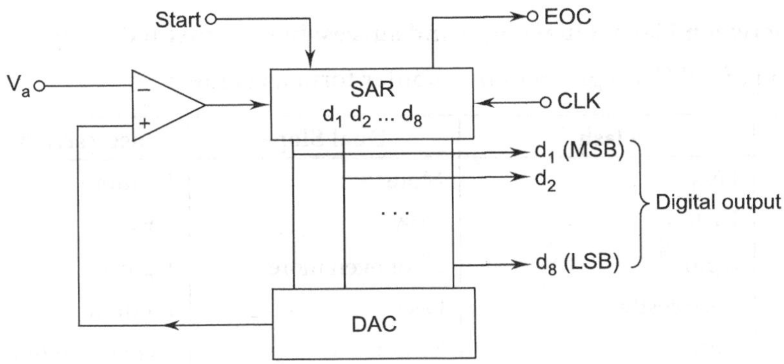

Fig. 3.130. Block diagram of successive approximation type A/D converter

Operation

The circuit operates as follows. With the arrival of the START command, the SAR sets the MSB $d_1 = 1$ with all other bits to zero so that the trial code is 1 0 0 0 0 0 0 0. The output V_d of the DAC is now compared with analog input V_a. If V_a is greater than the DAC output V_d, then 1 0 0 0 0 0 0 0 is less than the correct digital representation. The MSB is left at '1' and the next lower significant bit is made '1' and further tested.

However, if V_a is less than the DAC output, then 1 0 0 0 0 0 0 0 is greater than the correct digital representation. So reset MSB to '0' and go on to the next lower significant bit. This procedure is repeated for all subsequent bits, one at a time, until all bit positions have been tested. Whenever the DAC output crosses V_a, the comparator changes state and this can be taken as the end of conversion (EOC) command. Figure 3.131 shows a typical conversion sequence.

Correct digital representation	Successive approximation register output V_d at different stages in the conversion	Comparator output
1 1 0 1 0 1 0 0	1 0 0 0 0 0 0 0	1 (Initial output)
	1 1 0 0 0 0 0 0	1
	1 1 1 0 0 0 0 0	0
	1 1 0 1 0 0 0 0	1
	1 1 0 1 1 0 0 0	0
	1 1 0 1 0 1 0 0	1
	1 1 0 1 0 1 1 0	0
	1 1 0 1 0 1 0 1	0
	1 1 0 1 0 1 0 0	

Fig. 3.131. Successive approximation conversion sequence for a typical analog input

The time for one analog to digital conversion must depend on both the clock's period T and number of bits n. It is given as

$$T_c = T(n+1)$$

where T_c = conversion time

T = clock period

n = number of bits

Comparison between Flash, Dual slope and successive approximation type A/D Converters

The comparison of ADC's is given in the tabular form as below :

Parameter	Flash	Dual Slope	Successive Approximation
Accuracy	Less	More	Medium
Speed	Fastest	Slow	Fast
Resolution	Upto 2^8	2^{16} or even more	Upto 2^{16}
Cost	Very costly	Less	Medium
Input hold time	Very less	Maximum	Depend on number of bits. It is more than flash type but less than dual slope A/D converter
Applications	High speed fiber optic communication, digital storage oscilloscope, imaging, etc.	These are used when high accuracy and resolution is required and speed is not the important criteria	Used in data acquisition systems

Specifications of A/D Converter

The various performance parameters of A/D converter are :

Accuracy : Since both analog and digital systems such as comparator and D/A converter are used for the construction of A/D converter, the overall accuracy of an A/D converter depends on the errors from both analog and digital systems.

Quantization error : A digital error in an ADC is based on the resolution of the digital system. In A/D conversion, a continuous analog voltage is represented by an equivalent set of digital numbers. When the digital numbers are converted back to analog voltage by a DAC, the output is a staircase waveform, which is a discontinuous signal, composed of a number of discrete steps. The smallest digital step is due to the LSB and it can be made smaller only by increasing the number of bits in the detail representation. This error is called quantization error, or digitizing error and it is commonly ± 1 bit. If the comparator is adjusted, the quantization error can be made $\pm \frac{1}{2}$ LSB. For an 8-bit A/D converter, the quantization error is $\frac{1}{2^8} = 0.4$ per cent while for a 10 bit A/D converter, it is $\frac{1}{2^{10}} = 0.1$ per cent.

Analog error : Analog error in an ADC is mainly due to variations in the DC switching point of the comparator. The variations are mainly due to offset, gain and linearity of the operational amplifier used in the comparator. The other sources of analog error are the resistors in DAC, the reference voltage, ripple and noise.

Conversion time : The time required by an A/D converter to convert an analog input value into its equivalent digital data is called conversion time.

Resolution: The resolution of an ADC depends on the reference voltage (V_R) and the number of bits (n) in the digital output. This is given by

$$\text{Resolution} = \frac{V_R}{2^n}$$

Input voltage range : It is the range of voltage that an ADC can accept as its input.

Differential linearity : This is an important measure of ADC performance. It is defined as a measure of the variation in voltage step size that causes the converter to change from one state to the next. It is normally expressed as a percentage of average step size.

4

8085 MICROPROCESSOR

4.1 MICROCOMPUTERS

- Microcomputers are much small computers.
- They operate more slowly and they work with smaller data words.
- They can address a few thousands to a few million memory locations.
- They have almost all the features of minicomputer and main frames.
- An important feature of a microcomputer is that the CPU is usually a single integrated circuit called a microprocessor.

4.2 BLOCK DIAGRAM OF MICROCOMPUTER

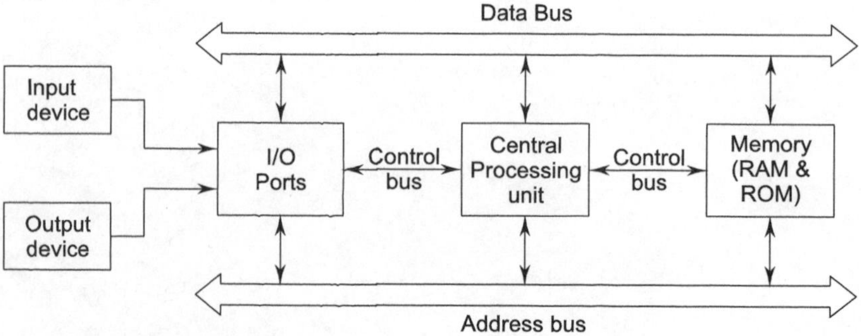

Fig. 4.1 Block Diagram of Microcomputer

The main parts of a microcomputer are :

1. Central Processing Unit (CPU)
2. Memory (ROM and RAM)
3. Input and output circuitry or I/O
4. Buses (Address bus, data bus and control bus).

1. **Central Processing Unit (CPU)** : The job of a CPU is to control the operations of the computer. The operation of CPU are :
   (*i*) **Fetching** : It fetches instructions from memory. They are in binary coded form.
   (*ii*) **Decoding** : It decodes these instructions into series of simple actions.
2. **Memory** : The memory section of a microcomputer consists of RAM (Random Access Memory) as well as ROM (Read Only Memory). The memory may also contain the floppy disks, magnetic hard disks or optical disks. The purpose of using memory is storage of sequence of instructions in the form of binary codes.
3. **Input/Output (I/O)** : The I/O section enables the computer to take the data from the user and / or send the data to the outside world. Devices such as keyboards, video display terminals, printers and modems are called as peripherals and they are connected to the I/O section.

   The actual physical devices which are used for interfacing the computer buses with the external systems are called as ports. They are of two types, namely input ports and output ports.

   The input port will allow the connection of data from the keyboard or an A/D converter or some other source to the computer. An output port is used for connecting the data from the computer to the outside world. The processed data is sent from the computer to a peripheral such as video terminal, or a printer or a D/A converter.
4. **Buses:** A bus is a group of wires that are grouped together to serve a single purpose. There are three types of buses namely address, data and control bus.

Address Bus: The bus over which the CPU sends out the address of the memory location is called as the address bus. The address bus may consist of 16, 20, 24 or 32 parallel signal lines. If there are N address lines, then it can directly address 2^N memory location. Address bus in unidirectional.

Data Bus: The data bus consists of 8, 16 or 32 parallel lines. The data bus is a bi-directional bus. That is the data can get transferred from CPU to memory and *vice-versa*. The data bus also connects the I/O ports and CPU. So the CPU can write data to or read it from the memory or I/O ports. The number of data lines used in the data bus is equal to the size of data word being written or read.

Control Bus: The control bus consists of 4 to 10 parallel signal lines. The control bus is used for sending control signals to the memory and I/O devices. Some of the control bus signals are memory read, memory write, I/O read, I/O write etc.

Software

Computer software is divided into two broad categories:

1. **System Software** : Collection of programs which are needed in the creation, preparation and execution of other programs.
2. **User Software** : These are the programs generated by various users of the system.

Operating System

Operating system is nothing but collection of system programs. It is an interface between the user and the machine.

Various components of operating system:

1. **I/O driver** : Whenever, user program or any other system program needs to use an I/O device, it will request operating system to use an I/O driver to perform the task.
2. **File Management** : It handles large collection of data that are stored in mass storage devices. The file management routines are used in conjunctions with I/O drivers for mass storage. It performs following functions :
   1. Access the file 2. Copies to file 3. Manipulates the file.
3. **Translators** : This program converts, assembly language or high level language program into machine language. If the program being translated is in assembly language; the translator is called an Assembler. But if it translates, higher-level language program, then it is referred to as a compiler or interpreter.
4. **Text-editors** : This program is for inputting or modifying the text that is stored on a mass storage device or that is to be stored.
5. **Library** : Collection of small programs (Procedure or subroutines), is normally referred to as library. System library is normally for general use. Even user can create their own libraries.
6. **Linker** : The program, which links different subroutines or library routines, is referred to as a linker or linkage editor.
7. **Loader** : The program, which transfers the program to be executed into memory is called as loader.

4.3 MICROPROCESSORS

- A microprocessor is a semiconductor chip that implements the central processor of a computer. It consists of an Arithmetic and Logic Unit (ALU) and a control unit.
- The central processing unit built into single chip is called as microprocessor.

4.4 FEATURES OF INTEL 8085

The features of 8085 include :

1. It is an 8-bits microprocessor.
2. It is a single chip. NMOS device with 6200 transistors.
3. Requires +5 V power supply.
4. The maximum clock frequency is 5 MHz. Normally it operate at 3 MHz.
5. It has an 8-bits data bus and 16-bits address bus.
6. It provides 74 instructions.

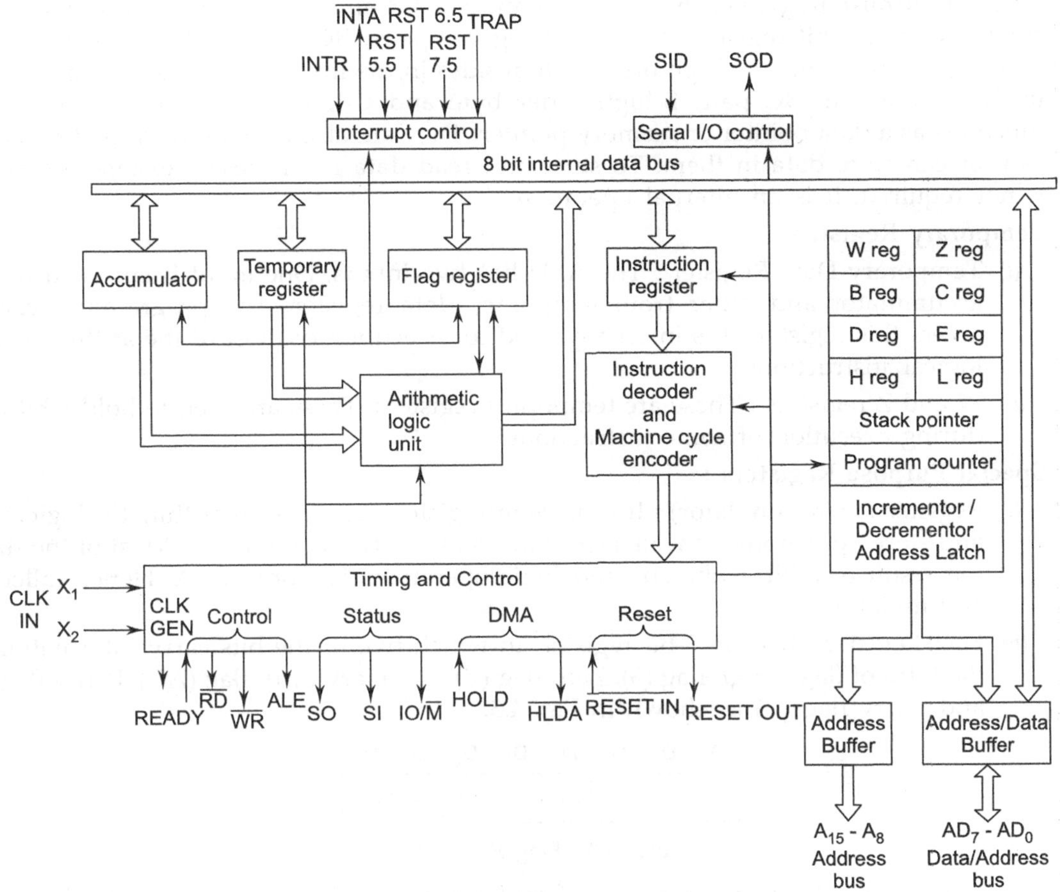

Fig. 4.2 Architecture of 8085

4.5 ARCHITECTURE OF 8085

Figure shows the architecture of 8085

It consists of various functional blocks as listed below :

- Arithmetic and logic unit
- Registers
- Instruction decoder and machine cycle decoder
- Address buffer
- Address / data buffer
- Incrementer / Decrementer Address Latch
- Serial I/O control
- Interrupt control
- Stack pointer and program counter
- Timing and control unit

Registers

It has eight addressable 8-bit registers : A, B, C, D, E, H, L, F and two 16-bit registers PC and SP. They are classified as :

(*a*) **General Purpose Registers :** B, C, D, E, H and L are 8-bit general purpose registers. They can be used as 8-bit register or as 16-bit register pairs, BC, DE and HL. When they are used as register pairs, the high order byte resides in the first register and low order byte in the second. [In BC pair, B-high order byte and C-low order byte]. HL pair also functions as a data pointer or memory pointer. They are also called scratch pad registers, as user can store data in them. To store and read data from these registers bus access is not required, it is an internal operation.

(*b*) **Temporary Registers :**

(*i*) Temporary Data Register : The ALU has two inputs. One input is supplied by the accumulator and other from temporary data register. The programmer cannot access this register. It is internally used for execution of most of the arithmetic and logical instructions.

(*ii*) W and Z registers : These are temporary registers. These are used to hold 8-bit data during execution of some instructions.

(*c*) **Special Purpose Registers :**

(*i*) Register A (Accumulator) : It is an 8-bit register. It is used in arithmetic, logic, load and store operations, as well as in, Input/Output (I/O) operations. Most of the times the result of 8-bit arithmetic and logic operations is stored in A. Hence called as accumulator.

(*ii*) Flag Register : It is an 8-bit register, in which five of the bits carry information in the form of flags : Sign flag (S), Zero flag (Z), Auxiliary carry flag (AC), Parity flag (P) and Carry flag (C) as shown in Fig. 4.3.

D_7	D_6	D_5	D_4	D_3	D_2	D_1	D_0
S	Z	X	AC	X	P	X	CY

Fig. 4.3 Flag Register

S-sign flag : After the execution of arithmetic or logical operations, if bit D_7 of the result is 1, the sign flag is set. If D_7 is 0, the number will be considered as positive number. If D_7 is 1, it will be viewed as negative number.

Z-zero flag : The zero flag sets if the result of operation in ALU is zero and flag resets if result is non-zero.

AC-Auxiliary carry flag : This flag is set if there is an overflow out of bits *i.e.,* carry from lower nibble to higher nibble (D_3 bit to D_4 bit). This flag is used for BCD operations and not available for programmer.

P-purity flag : Parity is defined by the number of ones present in the accumulator. After any operation if the result has an even number *i.e.,* even parity, the flag is set. If the parity is odd flag is reset.

CY-Carry flag : This flag is set if there is an overflow out of bit 7. The carry flag also serves as a borrow flag for subtraction.

(*iii*) Instruction Register : In processor operation, the processor first fetches the op code of instruction from memory. The CPU stores this op code in a register called the instruction register.

Arithmetic Logic Unit (ALU)

It performs arithmetic and logical functions on 8-bit variables. The arithmetic unit performs operation such as addition, subtraction, increment and decrement. The logical unit performs AND, OR, EX-OR, complement, rotate and clear.

Instruction Decoder

The processor first fetches the op code of instruction from memory and stores this op code in the instruction register. It is then sent to instruction decoder. The decoder decodes it and gives timing and control signals which controls register, data buffers, ALU and external peripheral signals.

Address Buffer

This is an 8-bit unidirectional buffer. It is used to drive external high order address bus ($A_{15} \cdot A_8$). It is also used to tristate the high order address bus under reset, hold, halt conditions.

Address/Data Buffer

This is an 8-bit bidirectional buffer. It is used to drive multiplexed address/data bus *i.e.*, low order address bus ($A_7 - A_0$) and data bus ($D_7 - D_0$). Due to address and data buffers the address and data buses can be tri-stated when they are not in use.

Incrementer/Decrementer Address Latch

This 16-bit register is used to increment or decrement the contents of program counter or stack pointer as a part of execution of instructions.

Program Counter (PC)

Program is a sequence of instructions. Microprocessor fetches these instructions from memory and executes sequentially. The program counter is a special purpose register which, at a given time stores the address of the next instruction to be fetched. Its acts as a pointer to the next instruction.

Stack Pointer (SP)

The stack is a reserved area of the memory in the RAM where temporary information may be stored. [Last in first out technique]. A 16-bit stack pointer is used to hold the top of stack.

Interrupt Control

The processor fetches, decodes and executes instructions in a sequence. Sometimes it is necessary to have processor to automatically execute one of a collection of special system. The occurrence of this special condition is referred as interrupt. The interrupt control block has five interrupt inputs RST 5.5, 6.5, 7.5, TRAP and INTR and one acknowledge signal $\overline{\text{INTA}}$.

Serial I/O Control

In situations like, data transmission over long distance and communication with cassette tapes or a CRT terminal, it is necessary to transmit data bit by bit to reduce the cost of cabling. The serial I/O control provides two lines, SOD and SID for serial communication.

Timing and Control Circuitry

The control circuitry and the operations in 8085 are synchronized with the help of clock signal. The control circuitry helps in control of fetching, decoding operations and generating appropriate signals for instruction execution. It also generates signals required to interface external devices to the processor, 8085.

4.6 PIN DEFINITIONS OF 8085

Figure 4.4(*a*) and (*b*) shows pin configuration and functional pin diagram of 8085 respectively.

Power Supply Signals

(*i*) V_{CC} : It requires a single +5 V power supply.
(*ii*) V_{SS} : Ground reference

Frequency Signals

(*i*) X_1 and X_2 : A tuned circuit like LC, RC or crystal is connected at these two pins. The internal clock generator divides oscillator frequency by 2, therefore to operate a system at 3 MHz, the crystal of tuned circuit must have a frequency of 6 MHz.
(*ii*) CLK OUT : This signal is used as a system clock for other devices. Its frequency is half the oscillator frequency.

Data Bus and Address Bus

(*i*) AD_0 to AD_7 : The 8-bit data bus ($D_0 - D_7$) is multiplexed with the lower half ($A_0 - A_7$) of the 16-bit address bus. During first part of the machine cycle T_1, lower 8-bits of memory address or I/O address appear on the bus. During remaining part (T_2 and T_3) of the machine cycle these lines are used as bi-directional data bus.
(*ii*) A_8 to A_{15} : The upper half of the 16-bit address appears on the address lines A_8 to A_{15}. These lines are used for the most significant 8-bits of the 16-bit address lines.

Serial I/O Signals

(*i*) SID (Serial I/P data) : Used to accept serial data bit by bit from the external device.
(*ii*) SOD (Serial O/P data) : Used to transmit serial data bit by bit to the external device.

Interrupt Signals

8085 has five hardware interrupt signals : RST 5.5, RST 6.5, RST 7.5, TRAP and INTR. The microprocessor recognizes interrupt requests on these lines at the end of the current instruction execution. The $\overline{\text{INTA}}$ signal (Interrupt Acknowledge) is used to indicate that the processor has acknowledged an INTR interrupt.

DMA Signal

(*i*) HOLD : This signal indicates that another master is requesting for the use of address bus, data bus and control bus.
(*ii*) HLDA : This active high signal is used to acknowledge HOLD request.

Control and Status Signals

(*i*) ALE (Address Latch Enable) : In 8085, AD_0 to AD_7 are multiplexed signals and the lower half of address ($A_0 - A_7$) is available only during T_1 of the machine cycle. This lower half is also necessary during T_2 and T_3 of machine cycle to access specific location in memory or I/O port. This means lower half of address must be latched. This latching is done by using external signal and ALE.

(*ii*) READY : It is used by the microprocessor to sense whether the peripheral is ready or not for data transfer.

(*iii*) $\overline{RD}$ and $\overline{WR}$: These signals control the data flow between processor and memory or I/O device. A low on $\overline{RD}$ indicates that the data must be read from the selected location or I/O port. Similarly a low on $\overline{WR}$ indicates that the data must be written into selected location or I/O device port.

(*iv*) IO/$\overline{M}$, S_0 and S_1 : IO/$\overline{M}$ indicates whether I/O or memory operation is carried out. S_1 and S_0 indicate the type of machine cycle in progress.

RESET Signals

(*i*) $\overline{\text{RESET IN}}$: A low on this pin sets the program counter to (OOOOH) resets the interrupt enable and HLDA flip-flops, tri-states all the buses.

(*ii*) RESET OUT : A active high on this pin indicates that processor is being reset. This signal is synchronized to the processor clock and it can be used to reset other devices connected in the system.

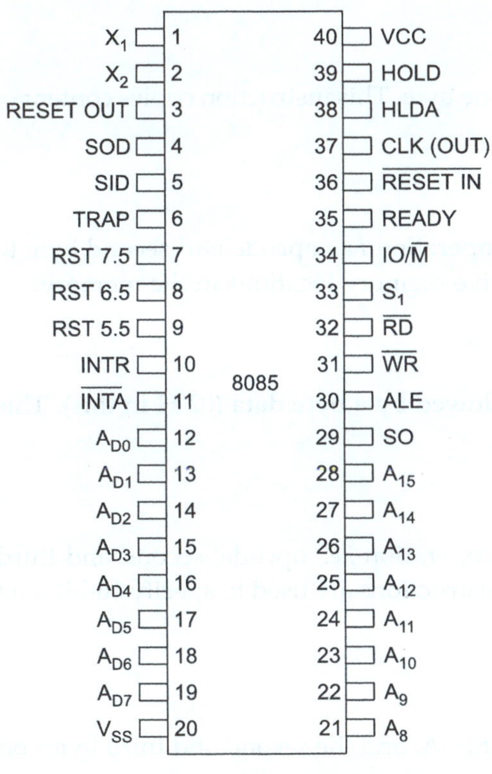

(*a*) Pin configuration

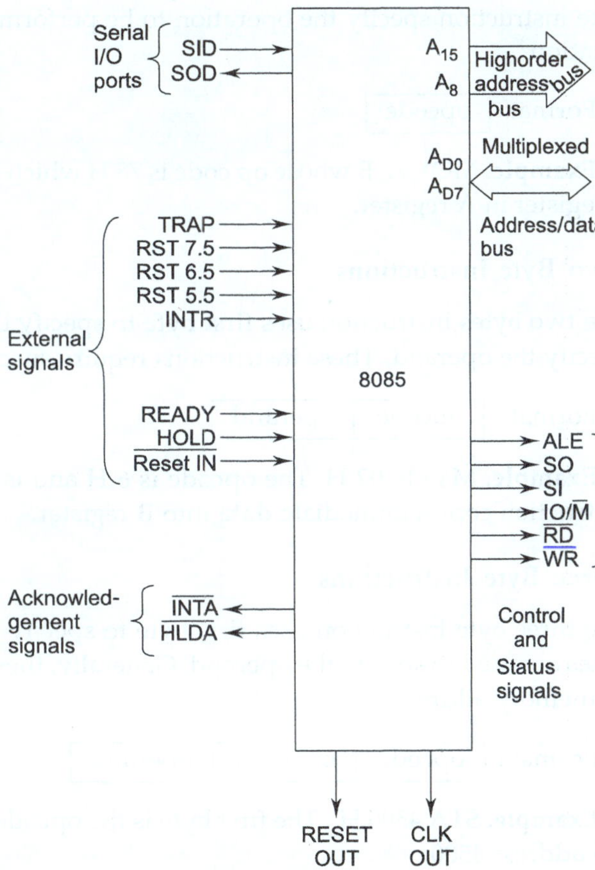

(*b*) Functional pin diagram

Fig. 4.4

4.7 INSTRUCTION FORMATS

- An instruction is a binary pattern designed inside a microprocessor to perform a specific function. The entire group of instructions called the instruction set, determines what functions the microprocessor can perform.
- Each instruction basically consists of two parts. The first part is called as opcode and the second is called as operand.

Types of Instruction

Depending on the number of bytes or bit patterns required to specify an operation and the data the instruction are of 3 types :

(*a*) One byte instructions
(*b*) Two byte instructions
(*c*) Three byte instructions

One Byte Instructions

A one byte instruction includes the opcode and the operand in the 8-bit only *i.e.,* one byte. One byte instruction specify the operation to be performed. These instructions, require one memory location

Format:

opcode

Example. MOV A, B whose op code is 78 H which is one byte. This instruction copies content of B register in A register.

Two Byte Instructions

The two bytes instruction uses first byte to specify the operation *i.e.,* opcode and second byte to specify the operand. These instructions require successive memory locations in the memory.

Format :

opcode	operand

Example. MVI B, 02 H. The opcode is 6 H and is followed by a byte data (02 H in this). This instruction copies immediate data into B register.

Three Byte Instructions

The three byte instruction uses first byte to specify the operation *i.e.,* opcode, second and third bytes are used to specify the operand. Generally, these instructions are used to specify 16-bit data or memory address.

Format :

opcode	operand	operand

Example. STA 4500 H : The first byte is the opcode for STA, and the second and third bytes on the address 4500.

4.8 ADDRESSING MODES

The various formats for specifying operands are called the addressing modes. For 8085, they are :

1. Immediate,
2. Register,
3. Direct,
4. Indirect,
5. Implied.

1. **Immediate addressing :** In an immediate addressing mode, 8 or 16-bit data can be specified as a part of instruction. In 8085, the instructions having 'I' letter fall under this category. 'I' indicate Immediate addressing.

   Example. MVI A, 20 H : Moves 8-bit immediate data to accumulator.

   LXI D, 203FH : moves 16-bit immediate data into DE.
2. **Register addressing :** This (mode) register pair (D = 20, E = 3 F) ; specifies the source operand ; destination operand or both to be contained in 8085 registers.

   Example. MOV A, B : Moves contents of B register to accumulator.

   ADD C : Adds C contents to accumulator content and results gets stored in accumulator.
3. **Direct addressing :** This mode specifies the 16-bit address of the operand within the instruction itself.

   Example. LDA 4000 H : Loads contents of memory location 4000 into the accumulator.
4. **Indirect addressing mode :** In indirect addressing mode, the memory address when the operand located is specified by the content of register pair.

   Example. MOV M, A : Story the contents of accumulator to the memory location pointed by HL register pair.
5. **Implied addressing mode :** In implied addressing mode, opcode specifies the address of the operands.

   Example. CMA : Complements constants of accumulator.

4.9 8085 INSTRUCTION SET

The 8085 instruction set can be categorised into five different groups based on the nature of functions the instruction carry out :

(*a*) Data transfer group

(*b*) Arithmetic group

(*c*) Logic group

(*d*) Branch group

(*e*) Stock, Input/Output and Machine Control Group.

Data Transfer Group

The data transfer group do not affect any flag in the flag register. This group load the given data into register, copy data from register to memory location, and copy data from register to register. This group consists of following set of instructions.

1. MVI r, data (8-bit)
2. MVI M, data (8-bit)
3. MOV rd, rs
4. MOV rd, M
5. MOV M, rs
6. STA addr

7. LDA addr
8. LXI rp, data (16-bit)
9. SHLD addr
10. LHLD addr
11. STAX rp
12. STAX rp
13. XCHG

1. MVI r, data (8) : Move 8-bit data immediately to register r.
   Operation : r ← 8-bit data (byte)
   Description : This instruction directly loads a specified register with an 8-bit data given within instruction. (r may be A, B, C, D, E, H and L).
   No. of bytes : 2 bytes.
   Addressing mode : Immediate addressing.
   Flags : Not affected.
2. MVI M, data (8) : Move 8-bit immediate data to memory whose address is in HL register pair.
   Description : This instruction directly loads an 8-bit data given within the instruction into a memory location.
   No. of bytes : 2 bytes
   Addressing mode : Immediate and indirect addressing
   Flags : Not affected.
3. MOV rd, rs : Move data from source register (rs) to destination register (rd)
   Description : This instruction copies data from the source register into destination register. rd and rs may be A, B, C, D, E, H and L.
   No. of bytes : 1 byte.
   Addressing mode : Register addressing
   Flags : Not affected.
4. MOV M, rs : Move data from source register (rs) to memory whose address is in HL register pair.
   Description : This instruction copies data from the source register into memory location pointed by HL register pair. rs is an general purpose register such as A, B, C, D, E, H and L.
   No. of bytes : 1 byte
   Addressing mode : Indirect addressing
   Flags : Not affected.
5. MOV rd, M : Move data from memory location specified by HL register pair to the destination register (rd).
   Description : This instruction copies data from memory location whose address is specified by HL register pair into destination register. rd may be A, B, C, D, E, H and L.
   No. of bytes : 1 byte
   Addressing mode : Indirect addressing
   Flags : Not affected.
6. LXI rp, data (16) : Load 16-bit immediate data to specified register pair.
   Description : This instruction loads immediate 16-bit data specified within the instruction into register pair or stack pointer.
   The rp is 16-bit register pair such as BC, DE, HL or stack pointer.
   No. of bytes : 3 bytes.

Addressing mode : Immediate addressing.

Flags : Not affected.

7. STA addr : Store the contents of A register at address given within the instruction.

   Description : This instruction stores the contents of A register into the memory location whose address is directly specified within the instruction.

   No. of bytes : 3 bytes

   Addressing mode : Direct addressing

   Flags : Not affected.

8. LDA addr : Load data into A register directly from the address given within the instruction.

   Description : This instruction copies the contents of the memory location whose address is given within the instruction into the accumulator.

   No. of bytes : 3 bytes

   Addressing mode : Direct addressing

   Flags : Not affected.

9. SHLD addr : Store HL register part in memory.

   Description : This instruction stores the contents of L register in the memory location given within the instruction and contents of H register at address next to it.

   No. of bytes : 3 bytes

   Addressing mode : Direct addressing

   Flags : Not affected

10. LHLD addr : Load HL register pair from memory.

    Description : This instruction copies the contents of the memory location given within the instruction into the L register and the contents of the next memory location into the H register.

    No. of bytes : 3 bytes.

    Addressing mode : Direct addressing

    Flags : Not affected

11. STAX rp : Stores the content of A register in memory location whose address is specified by BC or DE register pair.

    Description : This instruction copies the contents of accumulation into the memory location whose address is specified by the specified register pair. The rp is BC or DE register pair.

    No. of bytes : 1 byte

    Addressing mode : Register indirect addressing

    Flags : Not affected

12. LDAX rp : Load A register with the contents of memory location whose address is specified by BC or DE register pair.

    Description : This instruction copies the contents of memory location whose address is specified by the register pair into the accumulator. The rp is BC or DE register pointer.

    No. of bytes : 1 byte.

    Addressing mode : Register indirect addressing.

    Flags : Not affected.

13. XCHG : Exchange the contents of H with D and L with E.
    Description : This instruction exchanges the contents of the register H with that of D and of L with that of E. [H ↔ D and L ↔ E].
    No. of bytes : 1 byte
    Addressing mode : Register addressing
    Flags : Not affected

Arithmetic Group

Arithmetic group includes add, subtract, increment or decrement data in registers or in memory. In addition, there is one instruction which adjusts 8-bit data to form BCD numbers.

1. ADD r
2. ADD M
3. ADI data (8)
4. ADC r
5. ADC M
6. ACI data (8)
7. DAD rp
8. SUB r
9. SUB M
10. SUI data
11. SBB r
12. SBB M
13. SBI data
14. DAA
15. INR r
16. INR M
17. INX rp
18. DCR r
19. DCR M
20. DCX rp

1. Add r : Add register r to accumulator.
   Description : This instruction adds the contents of the specified register to the contents of accumulator and stores result in the accumulator (r may be A, B, C, D, E, H or L).
   No. of bytes : 1 byte
   Addressing mode : Register addressing
   Flags : All flags are affected
2. ADD M : Add data in memory to accumulator
   Description : This instruction adds the contents of the memory location pointed by HL register pair to the contents of accumulator and stops result in the accumulator.
   No. of bytes : 1 byte
   Addressing mode : Register indirect addressing.
   Flags : All flags are affected
3. ADI Data (8) : Add immediate 8-bit data to accumulator.
   Description : This instruction adds the 8-bit data given within the instruction to the contents of accumulator and stores result in accumulator.
   No. of bytes : 2 bytes.
   Addressing mode : Immediate addressing.
   Flags : All flags are affected.
4. ADC r : Add register r with carry to register A.
   Description : This instruction adds the contents of specified register to the contents of accumulator with carry. This means, if the carry flag is set by some previous operation,

it adds 1 and the contents of the specified register to the contents of accumulator, else it adds the contents of the register only.

No. of bytes : 1 byte

Addressing mode : Register addressing

Flags : All flags are affected

5. ADC M : Add data in memory to accumulator with carry.

   Description : This instruction adds the contents of the memory location pointed by HL register pair to the contents of accumulator with carry and stores the result in the accumulator.

   No. of bytes : 1 byte

   Addressing mode : Register indirect addressing

   Flags : All flags are affected

6. ACI data (8) : Add 8-bit immediate data to accumulator with carry.

   Description : This instruction adds 8-bit data given within the instruction to the contents of accumulator with carry and stores the result in the accumulator.

   No. of bytes : 2 bytes.

   Addressing Mode : Immediate addressing

   Flags : All flags are affected.

7. DAD rp : Add register pair rp to HL register pair.

   Description : This instruction adds the contents of the specified register pair to the contents of the HL register pair and stores the result in the HL register pair. The rp is 16-bit register pair. Such as BC, DE, HL or stack pointer.

   No. of bytes : 1 byte

   Addressing mode : Register addressing.

   Flags : Only carry flag is affected.

8. SUB r : Subtract specified register from accumulator.

   Description : This instruction subtracts the contents of the specified register from the contents of the accumulator and stores the result in the accumulator. r is an 8-bit general purpose register such as A, B, C, D, E, H and L.

   No. of bytes : 1 byte

   Addressing mode : Register addressing

   Flags : All flags are affected. If carry flag = 1, result is negative and 1s to 2s complement form and it carry flag = 0 result is positive.

9. SUB M : Subtract data in memory from accumulator.

   Description. This instruction subtracts the contents of the memory location pointed by HL register pair from the contents of accumulator and stores the results in the accumulator.

   No. of byte : 1 byte.

   Addressing mode : Register indirect addressing.

   Flags : All flags are affected.

10. SUI Data (8) : Subtract 8-bit immediate data from the accumulator.

    Description : This instruction subtracts an 8-bit data given within the instruction from the contents of the accumulator and story the result in the accumulator.

No. of bytes : 2 bytes.

Addressing mode : Immediate addressing.

Flags : All flags are attached.

11. SBB r : Subtract specified register and borrow flag (carry flag) from accumulator.

Description : This instruction subtracts the specified register contents and borrow flag from the accumulator contents. This means, if the borrow flag (carry) is set by some previous operation, it subtracts 1 and the contents of the specified register from the contents of the accumulator, else it subtracts the contents of the specified register only.

No. of bytes : 1 byte.

Addressing mode : Register addressing

Flags : All flags are attached.

12. SBB M : Subtract memory data and borrow flag from accumulator.

Description : This instruction subtract the contents of memory location pointed by HL register pair from the contents of accumulator and borrow flag and stores result in accumulator.

No. of bytes : 1 byte.

Addressing mode : Register indirect addressing

Flags : All flags are affected.

13. SBI data (8) : Subtract 8-bit immediate data and borrow flag from accumulator.

Description : This instruction subtracts 8-bit data given within the instruction and borrow flag from the contents of accumulator and stores the result in the accumulator.

No. of bytes : 2 byte.

Addressing mode : Immediate addressing.

Flags : All flags are affected.

14. DAA : Decimal Adjust Accumulator

This instruction adjusts accumulator to packed BCD (binary coded decimal) after adding two BCD numbers.

Instruction works as follows :

1. If the value of the low-order four bits ($D_3 - D_0$) in the accumulator its greater than 9 or if auxiliary carry flag is set, the instruction adds 6 (06) to the low-order four bits.
2. If the value of the high-order four bits ($D_7 - D_4$) in the accumulator is greater than 9 or it carry flag is set, the instruction adds 6 (60) to the high-order bits.

No. of bytes : 1 byte.

Flags : All flags are affected.

Example :

If A = 0 0 1 1 1 0 0 1 = 39 BCD

and C = 0 0 0 1 0 0 1 0 = 12 then

ADDC : Gives A = 0 1 0 0 1 0 1 1 = $4B_H$

DAA : adds 6 [0 1 1 0] because 1 0 1 1 > 9,

A = 0 1 0 1 0 0 0 1 = 51 [BCD]

15. INR r : Increment specified register :

Description : This instruction increments the contents of specified register by 1. The result is stored in the same register.

No. of bytes : 1 byte.

Addressing mode : Register addressing.

Flags : All flags are affected.

16. INR M : Increment data in memory

    Description : This instruction increments the contents of memory location pointed by HL register pair by 1. The result is stored in same memory location. HL act as memory pointer.

    No. of bytes : 1 byte.

    Addressing mode : Register indirect addressing.

    Flags : All flag except carry flag are affected.

17. INX rp : Increment specified register pair.

    Description : This instruction increment the contents of register pair by one. The result is stored in the same register pair. The rp is register pair such as BC, DE, HL or Stack Pointer (SP).

    No. of bytes : 1 byte.

    Addressing mode : Register addressing.

    Flags : No flags are attached.

18. DCR r : Decrement specified register

    Description : This instruction decrements the contents of the specified register by one. It stores the result in the same register. r is 8-bit general purpose register such as A, B, C, D, E, H and L.

    No. of bytes : 1 byte.

    Addressing mode : Register addressing.

    Flags : All flags except carry flag are affected.

19. DCR M : Decrement data in memory

    Description : This instruction decrements the content of memory location pointed by HL register pair by result is stored in same memory location.

    No. of bytes : 1 byte.

    Addressing mode : Register indirect addressing.

    Flags : All flags except carry flag are affected

20. DCX rp : Decrement specified register pair.

    Description : This instruction decrement the contents of register pair by one. The result is stored in the same register pair. rp is register pair such as BC, DE and HL or stack pointer

    No. of bytes : 1 byte.

    Addressing mode : Register addressing.

    Flags : No flags are affected.

Branch Group

This group includes :

- Jump instructions
- Call and return instructions
- Restart instructions.

In this section only jump instructions are explained. The call, return and restart instructions are associated with stack operation, hence discussed in stack group of instructions.

Jump : This group consists of the following set of instructions

1. JMP addr
2. J condition addr
3. PCHL

1. JMP addr : Jump unconditionally to the address.

   Description : This instruction loads the PC with the address given within the instruction and resumes the program execution from this location.

   No. of bytes : 3 bytes.

   Addressing mode : Immediate addressing.

   Flags : Not affected.

2. J condition addr : Jump conditionally to the address.

   Description : This instruction causes a jump to an address given in the instruction if the desired condition occurs in the program before the execution of the instruction.

   Possible Conditions for Jumps

Instruction Code	Description	Condition for Jump
JC	Jump on carry	CY = 1
JNC	Jump on not carry	CY = 0
JP	Jump on positive	S = 0
JM	Jump on minus	S = 1
JPE	Jump on parity even	P = 0
JPO	Jump on parity odd	P = 1
JZ	Jump on zero	Z = 1
JNZ	Jump on not zero	Z = 0

   No. of bytes : 3 bytes.

   Addressing mode : Immediate addressing.

   Flags : Not affected, only checked.

3. PCHL : Load HL contents into program counter.

   Description : This instruction loads the contents of HL register pair into the program counter. Thus, the program control is transferred to the location whose address is in HL register pair.

   No. of bytes : 1 byte

   Addressing mode : Register addressing

   Flags : Not affected.

Logic Group

This group perform operations such as AND, OR and XOR, compare data between registers or between register and memory, rotate and complement data in registers.

This group consists of the following set of instructions :

1. ANA r	2. ANA M
3. ANI data	4. XRA r
5. XRA M	6. XRI data
7. ORA r	8. ORA M
9. ORI data	10. CMP r
11. CMP M	12. CPI data
13. STC	14. CMC
15. CMA	16. RLC
17. RRC	18. RAL
19. RAR	

1. ANA r : AND specified register with accumulator.

   Description : This instruction logically AND's the contents of the specified register with the content of accumulator and stores the result in the accumulator. Each bit in the accumulator is logically AND cd with the corresponding bit to register r *i.e.,* D_0 in bit in A with D_0 in r, D_1 in A with D_1 in r upto D_7 bit. r is a 8-bit general purpose register such as A, B, C, D, E, H and L.

   No. of bytes : 1 byte

   Addressing mode : Register addressing

   Flags : All flags are affected with Cy flag = 0 and auxiliary carry flag = 1.

2. ANA M : AND data in memory with accumulator.

   Description : This instruction logically ANDs the content of memory location pointed by HL pair with contents of accumulator and results get stored in accumulator.

   No. of bytes : 1 byte.

   Addressing mode : Register indirect addressing

   Flags : All flags are affected with Cy flag = 1 and auxiliary carry flag = 1.

3. ANI data : AND 8-bit immediate data with accumulator.

   Description : This instruction logically ANDs the 8-bit data given in the instruction with the contents of the accumulator and stores the result in the accumulator.

   No. of byte : 2 bytes

   Addressing mode : Immediate addressing

   Flags : All flags are affected with Cy flag = 0 and auxiliary carry flag = 1.

4. XRA r : XOR specified register with accumulator.

   Description : This instruction logically XORs the contents of the specified register with the contents of accumulator and stores the result in the accumulator.

   No. of bytes : 1 byte.

   Addressing mode : Register addressing.

   Flags : All flags are affected with carry flag = 0 and auxiliary carry flag = 0.

5. XRA M : XOR data in memory with accumulator.

   Description : This instruction logically XOR's the contents of memory location pointed by HL register pair with the contents of accumulator.

   No. of bytes : 1 byte.

   Addressing mode : Register indirect addressing.

   Flags : All flags are affected with carry flag = 0 and auxiliary flag = 0.

6. XRI data : XOR 8-bit immediate data with accumulator.

   Description : This instruction logically XORs the 8-bit data given in the instruction with the contents of accumulator and stores the result in the accumulator.

   No. of bytes : 2 bytes.

   Addressing mode : Immediate addressing

   Flags : All flags are affected with carry flag = 0 and auxiliary carry flag = 0.

7. ORA r : OR specified register with accumulator.

   Description : This instruction logically ORs the contents of specified register with the contents of accumulator and stores the result in the accumulator. Each bit in the accumulator is OR cd with corresponding bit in register r.

   No. of bytes : 1 byte.

   Addressing mode : Register addressing.

   Flags : All flags are affected with carry flag = 0 and auxiliary carry flag = 0.

8. ORA M : OR data in memory with accumulator.

   Description : This instruction logically ORs the content of memory location pointed by HL register pair with the contents of accumulator and result is stored in accumulator.

   No. of bytes : 1 byte.

   Addressing mode : Register indirect addressing.

   Flags : All flags are affected with carry flag = 0 and auxiliary carry flag = 0.

9. ORI data : OR 8-bit immediate data with accumulator.

   Description : This instruction logically ORs the 8-bit data given in the instruction with the contents of the accumulator and stores the result in the accumulator.

   No. of bytes : 2 bytes.

   Addressing mode : Immediate addressing.

   Flags : All flags are affected with carry flag = 0 and auxiliary carry flag = 0.

10. CMP r : Compare specified register with accumulator.

    Description : This instruction subtracts the contents of the specified register from the contents of the accumulator and sets the condition flags as a result of the subtraction. It sets zero flag if A = r and sets carry flag if A < r.

    The result of the subtraction is not stored in the accumulator, so accumulator contents remain unchanged.

    No. of bytes : 1 byte.

    Addessing mode : Register addressing.

    Flags : All flags are affected

    $$A > r : CY = 0, Z = 0$$
    $$A = r : CY = 0, Z = 1$$
    $$A < r : CY = 1, Z = 0$$

11. CMP M : Compare data in memory with accumulator.

    Description : This instruction subtracts the contents of the memory location specified by HL pair from the contents of the accumulator and sets the condition flags as a result of subtraction. It sets zero flag if A = M and sets carry flags if A < M.

    The result of the subtraction is not stored in the accumulator. So, accumulator contains remain unchanged.

No. of bytes : 1 byte.

Addressing mode : Register indirect addressing

Flags : All flags are affected.

$$A > M : CY = 0, Z = 0$$
$$A = M : CY = 0, Z = 1$$
$$A < M : CY = 1, Z = 0$$

12. CPI data : Compare immediate 8-bit data with accumulator.

Description : This instruction subtracts the 8-bit data given in the instruction from the contents of the accumulator and sets the condition flags as a result of subtraction. It sets zero flag if A = data and sets carry flag if A < data. Accumulator content remain unchanged.

No. of bytes : 2 bytes

Addressing mode : Immediate addressing

Flags : All flags are affected

$$A > \text{data} : CY = 0, Z = 0$$
$$A = \text{data} : CY = 0, Z = 1$$
$$A < \text{data} : CY = 1, Z = 0.$$

13. STC : Set carry

Description : This instruction sets carry flag = 1.

No. of bytes : 1 byte

Flags : Carry flag is set, all other flags are unaffected.

14. CMC : Complement carry

Description : This instruction complements the carry flag

No. of Bytes : 1 byte.

Flags : Carry flag is complemented, all flags are unaffected.

15. CMA : Complement A

Description : This instruction complements each bit of the accumulator.

No. of bytes : 1 byte

Addressing mode : Implied addressing

Flags : Flags are not affected.

16. RLC : Rotate Accumulator Left.

Operation :

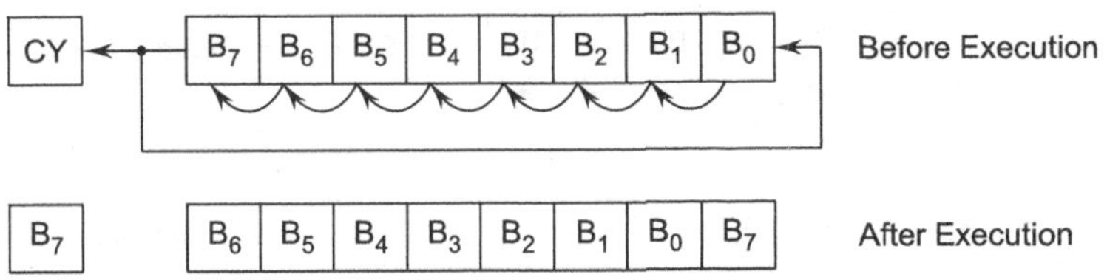

Description : Ths instruction rotates the contents of the accumulator left by one position. Bit B_7 is placed in B_0 as well as in CY.

No. of bytes : 1 byte

Addressing mode : Implied addressing

Flags : Carry flag is affected. All other flags are unaffected.

17. RRC : Rotate accumulator right.

Operation :

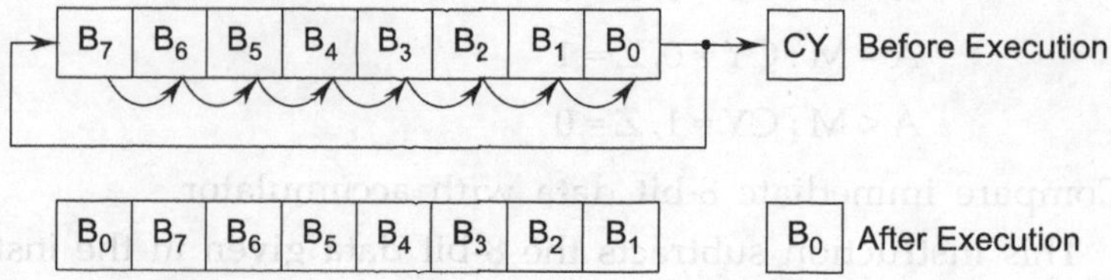

Description : This instruction rotates the contents of the accumulator right by one position. Bit B_0 is placed in B_7 as well as in CY.

No. of bytes : 1 byte

Addressing mode : Implied addressing.

Flags : Carry flag is affected. All other flags are unaffected.

18. RAL : Rotate accumulator left through carry

Description : This instruction rotates the contents of the accumulator left by one position. Bit B_7 is placed in CY and CY is placed in B_0.

No. of bytes : 1 byte

Addressing mode : Implied addressing

Flags : Carry flag is affected. All other flags are unaffected.

Operation :

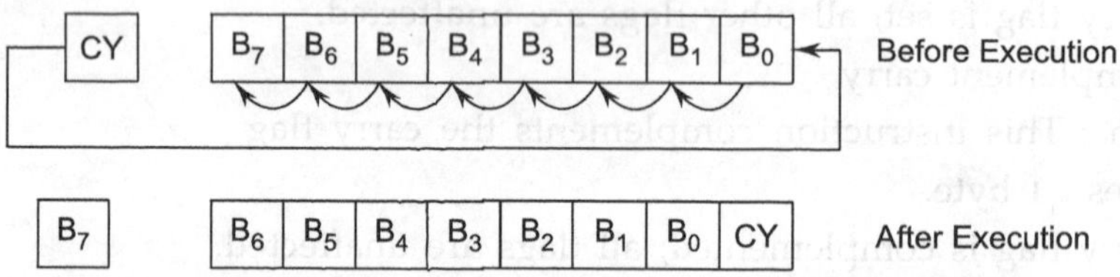

19. RAR : Rotate accumulator right through carry.

Description : This instruction rotates the contents of the accumulator right by one position bit B_0 is placed in CY and CY is placed in B_7.

No. of bytes : 1 byte.

Addressing mode : Implied addressing.

Flags : Carry flag is affected. All flags are unaffected.

Operation :

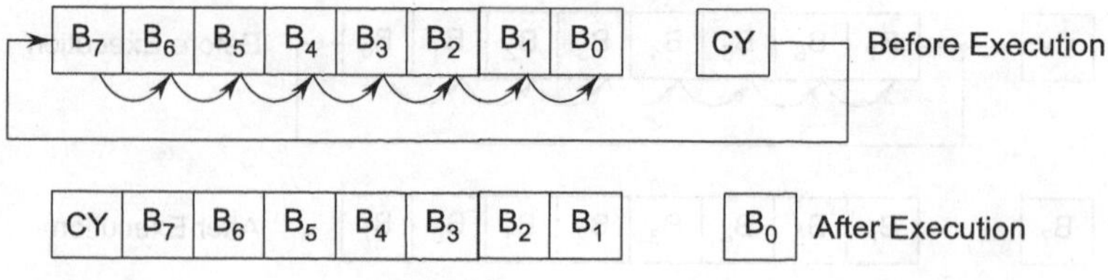

Stack Operations, Input/Output and Machine Control Group

The stack is a part of read/write memory that is used for temporary storage of binary information during the execution of the program. The binary information is basically the immediate results and the return address in case of the subroutine calls.

Need for stack : For the application programs, the internal memory of the microprocessor is not sufficient to store the intermediate results. These intermediate results can be stored temporarily on the stack and can be referred back when required.

What is a subroutine ?

A subroutine is a group of instructions, performs a particular subtask which is executed number of times. The use of subroutines is a very important technique in designing software for microprocessor systems because it eliminates the need to write subtasks repeatedly.

Stack Operations

This group consists of the following set of instructions :

1. PUSH and POP instructions
2. CALL and RETURN instructions
3. RESTART instructions
4. XTHL
5. SPHL.

1. PUSH rp : Stores the contents of specified register pair on the stack.

   Description : This instruction decrements stack pointer by one and copies the higher byte of the register pair into the memory location pointed by the stack pointer. It then decrements the stack pointer again by one and copies the lower byte of the register pair into the memory location pointed by the stack pointer.

   No. of bytes : 1 byte.

   Addressing mode : Register indirect addressing

   Flags : Flags are not affected.

2. PUSH PSW : Stores the contents of program status word and accumulator on the stack.

   Description : PSW is program status word. The A register acts as higher order register and flag register as low order register. When this instruction is executed the contents of the specified register pair are copied onto stack in this sequence.

   1. The stack pointer is decremented by one and the contents of higher order register of register pair are copied to that location.
   2. The stack pointer is again decremented by 1 and the contents of lower order register of register pair are copied at that location.

   No. of bytes : 1 byte.

   Addressing mode : Register indirect addressing mode.

   Flags : No flags are modified.

   Operation :
   $$SP = SP - 1$$
   $$(SP) = A$$
   $$SP = SP - 1$$
   $$(SP) = F.$$

3. POP rp : Loads specified register pair with 16-bit data from the stack.

   Description : This instruction copies the contents of memory location pointed by the stack pointer into the lower order byte of the specified register pair and increments the stack pointer by one. It then copies the contents of memory location pointed by stack pointer into the higher byte of the specified register pair and increments the stack pointer again by one.

   No. of bytes : 1 byte.

   Addressing mode : Register indirect addressing.

   Flags : Flags are not affected.

4. POP PSW : Loads program status word and accumulator with 16-bit from the stack.

   Description : PSW is program status word. The A register acts as higher order register and flag register as low order register.

   1. When this instruction is executed, the contents of the memory location pointed by the Stack Pointer (SP) register, are copied into flag register.
   2. The SP is incremented by one and contents of that memory location are copied to the register A.
   3. The SP is again incremented by 1.
   4. The process of POP is exactly opposite to PUSH. So the contents stored by PUSH are taken back using POP instructions.

   No. of bytes : 1 byte.

   Addressing mode : Register indirect addressing mode.

   Flags : No flags are modified.

5. CALL addr : Call unconditionally a subroutine whose starting address is given within the instruction.

   Description : The CALL instruction is used to transfer program control to a subprogram or subroutine. This instruction pushes the current PC contents onto the stack and loads the given address into the PC. Thus the program control is transferred to the address given in the instruction stack pointer is decremented by two. When the subroutine is called, the program control is transferred to subroutine. After execution the program control is transferred again to the calling program.

   No. of bytes : 3 bytes.

   Addressing mode : Immediate register indirect addressing.

   Flags : Flags are not affected.

6. RET : Return from the subroutine unconditionally.

   Description : This instruction pops the return address (address of the instruction next to CALL in the main program) from the stack and loads program counter with this return address. Thus transfers program control the instruction next to CALL in the main program.

   No. of bytes : 1 byte.

   Addressing mode : Register indirect addressing

   Flags : Flags are not affected.

7. Conditional call : Call the subroutine conditionally at given addr.

   Description : This instruction call the subroutine at the given address if a specified condition is satisfied. Before call it stores the address of instruction next to the call on the stack and decrements stack pointer by two. The following table shows the possible condition for calls.

Table 4.1. Conditions for CALL

Instruction Code	Description	Condition for Call
CC	Call on carry	CY = 1
CNC	Call on not carry	CY = 0
CP	Call on positive	S = 0
CM	Call on minus	S = 1
CPE	Call on parity even	P = 1
CPO	Call on parity odd	P = 0
CZ	Call on zero	Z = 1
CNZ	Call on not zero	Z = 0.

No. of bytes : 3 bytes.

Addressing mode : Immediate register indirect addressing.

Flags : Flags are not affected. Flags are checked.

8. R Condition : Return from the subroutine conditionally.

Description : This instruction returns the control to the main program if the specified condition is satisfied. The following table shows the possible conditions for return.

Table 4.2. Conditions for return

Instruction Code	Description	Condition for RET
RC	Return on carry	CY = 1
RNC	Return on not carry	CY = 0
RP	Return on positive	S = 0
RM	Return on minus	S = 1
RPE	Return on parity even	P = 1
RPO	Return on parity odd	P = 0
RZ	Return on zero	Z = 1
RNZ	Return on not zero	Z = 0

No. of bytes : 1 byte.

Addressing mode : Register indirect addressing.

Flags : Flags are not affected.

9. RESTART : This group consists of restart instructions.

RST n : Restart n (0 to 7)

Description : This instruction transfers the program control to the specific memory address as shown in table. This instruction is like a fixed address CALL instruction. These fixed addresses are also referred to as vector address. The processor multiply the RST number by 8 to calculate these vector addresses. Before transferring the program control to the instruction following the vector address RST instruction saves the current PC contents on the stack.

Table 4.3. Vector addresses for RST instructions

Instruction Code	Vector Address
RST 0	0 × 8 = 0 0 0 0 H
RST 1	1 × 8 = 0 0 0 8 H
RST 2	2 × 8 = 0 0 1 0 H
RST 3	3 × 8 = 0 0 1 8 H

RST 4	$4 \times 8 = 0\,0\,2\,0$ H
RST 5	$5 \times 8 = 0\,0\,2\,8$ H
RST 6	$6 \times 8 = 0\,0\,3\,0$ H
RST 7	$7 \times 8 = 0\,0\,3\,8$ H.

No. of bytes : 1 byte

Addressing mode : Register indirect addressing.

Flags : Flags are not affected.

10. XTHL : Exchange top of stack with H and L.

    Description : This instruction exchanges the contents of memory location pointed by the stack pointer with the contents of L register and the contents of the next memory location with the contents of H register. It does not modify stack pointer.

    No. of bytes : 1 byte.

    Addressing mode : Register indirect addressing

    Flags : Flags are not affected.

11. SPHL : Move data from HL to stack pointer.

    Description : This instruction copies the contents of HL register pair into the stack pointer. The contents of H register are copied to higher order byte of stack pointer and contents of L register are copied to the lower byte of SP.

    No. of bytes : 1 byte.

    Addressing mode : Register addressing.

    Flags : Register addressing.

Input/Output

This group consists of following set of instructions :

1. IN addr (8-bit) : Copy data from input port into accumulator.

   Description : This instruction copies the data at the port whose address is specified in the instruction into the accumulator.

   No. of bytes : 2 bytes.

   Addressing mode : Direct addressing.

   Flags : Flags are not affected.

2. OUT addr (8-bit) : Send data to the output port.

   Description : This instruction sends the contents of accumulator to the output port whose address is specified within the instruction.

   No. of bytes : 2 bytes.

   Addressing mode : Direct addressing.

   Flags : Flags are not affected.

Machine Control

This group consists of following set of instructions

1. EI
2. DI
3. NOP
4. HLT
5. SIM
6. RIM

1. EI : Enable Interrupt :

   Description : This instruction sets the interrupt enable flip-flop to enable interrupts when the microprocessor is reset or after interrupt acknowledge, the interrupt enable flip-flop is reset.

   No. of bytes : 1 byte.

   Addressing mode : None

   Flags : Flags are not affected.

2. DI : Disable Interrupts

   Description : This instruction resets the interrupt enable flip-flop to disable interrupts. This instruction disables all interrupts except TRAP since TRAP is non-maskable interrupt.

   No. of bytes : 1 byte

   Addressing mode : None

   Flags : Flags are not affected

3. NOP : No operation

   Description : No operation is performed

   No. of bytes : 1 byte

   Addressing mode : None

   Flags : Not affected

4. HLT : Halt the processor

   Description : This instruction halts the processor. It can be restarted by a valid interrupt or applying RESET signal.

   No. of Bytes : 1 byte

   Addressing Mode : None

   Flags : Not affected

5. SIM : Set Interrupt Mask

   Description : This instruction masks the interrupts as desired. It also sends out serial data through the SOD pin. For this instruction command byte must be loaded into accumulator.

   The pattern for command byte is

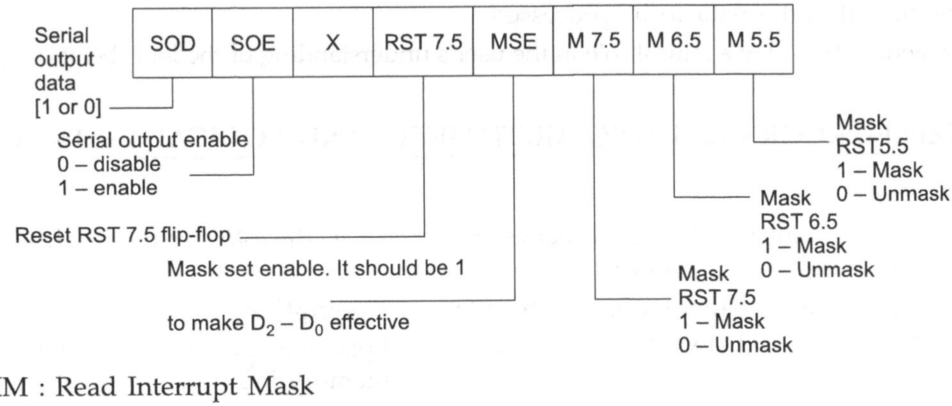

6. RIM : Read Interrupt Mask

   Description : This instruction copies the status of the interrupt into the accumulator. It also reads the serial data through the SID pin. The pattern of the stator byte is :

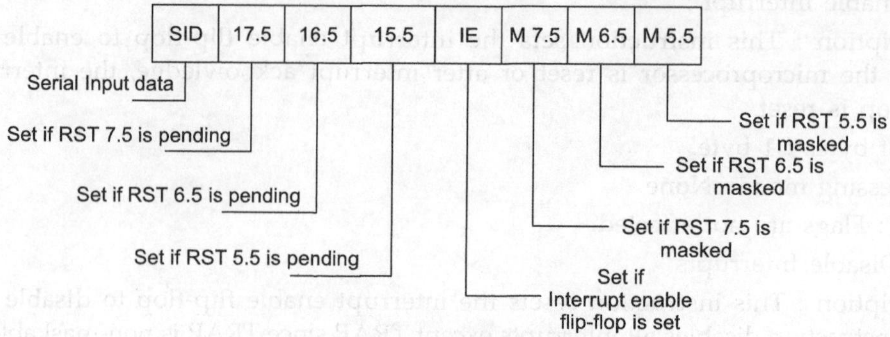

4.10 PROGRAMS USING MICROPROCESSOR 8085

A program is a sequence of instructions to be executed for performing the task of a user. The program and data are fed into the microprocessor via input device and the program is stored in the memory of the microprocessor. Microprocessor has ALU which performs the necessary arithmetic and logic operations and the result is stored in memory which has to be displayed in output device. Micorprocessor instruction are written in Hex lode or by using Mnemonics. A source program written in the Mnemonic code is called assembly language and the software which translate this source program into machine language is called an assembler.

To write a program for performing a task, the following steps are required.

Step 1. Write as algorithm to solve the given problem. Algorithm is defined as a precise statement of the procedures required for solving a problem.

Step 2. Represent the steps in the algorithm and the sequence in the pictorial form called flow chart.

Step 3. Convert the blocks in the flow chart into 8085 operations which is called an Assembly Language Program.

Assembly Language Program has five sections :

1. Memory address : 16-bit addresses of the user to store the program.
2. Machine code : Hexadecimal numbers (instruction codes) to be entered.
3. Opcode : Opcodes are the symbols to indicate the operation of the function.
4. Operand : It is the data to be processed.
5. Comments : This play a critical role in the user's understanding of the logic behind a program.

4.11 SIMPLE PROGRAMS USING ARITHMETIC AND LOGICAL INSTRUCTIONS

8-bit addition

Example 4.1. *Store the data 25 H and 37 H in memory locations 4000 and 4001 respectively. Add them and store the result in the memory location 4600.*

Memory location	Mnemonics	Opcode	Description
4200	LDA 4000	3 A	Load Accumulator with first data from memory location
4201		00	
4202		42	

4203	MOV B, A	47	Move data from accumulator to B region.
4204	LDA 4201	3 A	Load accumulator with second data from the memory location
4205		01	
4206		42	
4207	ADDB	80	Add the content of B register with accumulator
4208	STA 4600	32	Store the result to memory location from accumulator
4209		00	
420A		46	
420B	HLT	76	Stop execution.

Example 4.2. *Add two numbers 25 H and 37 H; store the results in the memory locations 4600 and 4601.*

Memory Location	Label	Mnemonics	Opcode	Description
4001		LXI H, 4200	21	Load HL register pair with the address of the first number
4002			00	
4003			42	
4004		MVI C, 00	0 E	Set the reg. C to zero
4004			00	
4005		MOV A, M	7 E	Move the data to the accumulator
4006		INX H	23	Increment the HL pair
4007		ADD M	86	Add the two numbers
4008		JNC Loop	02	Check for carry
4009			0C	
400A			40	
400B		INR C	0C	Increment the reg. C
400C	Loop	STA 4600	32	Store the result to the memory location
400D			00	
400E			46	
400F		MOV A, C	79	Move the content of reg. C to accumulator
4010		STA 4601	32	Store the result to the next memory location
4011			01	
4012			46	
4013		HLT	76	Stop execution.

16-bit addition :

Example 4.3. *Store the data 2424 H and 2011 H in memory locations 4200 to 4203 respectively. Add them and store the result in memory locations 4500, 4501 and 4502 H.*

Memory Location	Label	Mnemonics	Opcode	Description
4001		LHLD 4200	2 A	Load HL pair with 16-bit data

4002			00	
4003			42	
4004		XCHG	EB	Move the data into DE pair
4005		LHLD 4202	2 A	Load HL pair with next 16-bit data
4006			02	
4007			42	
4008		MVI A, 00	3 E	Clear accumulator
4009			00	
400 A		DADD	19	Add two datas (16-bit addition)
400 B		JNC LOOP	D2	Check for carry
400 C			OF	
400 D			40	
400 E		INR A	3 C	Increment A reg.
400 F	Loop	SHLD 4500	22	Store the HL pair data to memory location
4010			00	
4011			45	
4012		STA 4502	32	Store the accumulator content to memory location
4013			02	
4014			45	
4015		HLT	76	Stop execution.

Decimal Addition :

Example 4.4. *Add the decimal numbers 84 and 57 and store the results in memory locations 4500 and 4501.*

Memory Location	Label	Mnemonics	Opcode	Description
4000		MVI C, 00	0 E	Set C reg. to zero
4001			00	
4002		MVI A, 84	3 E	Move the data into Acc.
4003			84	
4004		MVI B, 57	06	Move the data into B reg.
4005			57	
4006		ADD B	80	Add two datas
4007		DAA	27	Decimal Adjust
4008		JNC Loop	D2	Check for carry
4009			0 D	
400 A			40	
400 B		INR C	0 C	Increment C register
400 C	Loop	STA 4500	32	Store the data from Acc to memory location
400 D			00	
400 E			45	

400 F	MOV A, C	79	Move the content of C reg. to Acc
4010	STA 4501	32	Store the result to the memory location
4011		01	
4012		45	
4013	HLT	76	Stop execution.

8-bit Subtraction :

Example 4.5. *Store the data 25 H and 17 H in memory locations 4200 and 4201 respectively. Add them and store the results in 4500 and 4501 H.*

Memory Location	Label	Mnemonics	Opcode	Description
4000		LDA 4201	3A	Load the first data in the memory location given
4001			01	
4002			42	
4003		MOV B, A	47	Move the data to B region.
4004		LDA 4200	3 A	Load the second data to the given memory location
4005			00	
4006			42	
4007		MVI C, 00	0 E	Clear C register
4008			00	
4009		SUB B	90	Subtract the contents
400 A		JNC Loop	D2	Check for carry
400 B			0 E	
400 C			40	
400 D		INR C	0 C	Increment carry register
400 E	Loop	STA 4500	32	Store the result to the memory location
400 F			00	
4010			45	
4011		MOV A, C	79	Move C reg. content to accumulator
4012		STA 4501	32	Store the result to the memory location
4013			01	
4014			45	Stop Execution
4015		HLT	76	

2's Complement :

Example 4.6. *Store the data 24 H in memory location 4200. Get the 2s complement of the number and store result in 4500 H.*

Memory Location	Label	Mnemonics	Opcode	Description
4000		LDA 4200	3 A	Load the data in the given memory location
4001			00	

4002			42	
4003		CMA	2 F	Get one's complement
4004		ADI 01	C6	Add 01 to the above (*i.e.,* 2s complement)
4005				
4006		STA 4500	32	Store the result
4007			00	
4008			45	
4009		HLT	76	Stop execution.

16-bit subtraction :

Example 4.7. *Subtract 5533 H from 8877 H and store the results in memory locations 4500 and 4501 H.*

Memory Location	Label	Mnemonics	Opcode	Description
4000		LDA 4203	3 A	Load the lower byte of the second data
4001			03	
4002			42	
4003		MOV B, A	47	Move it to B reg.
4004		LDA 4201	3 A	Load the lower byte of the first data
4005			01	
4006			42	
4007		SUB B	90	Subtract the contents
4008		STA 4501	32	Store the lower byte result.
4009			01	
400 A			45	
400 B		LDA 4202	3 A	Load the higher byte of the second data
400 C			02	
400 D			42	
400 E		MOV B, A	47	Move it to B reg.
400 F		LDA 4200	3 A	Load the higher byte of the first data
4010			00	
4011			42	
4012		SBB B	98	Subtract the contents along with previous borrow
4013		STA 4500	32	Store the higher byte of result.
4014			00	
4015			45	
4016		HLT	76	Stop execution

8-bit Multiplication :

Example 4.8. *Store the data 08 H and 03 H in memory locations 4200 and 4201 respectively. Multiply the contents and store the results.*

Memory Location	Label	Mnemonics	Opcode	Description
4000		LDA 4200	3 A	Load the multiplicand in the memory location

4001			00	
4002			42	
4003		MOV B, A	47	Move it to B reg.
4004		LDA 4201	3 A	Load the multiplier (Here it is the count - How many times we have to add)
4005			01	
4006			42	
4007		MOV C, A	4 F	Move it to C reg.
4008		MVI A, 00	3 E	Clear accumulator (for initial step we have
4009			00	to add multiplicand with CO)
400 A	LOOP X	ADD B	80	Add the contents
400 B		DCR C	0 D	Decrement count
400 C		JNZ LOOP X	CZ	Check for zero
400 D			0 A	
400 E			40	
400 F		STA 5000	32	Store the result
4010			00	
4011			50	
4012		HLT	76	Stop execution.

Logic Operation :

Example 4.9. *Mask the higher nibble or most significant four bits and store it in the memory location 4500.*

Memory Location	Mnemonics	Opcode	Description
4000	LDA 4200	3 A	Load the data into Acc
4001		00	
4002		42	
4003	ANI OF	E 6	AND immediate of with Acc
4004		0 F	
4005	STA 4500	32	Store the data to the memory from Acc
4006		00	
4007		45	
4008	HLT	76	Stop execution

8-bit division :

Example 4.10. *Store the data 08 H and 03 H in memory locations 4200 and 4201. Divide 08 H by 03 H and store the results in 4500 H and 4501 H.*

Memory Location	Label	Mnemonics	Opcode	Description
4000		MVI C, 00	0 E	Clear C reg.
4001			00	
4002		LDA 4201	3 A	Load the divisor in memory location
4003			01	
4004			42	

4005		MOV B, A	47	Move to B reg.
4006		LDA 4200	3 A	Load the dividend in memory location
4007			00	
4008			42	
4009	LOOP 2	CMP B	B 8	Compare the contents
400 A		JC LOOP 1	DA, 12, 40	Check for carry
400 D		SUB B	90	Subtract (A – B)
400 E		INR C	0 C	Increment quotient
400 F		JMP LOOP 2	C 3	Jump unconditionally
4010			09	
4011			40	
4012	LOOP 1	STA 4500	32	Store the remainder
4013			00	
4014			45	
4015		MOV A, C	79	Move C content to Acc.
4016		STA 4501	32, 01, 45	Store the quotient
4019		HLT	76	Stop execution.

Example 4.11. *Set the 1st and 7th bit of the data in accumulator and store the results in memory location 4500.*

Memory Location	Mnemonics	Opcode	Description
4000	MVI A, 00	3 E, 00	Set Acc to zero
4002	ORI F2	F6, F2	OR immediate F2 with Acc
4004	STA 4500	32, 00, 45	Store the result to memory
4007	HLT	76	Stop execution

Arry addition :

Example 4.12. *Add a series of 8-bit number (5 datas) and store the result in memory location 4600 H.*

Memory Location	Label	Mnemonics	Opcode	Description
4000		LXI H, 4200	21	Load the HL register pair with 16-bit address
4001			00	
4002			42	
4003		MOV C, M	4 E	Move the HL pair data to C reg.
4004		MVI A, 00	3 E	Clear accumulator
4005			00	
4006	LOOP	INX H	23	Increment HL pair
4007		ADD M	86	Add the first two numbers
4008		DCR C	0 D	Decrement the C register
4009		JNZ LOOP	C2	Check for non-zero
400 A			06	
400 B			40	

400 C		STA 4600	32	Store the data to memory
400 D			00	
400 E			46	
400 F		HLT	76	Stop execution.

Block Transfer :

Example 4.13. *Transfer a set of data from one set of memory location to another memory area.*

Memory Location	Label	Mnemonics	Opcode	Description
4000		MVI C, 04	0 E 04	Load count - 1 to C reg.
4001		LXI H, 5000	21 00 50	Initialize the address of first data to be transferred.
4005		LXI D, 5500	11 00 55	Initialize the destination address
4008	LOOP	MOV A, M	7 E	Move M content to Acc.
4009		STAX D	12	Store Acc to destination address
400 A		INX H	23	Increment HL pair (address)
400 B		INX D	13	Increment DE pair (address)
400 C		DCR C	0 D	Decrement count
400 D		JNZ LOOP	C2, 08, 40	Check for zero
4010		HLT	76	Stop execution

Finding Smallest/Largest data in an array :

Example 4.14. *Find the largest / smallest number from the given 5 datas and store the result in memory location 5000 H.*

Memory Location	Label	Mnemonics	Opcode	Description
4000		LXI H, 4200	21	Load HL Pair with 16-bit address
4001			00	
4002			42	
4003		MOV B, M	46	Move data from M to B
4004		INX H	23	Increment HL pair
4005		DCR B	05	Decrement B reg.
4006		MOV A, M	7 E	Move data from memory to Acc
4007	LOOP 2	INX H	23	Increment HL pair
4008		CMP M	BE	Compare M with Acc
4009		JC LOOP 1	DA	Check for carry
400 A			14	
400 B			40	
400 C	LOOP 3	DCR B	05	Decrement B reg.
400 D		JNZ LOOP 2	C2	Check for non-zero
400 E			07	
400 F			40	
4010		STA 5000	32	Store data from Acc to memory

4011			00	
4012			50	
4013		HLT	76	Stop execution
4014	LOOP 1	MOV A, M	7 E	Move data from memory to Acc
4015		JMP LOOP 3	C 3	Branch to memory location
4016			0 C	
4017			40	

Note. For largest data it is **JC LOOP 1** and smallest **JNC LOOP 1.**

Sorting :

Example 4.15. *Arrange the given datas in ascending order and store the results in an array.*

Memory Location	Label	Mnemonics	Opcode	Description
4000		LXI H, 5000	21	Load HL pair with 16-bit address (first data)
4001			00	
4002			50	
4003		MOV B, M	46	Move data from M to B
4004		MOV E, B	58	Move data from B to E
4005		MOV C, B	48	Move data from B to C
4006	LOOP 2	LXI H, 5001	21	Load HL pair with next data
4007			01	
4008			50	
4009		DCR C	0 D	Decrement C. reg.
400 A		JNZ LOOP 1	C 2	Check for non-zero
400 B			0 E	
400 C			40	
400 D		HLT	76	Stop execution
400 E	LOOP 1	MOV B, E	43	Move data from E to B
400 F	LOOP 4	DCR B	05	decrement B reg.
4010		JZ LOOP 2	CA	Check for zero
4011			06	
4012			40	
4013		MOV A, M	7 E	Move the data from memory to Acc.
4014		INX H	23	Increment HL pairs
4015		CMP M	BE	Compare M with Acc
4016		JNC LOOP 3	D2	Check for no carry
4017			1F	
4018			40	
4019		JZ LOOP 3	CA	Check for zero
401 A			1 F	
401 B			40	

401 C		JMP LOOP 4	C 3	Jump unconditionally
401 D			0 F, 40	
401 F	LOOP 3	MOV D, A	57	Exchange the contents (swapping)
4020		MOV A, M	7 E	
4021		MOV M, D	72	
4022		DCX H	2 B	Decrement HL pair
4023		MOV M, A	77	Move A to memory
4024		INX H	23	Increment HC pairs
4025		JMP LOOP 4	C 3	Unconditional jump
4026			0 F	
4027			40	

Note. To arrange data in descending order, change JNC to JC in 4016.

4.12 REVIEW QUESTIONS

Part A

1. What are the registers available in 8085 microprocessor?
2. List the interrupts available in 8085 microprocessor.
3. What are the addressing modes in 8085?
4. What are the RIM and SIM instructions and where they are used?
5. Compare SUB and CMP instructions.
6. What is the size of data bus and address bus of 8085?
7. What are processor control instructions? Give an example.
8. Write an assembly language program with check to see if the number is even or odd and if it is odd returns a 0 in B register, returns a 1.
9. What is the need for ALE in 8085?
10. What is the need for multiplying of buses in 8085?
11. List out the hardware interrupts of 8085 with its addresses in priority order.
12. What are status signals in 8085?
13. Define T state and instruction cycle.
14. What are the functions of program counter and stack pointer?
15. What are PUSH and POP instructions?

Part B

1. Explain 8085 architecture with neat sketch.
2. Draw and explain bus architecture of 8085.
3. Explain the addressing modes of 8085.
4. Discuss the interrupt system of 8085.
5. Write an assembly language for 8-bit multiplication.

5

Interfacing and Applications of Microprocessor

5.1 INTERFACING OF INPUT AND OUTPUT DEVICES

The transfer of data between keyboard and microprocessor, and microprocessor and display device is called input/output data transfer or I/O data transfer. The transfer of data between microprocessor and input device is done with the help of input port. The data transfer between microprocessor and output device is done with the help of output port.

Input Port

It is used to read data from the input device such as keyboard. The simplest form of input port is a buffer. The input device is connected to the microprocessor through buffer as shown in Fig. 5.1. The buffer is a tristate buffer and its output is available only when enable signal is active. When microprocessor wants to read data from i/p device, the control signals from the microprocessor activates the buffer by asserting enable input of the buffer.

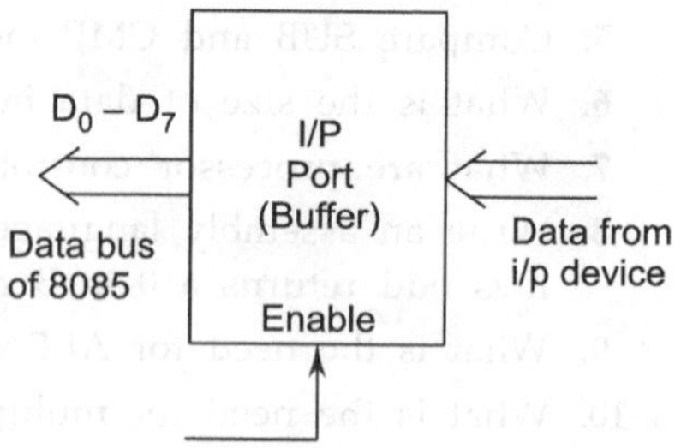

Fig. 5.1

Output Port

It is used to send data to the output device such as display from the microprocessor. The simplest form of output port is a latch. The output device is connected to the microprocessor through latch as shown in Fig. 5.2. When microprocessor wants to send data to the output device, it puts the data on the data bus and activates the clock signal of the latch, latching the data from the data bus at the o/p of the latch. It is then available at the output of latch for the o/p device.

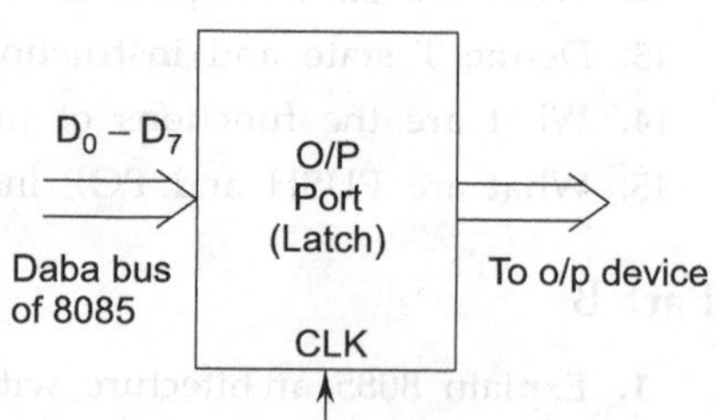

Fig. 5.2

5.2 I/O INTERFACING TECHNIQUES

I/O devices can be interfaced to an 8085 A system in two ways :

1. I/O mapped I/O
2. Memory mapped I/O

1. I/O Mapped I/O

In I/O mapped I/O, the 8085 uses $IO/\overline{M}$ signal to distinguish between I/O read/write and memory read/write operations. The 8085 has separate instructions IN and OUT for I/O data transfer. When 8085 executes IN or OUT instruction, it places device address on the demultiplexed low order address bus as well as the high order address bus, *i.e.*, the higher order address bus duplicates the contents of demultiplexed low-order address bus, when 8085 executes IN or OUT instruction.

A_{15}	A_{14}	A_{13}	A_{12}	A_{11}	A_{10}	A_9	A_8	A_7	A_6	A_5	A_4	A_3	A_2	A_1	A_0
0	1	1	0	0	0	0	0	0	1	1	0	0	0	0	0
								Device address = 60_H							

IN and OUT Instructions

IN : It is used to read 8-bit data from I/O device into the accumulator. This two byte instruction has, the first byte as an opcode and the second byte specifies the device address or a port number.

OUT : It is used to send 8 bit data from accumulator to the output device. This two byte instruction has the first byte as an opcode and the second byte specifies the device address or a port number.

I/O Device Selection

8085 gives 8 bit I/O address in I/O mapped I/O. This means it can select one of the 256 I/O ports. To select an appropriate I/O device, it has to do following:

1. Decode the address to generate unique signal corresponding to the device address on the bus.
2. When device address signal and control signal $(\overline{IOR}\text{ or }\overline{IOW})$ both are low, generate device select signal.
3. Use device select signal to activate the I/O port.

To generate device select signal (Y_0) low,

$$A_2\,A_1\,A_0 = 000_H$$
$$A_3\,A_4\,A_5\,A_6 = 0000_H \text{ (makes } \overline{G}_1 \,\&\, \overline{G}_2 \text{ low)}$$
$$A_7 = 1_H \text{ (makes G high).}$$

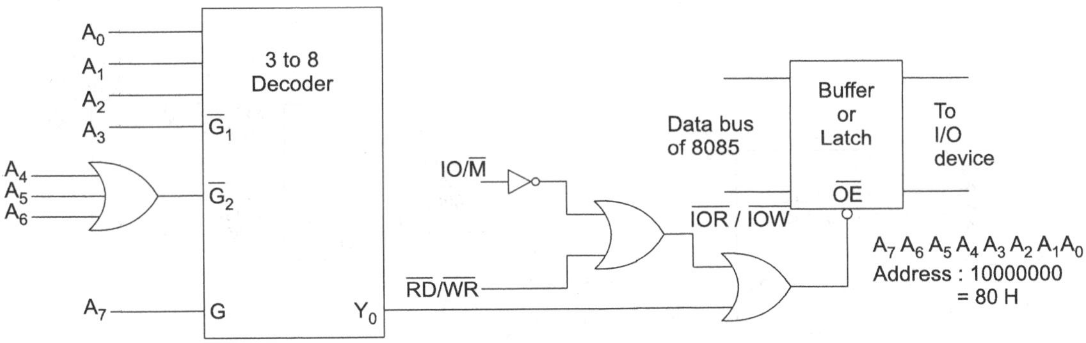

Fig. 5.3 Decoding circuit for I/O

Interfacing Input Device

The microprocessor 8085 accepts 8 bit data from the input device such as keyboard, sensors, transducers, etc. Figure 5.4 shows the circuit diagram to interface input port (buffer) which is used to read the status of 8 switches. The address for this input device is 80 H as device select signal goes low when address is 80 H.

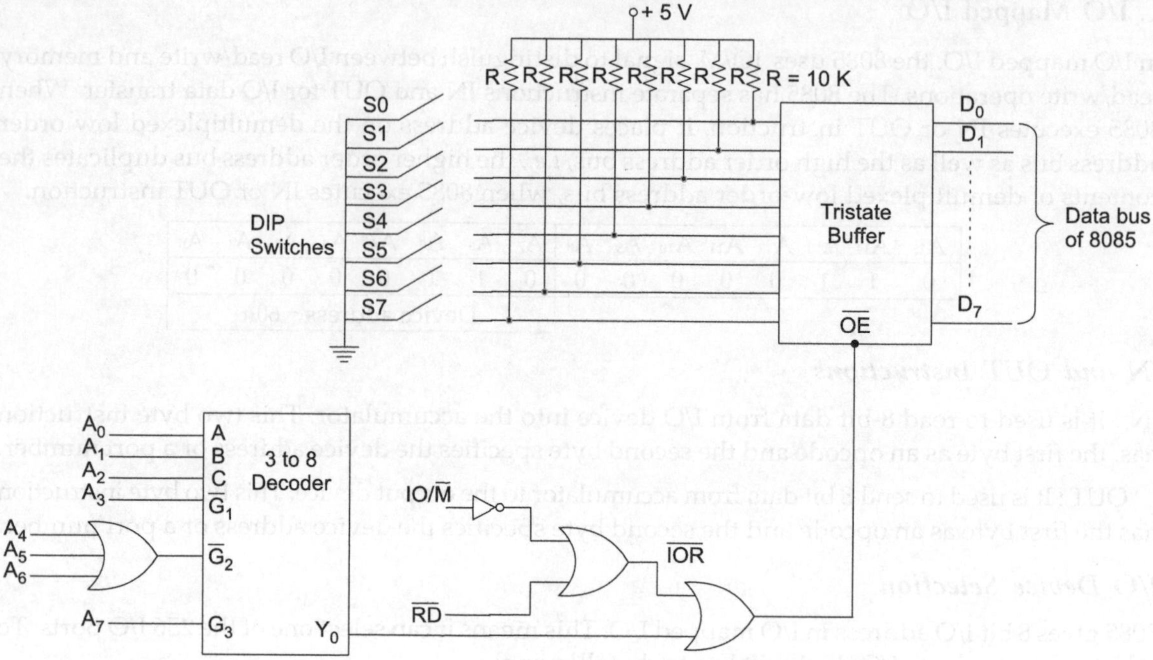

Fig. 5.4 Circuit diagram to interface input port

When, the switch is in released position, the status of line is high otherwise status is low. With this information microprocessor can check a particular key is pressed or not.

Interfacing Output Device

The microprocessor 8085 sends 8 bit data to the output device such as 7 segment displays, LEDs, printer etc. Figure 5.5 shows the circuit diagram to interface output port (latch) which is used to send the signal for glowing the LEDs. LED will glow when output pin status is low.

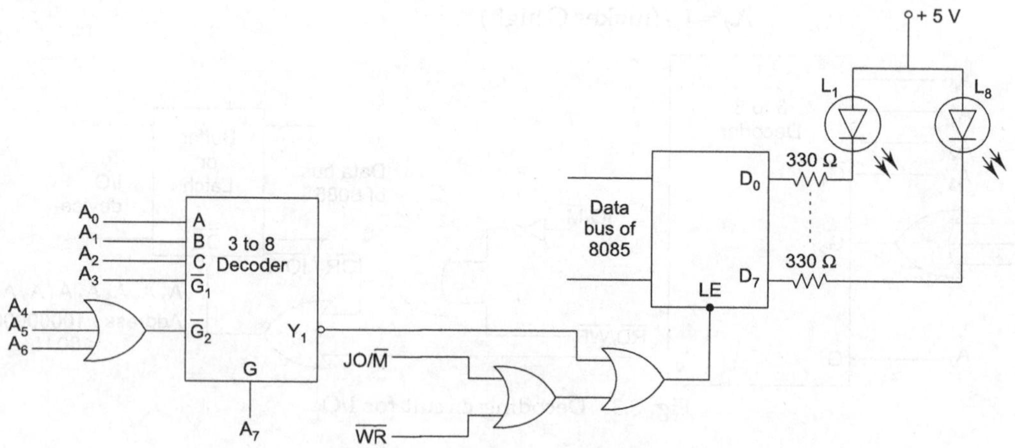

Fig. 5.5 Circuit diagram to interface output port

I/O Interfacing Using I/O Mapped I/O

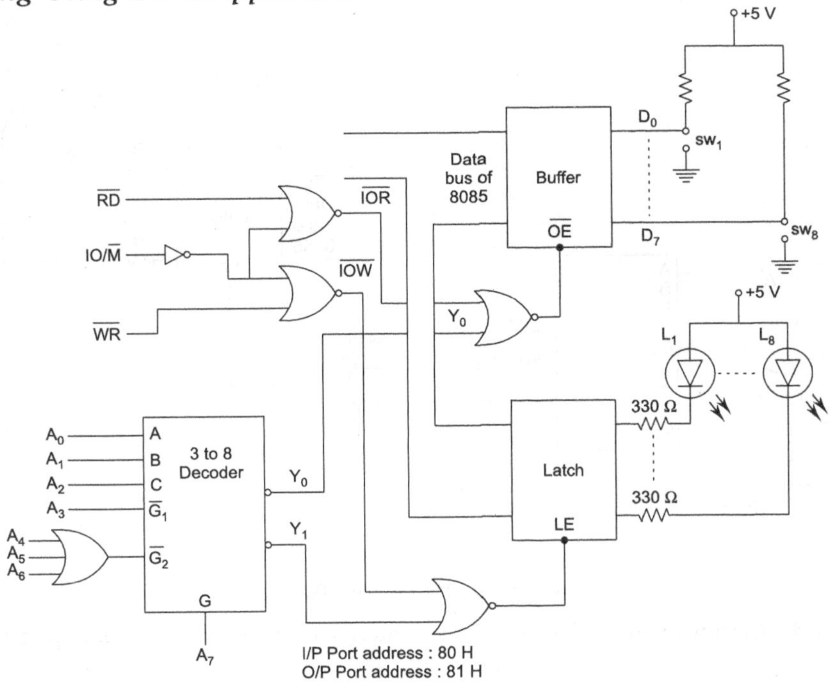

Fig. 5.6 I/O interfacing

2. Memory Mapped I/O

In memory mapped I/O, the I/O devices are assigned and identified by 16 bit addresses. The memory related instruction transfer the data between an I/O device and the microprocessor, as long as I/O port is assigned to the memory address space rather than to the I/O address space. Thus I/O device becomes a part of the system's memory map and hence its name. In memory mapped I/O every instruction that refers to memory location can control I/O.

Interfacing Input Device with Memory Mapped I/O

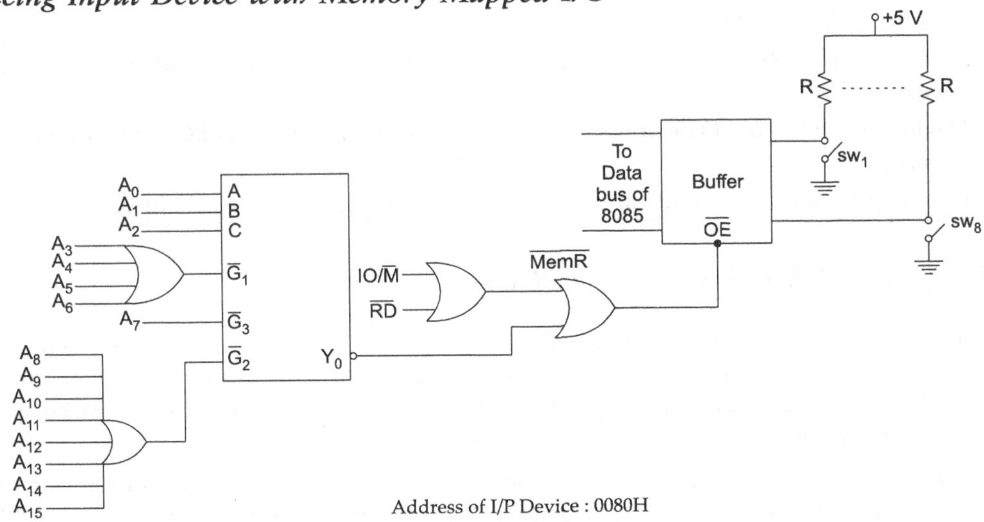

Fig. 5.7 Circuit diagram for I/P interfacing with memory mapped I/O

Interfacing Output Device using Memory Mapped I/O

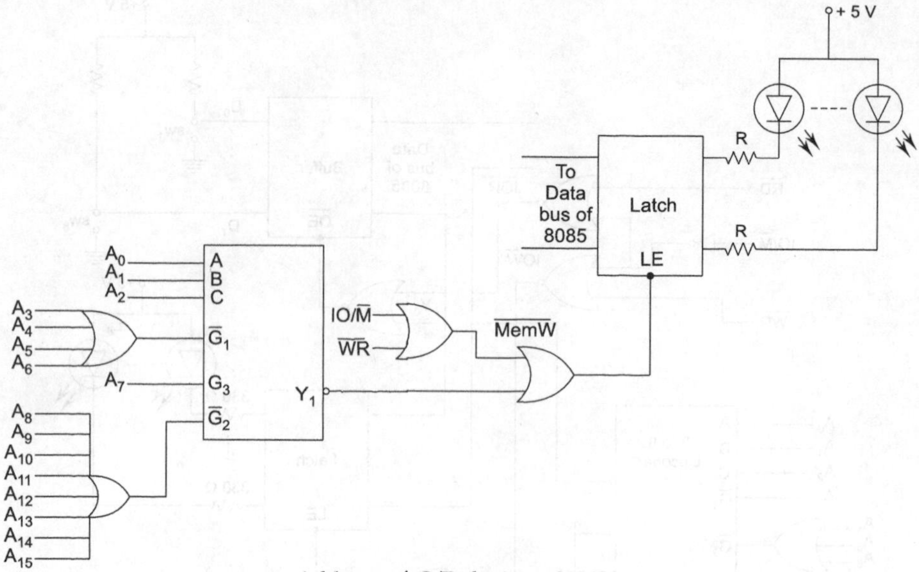

Address of O/P device : 0081H

Fig. 5.8 Circuit diagram for O/P device interfacing using memory mapped I/O

5.3 COMPARISON BETWEEN I/O MAPPED I/O AND MEMORY MAPPED I/O

S.No.	Memory Mapped I/O	I/O Mapped I/O
1.	In this, device address is 16 bit. Thus, A_0 to A_{15} lines are used to generate device address.	In this, device address is 8 bit. Thus, A_0 to A_7 or A_8 to A_{15} are used to generate device address.
2.	$\overline{\text{Mem R}}$ and $\overline{\text{Mem W}}$ control signals are used to control read and write I/O operations.	$\overline{\text{IOR}}$ and $\overline{\text{IOW}}$ control signals are used to control read and write I/O operations.
3.	Instructions available are LDA addr, STA addr, LDAXrp, STAXrp, MOV M, R, MOV R, M, ADDM, CMP M etc.	Instructions available are IN and OUT.
4.	Data transfer is between any register and I/O device.	Data transfer is between accumulator and I/O device.
5.	Maximum number of I/O devices are 65536 (theoretically)	Maximum number of I/O devices are 256.
6.	Decoding requires more hardware.	Decoding requires less hardware.

5.4 APPLICATIONS OF MICROPROCESSOR

With the advent of microprocessor, we can add intelligence and computer power to any instrument, machine process or system. The level of intelligence and computational power that can be added to a microprocessor based system depends on the memory capacity, software support and the interfacing arrangements.

To interface input/output devices to the microprocessor, some general and special purpose single chip interfacing devices are required. Some of the general purpose devices are I/O ports,

Programmable Peripheral Interface-PPI (8255). DMA controller (8237/57), interrupt controller (8259), communication interface (8251) and programmable counter/timer (8254). The most widely used programmable interfacing device is 8255.

5.5 TEMPERATURE CONTROL USING 8085

Microprocessor based systems are widely used in industries for the measurement display and monitoring of physical quantities like temperature, pressure, speed etc.

Block Diagram of Microprocessor Based Temperature Control

Transducer : For the measurement of physical quantities transducers are used. They convert them to electrical quantities. Here, for measuring temperature, sensors like thermocouple, thermistor, sensistor can be used.

Amplifier : If the electrical signal from transducer is small, it cannot be visualised or processed. Hence, it is amplified using amplifiers.

ADC : The electrical signal from transducer is an analog signal which a microprocessor cannot process. Hence, an analog to digital converter is used.

DAC : The signal from microprocessor will be digital signal which is going to control the analog elements like heater, cooler etc. For that the digital signal from the microprocessor has to be converted to analog by a DAC.

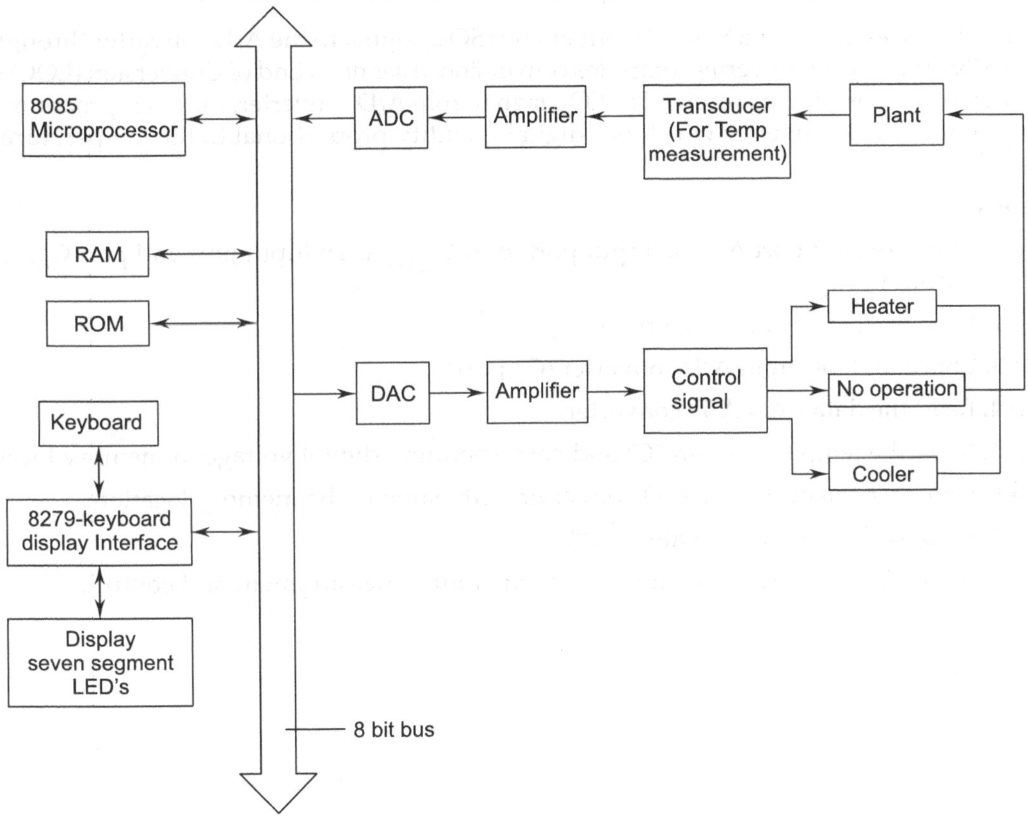

Fig. 5.9 8085 based temperature control system

Working : The plant may be an industry or an equipment or a furnace for which the temperature has to be monitored and maintained continuously at a particular temperature. The transducer measures current temperature and it is amplified by the amplifier which gets converted to hexadecimal digital value in ADC. This value is compared with the already set desired value in the microprocessor and the result of the comparison is as an error signal or null. The error may be positive meaning the temperature has to be increased by a heater arrangement. It may be negative meaning the temperature has to be reduced by some cooling set up. If there is no error, then no process has to be done and again the sensing activity has to be done repeatedly for continuous temperature maintenance.

For interfacing temperature control system with microprocessor, 8255 (PPI) and suitable ADC are connected between microprocessor and sensor output. Figure 5.10 shows the interface diagram of successive approximation A/D converter to the microprocessor through 8255 (PPI).

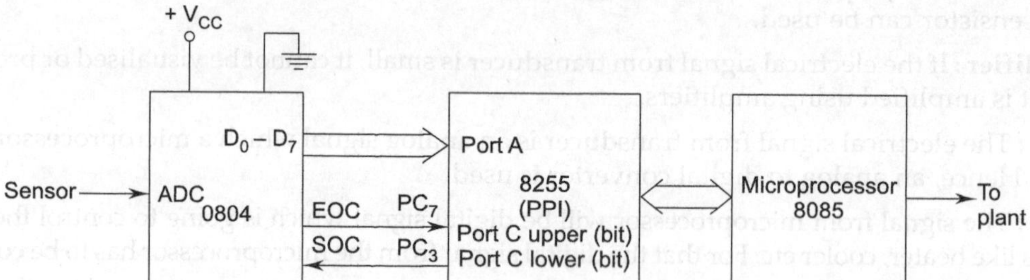

Fig. 5.10 Interfacing A/D converter with microprocessor

The microprocessor sends a Start of Conversion (SOC) signal to the A/D converter through Port C_{lower} of 8255. When A/D converter computes conversion, it sends as End of Conversion (EOC) signal to the microprocessor. Having received as EOC signal from A/D converter, the microprocessor reads the output of an A/D converter which is a digital quantity proportional to the temperature to be measured.

Algorithm:

Step 1. Initialize 8255 port A as an input port, port C_{upper} as an input port and port C_{lower} as an output port.

Step 2. Send SOC to an A/D converter (C_L port).

Step 3. Check for EOC from A/D converter (C_U port)

Step 4. Read the data from A/D converter

Step 5. Store the temperature (in °C) and corresponding digital voltage in memory locations.

Step 6. Compare the output of A/D converter with value in the memory location.

Step 7. Display the equivalent value (in °C)

Step 8. Repeat steps 2 to 7 for continuous temperature measurement and control.

Flow chart:

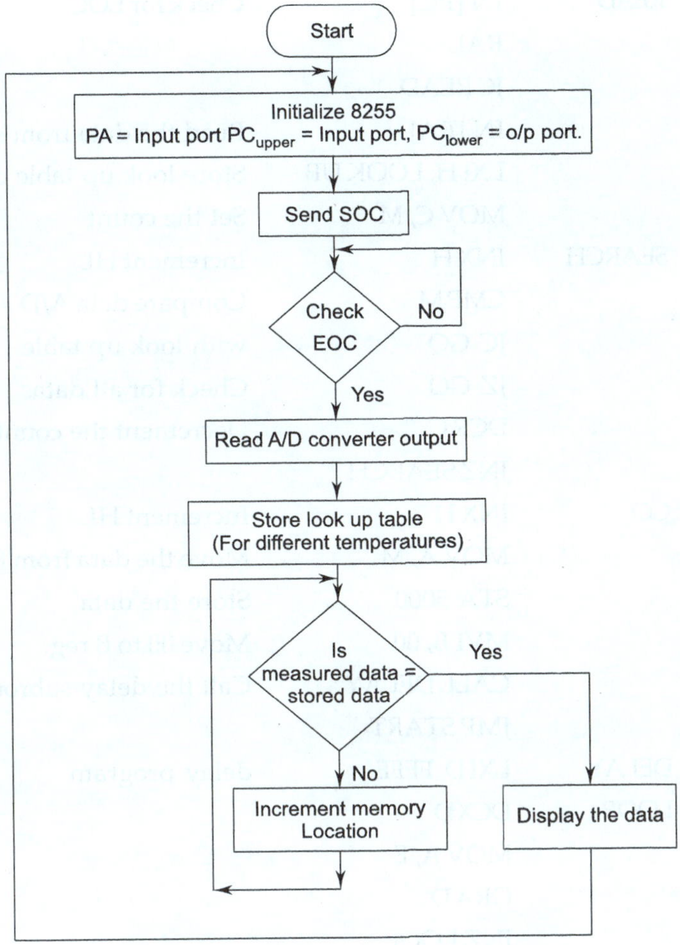

Fig. 5.11 Flow chart for temperature control system using 8085

LOOK UP

4500 0B	38	44	51	62	70	7F	8B	9E	AA	C1	C4	– Digital Hex Value
	50	55	60	65	70	75	80	85	90	95	99	– Analog Hex Value.

Program:

Memory Location	Label	Mnemonics	Description
4000		MVI A, 98	Initialize 8255
4002		OUT [CR]	Port A - i/p port
			PC_L - o/p port
4004	START	MVI A, 00	PC_H - i/p port
4006		OUT [PC]	Send pulse to converter
4008		MVI A, 08	

400A		OUT [PC]	
400C	READ	IN [PC]	Check for EOC
400E		RAL	
400F		JC READ	
4012		IN [PA]	Read the data from A/D converter
4014		LXI H, LOOK UP	Store look up table datas
4017		MOV C, M	Set the count
4018	SEARCH	INX H	Increment HL
4019		CMP M	Compare data A/D
401A		JC GO	with look up table
401D		JZ GO	Check for all datas
4020		DCR C	Decrement the counter
4021		JNZ SEARCH	
4024	GO	INX H	Increment HL
4025		MOV A, M	Move the data from HL to AC
4026		STA 5000	Store the data
4029		MVI B, 00	Move 00 to B reg.
402 B		CALL DELAY	Call the delay subroutine
402 E		JMP START	
4031	DELAY	LXI D, FFFE	delay program
4034	LOOP	DCX D	
4035		MOV A, E	
4036		ORA D	
4037		JNZ LOOP	

5.6 STEPPER MOTOR CONTROL USING 8085

A stepper motor rotates in steps in response to digital pulse input. The shaft of the motor rotates in equal increments when a train of pulse is applied. To control the direction and number of steps appropriate pulses are applied to the stator windings of the motor.

There are two types of stepper motors :

1. Permanent magnet type
2. Variable reluctance type

Figure 5.12 shows a permanent magnet type stepper motor with 4 pole stator and a rotor with six permanent poles.

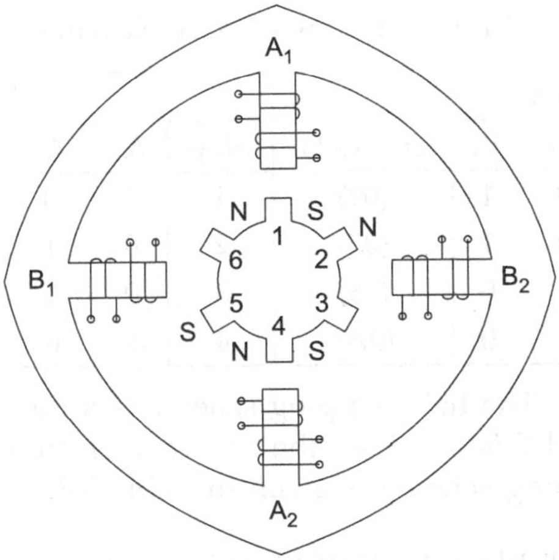

Fig. 5.12 Construction of a stepper motor

The stator is made of laminated soft iron. The stator windings are energized by pulses. The motor shown in Fig. 5.12 has four phase excitation as there are four poles on the stator. Each pole has two coils wound in the opposite sense so that the pole can be made either a north pole or a south pole as desired by applying appropriate pulse to one of the coils. If the pole A_1 is made a north pole, the pole A_2 is made a south pole. The permanent south pole No. 1 of the rotor will stand just below the A_1 of the stator. To give a clockwise motion the supply of pole A_1 and A_2 is switched OFF and B_1 and B_2 are energized. The pole B_1 is made a south pole and B_2 a north pole.

Now the permanent north pole No. 2 of the rotor comes just below the pole B_1. In the next step the pole A_2 is made a north pole and A_1 a south pole. After this B_2 is made a south pole and B_1 a north pole. Again the pole A_1 is made a north pole and A_2 a south pole and the whole sequence is repeated. In this order poles are energized to give a clockwise rotation. To rotate the rotor anticlockwise after making A_1 pole a north pole and A_2 a south pole, B_2 is made a south pole and B_1 a north pole.

Generally there are 3 different schemes available for stepping a stepper motor. They are (*i*) wave scheme (*ii*) two-phase scheme and (*iii*) half-stepping or mixed scheme.

(*i*) **Wave Scheme (Unipolar)**: In this, the stepper motor windings A_1, B_1, A_2, B_2 are cyclically excited with a DC current to run the motor in the clockwise direction. By reversing the phase sequence as A_1, B_2, A_2, B_1, the motor can be made to run in the anticlockwise direction. It is given in Table 5.1.

Table 5.1. Unipolar wave scheme

Anticlockwise						Clockwise					
Step	A_1	A_2	B_1	B_2	Hex data	Step	A_1	A_2	B_1	B_2	Hex data
1	1	0	0	0	(80)	1	1	0	0	0	(80)
2	0	0	0	1	(01)	2	0	0	1	0	(02)
3	0	1	0	0	(04)	3	0	1	0	0	(04)
4	0	0	1	0	(02)	4	0	0	0	1	(01)

(*ii*) **Two Phase Scheme:** In this scheme, any two adjacent windings are energized.

Table 5.2. Two Phase Scheme

Anticlockwise						Clockwise					
Step	A_1	A_2	B_1	B_2	Hex data	Step	A_1	A_2	B_1	B_2	Hex data
1	1	0	0	1	(09)	1	1	0	1	0	(0A)
2	0	1	0	1	(05)	2	0	1	1	0	(06)
3	0	1	1	0	(06)	3	0	1	0	1	(05)
4	1	0	1	0	(0A)	4	1	0	0	1	(09)

(*iii*) **Half-Stepping Scheme:** The half stepping scheme is a mixture of the wave scheme and the 2-phase scheme and this increases the accuracy of the stepper motor. The switching sequence for half stepping scheme is given in Table 5.3.

Table 5.3. Half-Stepping Scheme

Step	A_1	A_2	B_1	B_2	Hex data
1	1	0	0	0	(08)
2	1	0	1	0	(0A)
3	0	0	1	0	(02)
4	0	1	1	0	(06)
5	0	1	0	0	(04)
6	0	1	0	1	(05)
7	0	0	0	1	(01)
8	1	0	0	1	(09)

Block Diagram of Stepper Motor Interfacing with 8085

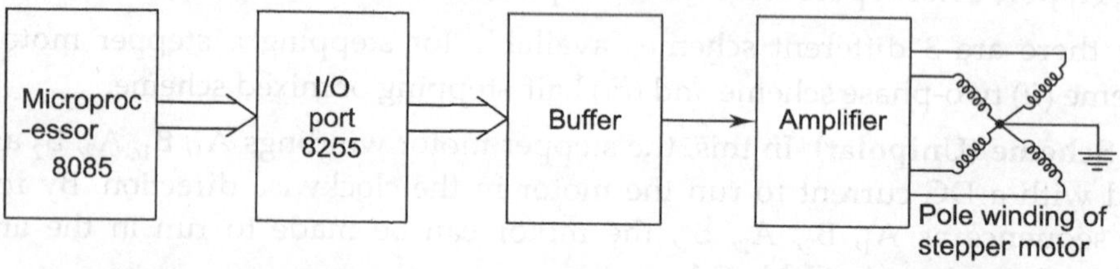

Fig. 5.13 Block diagram of stepper motor interfacing with 8085

The 12 V DC supply is used to energize the poles. Microprocessor has to send pulses, for which the windings gets excited. A delay subroutine has to be incorporated in the program. After energizing one set of pole windings some delay is provided, then the power supply is switched on to the other set of pole windings. This delay time govern the speed of the motor.

For a motor with 4 stator poles and 3 pairs of rotor poles (six) there are 12 possible stable positions in which a south pole of the rotor can lock with a north pole of the stator. The step size can be calculated from

$$\text{Step size} = \frac{360°}{N_s \times N_r}.$$

where N_s is the number of states poles.

N_r is the number of pairs of rotor poles.

Algorithm:

Step 1. Initialize PPI 8255, with port A as output port.

Step 2. Store the switching sequence in the memory.

Step 3. Set the counter to store the number of datas.

Step 4. Send the first data to the stepper motor through port A.

Step 5. Call subroutine for delay.

Step 6. Repeat steps 4 and 5 till the counter is zero.

Step 7. Repeat steps 2 to 6 for continuous rotation of stepper motor.

Flow chart:

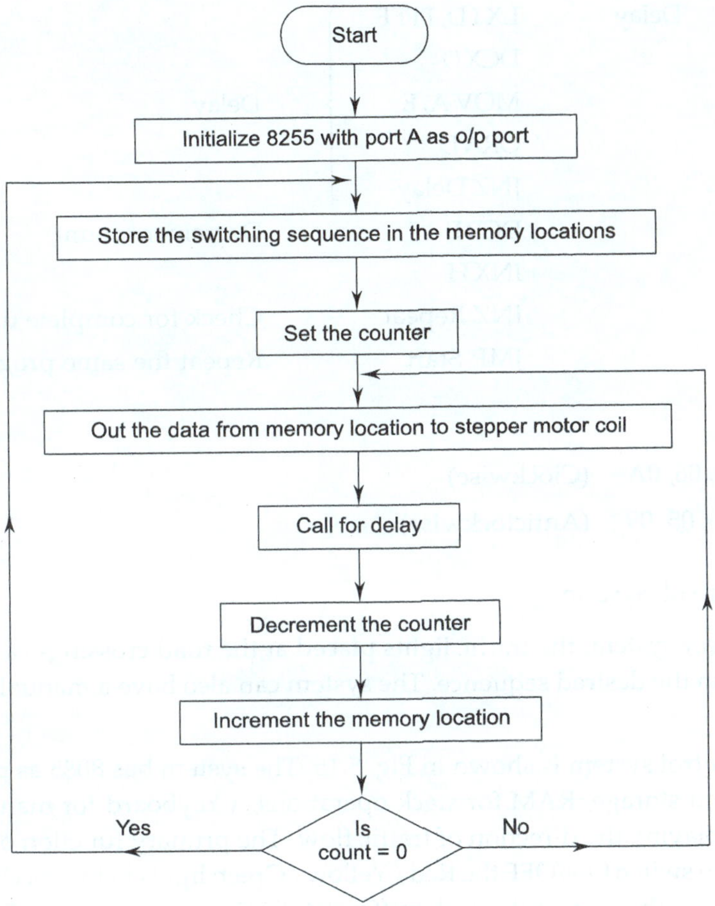

Fig. 5.14. Flow chart for stepper motor control using 8085

Program:

Memory Location	Label	Mnemonics	Description
4000		MVI A, 80	Initialize 8255
4002		OUT [CR]	
4004	Start	LXI H, LOOK UP	Store the datas in memory locations
4007		MVI B, 04	Initialize counter
4009			
400 A	Repeat	MOV A, M	
		OUT [PA]	Out the data through port A
400 C	Delay	LXI D, FFFF	Delay
400 F		DCXD	
4010		MOV A, E	
4011		ORA D	
4012		JNZ Delay	
4015		DCR B	Decrement count
4016		INX H	
4017		JNZ Repeat	Check for complete data transfer
401 A		JMP Start	Repeat the same program

LOOK UP

4500	09, 05, 06, 0A	(Clockwise)
4500	0A, 06, 05, 09	(Anticlockwise)

Traffic Light Control System

Using microprocessor system, the traffic lights placed at the road crossings can be automatically switched ON/OFF in the desired sequence. The system can also have a manual control option for emergency actions.

A traffic light control system is shown in Fig. 5.15. The system has 8085 as processor, EPROM memory for program storage, RAM for stack operations, a keyboard for manual control, 7 segment LED's for displaying the direction of traffic flow. The primary function of the microprocessor in the system is to switch ON/OFF the Red / Yellow / Green lights in the specified sequence. The LED's are interfaced to the system through buffer (74LS245) and ports of 8255.

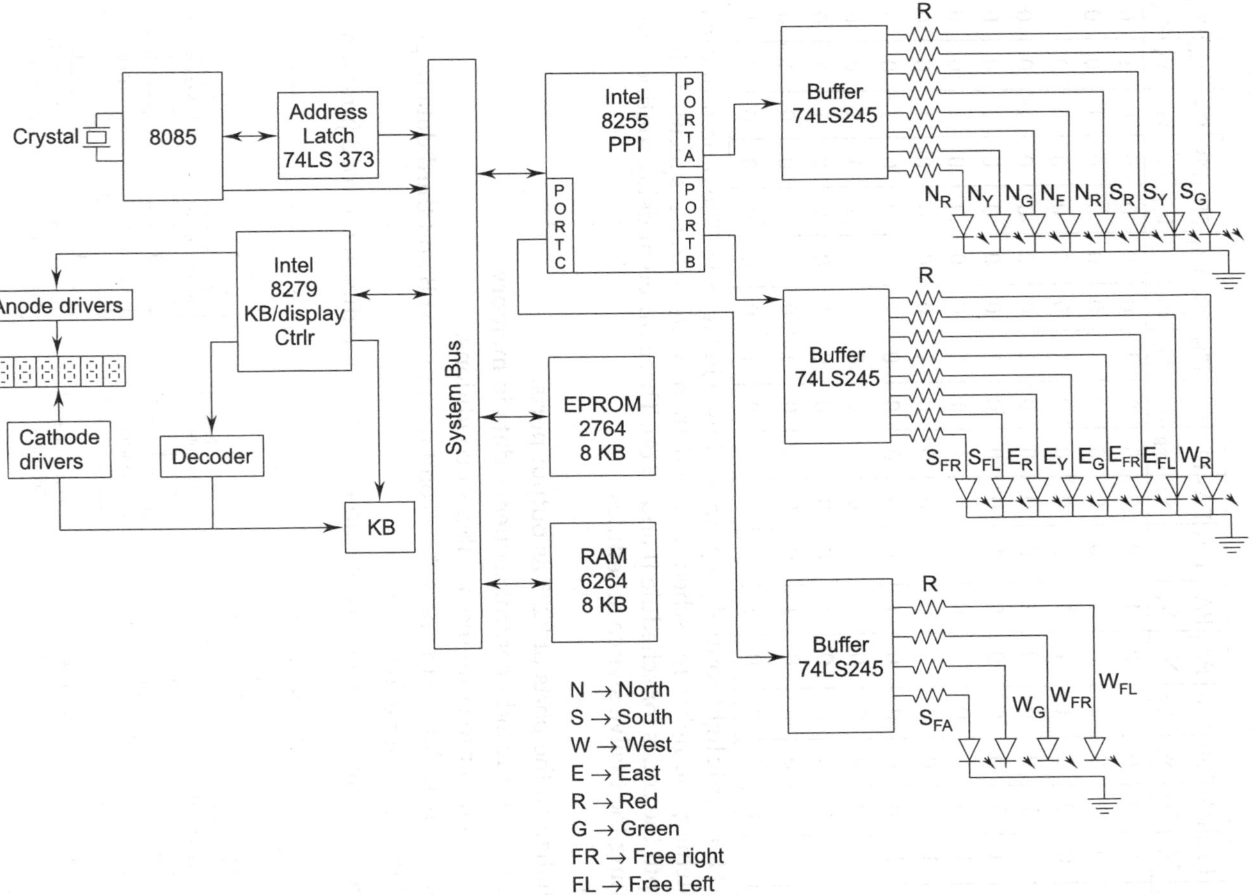

Fig. 5.15 8085 based traffic light control system

Table 5.4. Switching Schedule for Traffic Lights (1 → ON, 0 → OFF)

Switching Schedule	PA_0 N_R	PA_1 N_Y	PA_2 N_G	PA_3 N_{FR}	PA_4 N_{FL}	PA_5 S_R	PA_6 S_Y	PA_7 S_G	PB_0 S_{FR}	PB_1 S_{FL}	PB_2 E_R	PB_3 E_Y	PB_4 E_G	PB_5 E_{FR}	PB_6 E_{FL}	PB_7 W_R	PC_0 W_Y	PC_1 W_G	PC_2 W_{FR}	PC_3 W_{FL}
Schedule I	1	0	0	0	0	1	0	0	0	0	1	0	0	0	0	0	1	0	0	0
Schedule II	0	1	0	0	0	1	0	0	0	0	1	0	0	0	0	1	0	0	0	0
Schedule III	0	0	1	1	1	1	0	0	0	0	1	0	0	0	0	1	0	0	0	0
Schedule IV	0	1	0	0	0	1	0	0	0	0	1	0	0	0	0	1	0	0	0	0
Schedule V	1	0	0	0	0	0	1	0	0	0	1	0	0	0	0	1	0	0	0	0
Schedule VI	1	0	0	0	0	0	0	1	1	1	1	0	0	0	0	1	0	0	0	0
Schedule VII	1	0	0	0	0	0	1	0	0	0	1	0	0	0	0	1	0	0	0	0
Schedule VIII	1	0	0	0	0	1	0	0	0	0	0	1	0	0	0	1	0	0	0	0
Schedule IX	1	0	0	0	0	1	0	0	0	0	0	0	1	1	1	1	0	0	0	0
Schedule X	1	0	0	0	0	1	0	0	0	0	0	1	0	0	0	1	0	0	0	0
Schedule XI	1	0	0	0	0	1	0	0	0	0	1	0	0	0	0	0	1	0	0	0
Schedule XII	1	0	0	0	0	1	0	0	0	0	1	0	0	0	0	0	0	1	1	1

A flow chart for traffic light control program is shown in Fig. 5.16. The processor can output the codes for switching the lights for schedule I and then waits. After a specified time delay the processor output the codes for schedule II and so on. Three different delay routines are sufficient for implementing the twelve switch schedules.

Algorithm:

Step 1. Initialize all the ports of 8255 as output ports.

Step 2. Store the traffic light control sequence data in memory.

Step 3. Set the counter to store the number of conditions.

Step 4. Send the first data to the port A, second data to port B and third data to port C.

Step 5. Call the subroutine for delay.

Step 6. Repeat the steps 2 to 6 for continuous control of lights in the road junction.

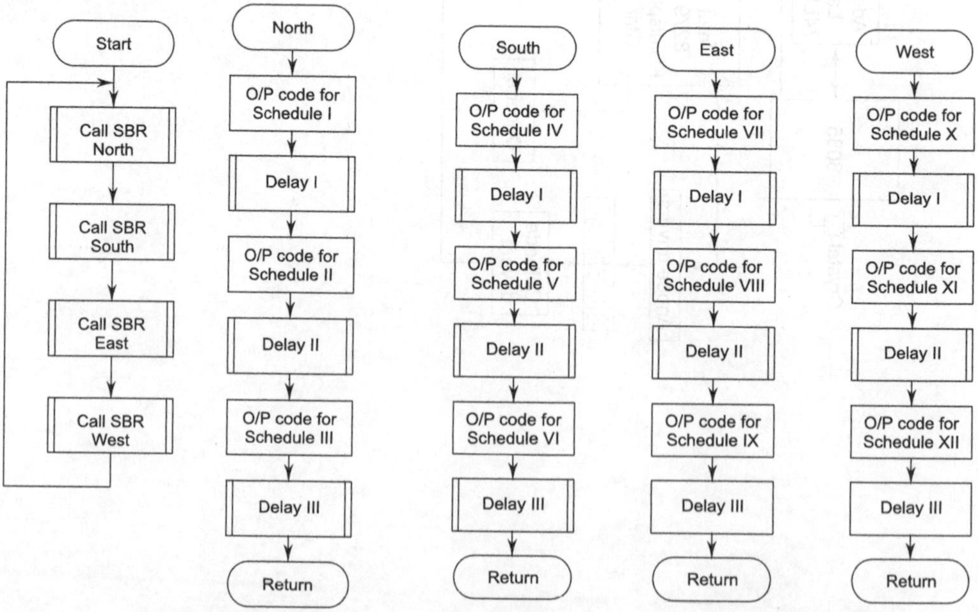

Fig. 5.16 Flow chart for traffic light control using 8085

Flow chart:

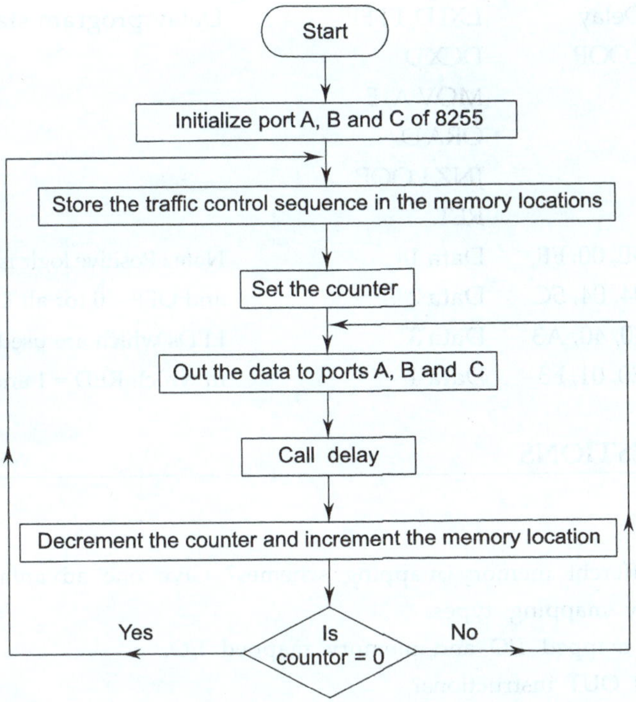

Fig. 5.17 Flow chart representing traffic control system using 8085

Program:

Memory Location	Label	Mnemonics	Description
4000		MVI A, 80	Initialize 8255
4002		OUT [CR]	Ports A, B & C
4004	Begin	LXI H,	Store the datas in memory locations
4007		MVI B, 04	Set counter
4009		MOV A, M	
400 A		OUT [PA]	Out the data through port A
400 C		INX H	
400 D		MOV A, M	
400 E		OUT [PB]	Out the data through port B
4010		INX H	
4011		MOV A, M	
4012		OUT [PC]	Out the data through port C
4014		CALL Delay	Call delay program
4017		INX H	
4018		DCR B	Decrement count
4019		JNZ Next	Check for complete data transfer

401 C		JMP Begin	Repeat the same program
401 F	Delay	LXI D, FFFF	Delay program starts here
4022	LOOP	DCX D	
4023		MOV A, E	
4024		ORA D	
4025		JNZ LOOP	
4028		RET	
4500	3E, 00, FF	Data 1	
4503	04, 04, 5C	Data 2	
4506	40, 40, A3	Data 3	
4509	E0, 01, F3	Data 4	

Note : Positive logic is used in LEDs *i.e.,* ON = 1 and OFF = 0 for all LEDs, except dual colour LEDs which are used for pedestrian crossing, in which RED = 1 and Green 0

5.7 REVIEW QUESTIONS

Part A

1. What are the different memory mapping schemes? Give one advantage for each.
2. Compare memory mapping types.
3. Differentiate I/O mapped I/O and memory mapped I/O.
4. What are IN and OUT instructions?
5. Explain about RIM and SIM instructions.
6. Define the term interfacing.
7. What is the role of microprocessor in stepper motor speed control?
8. What is the role of microprocessor in a traffic light controller?
9. How does 8085 control temperature variations of a particular system?
10. How many devices can be connected with 8085 in I/O mapped I/O?

Part B

1. Discuss in detail about I/O mapped I/O interfacing.
2. Explain briefly about memory mapped I/O technique with a neat example.
3. Interface a keyboard and 8 number of LED's with 8085 in I/O mapped I/O technique.
4. With neat block diagram explain temperature control system.
5. How does 8085 control speed of 4 phase stepper motor?

INDEX